中国建造

BUILT BY CHINA

中信大厦建造纪实

CONSTRUCTION RECORD OF CITIC TOWER

王伍仁 主编　WANG WUREN EDITOR IN CHIEF

中信和业｜中建三局｜中建安装 编

EDITED BY CITIC HEYE INVESTMENT CO., LTD. | CHINA CONSTRUCTION THIRD ENGINEERING BUREAU CO., LTD. | CHINA CONSTRUCTION INDUSTRIAL & ENERGY ENGINEERING GROUP CO.,LTD.

中国建筑工业出版社

CHINA ARCHITECTURE & BUILDING PRESS

中 国 建 造 系 列 丛 书 之 四
《 中 信 大 厦 建 造 纪 实 》 编 委 会

主 编 单 位 ： 中信和业投资有限公司

中国建筑股份有限公司 / 中建三局集团有限公司 联合体

中建安装集团有限公司

编 委 会 主 任 ： 常 振 明

编 委 会 副 主 任 ： 王　炯、王 伍 仁、梁 传 新

王 祥 明、陈 华 元、田　强

编 委 会 委 员 ： 丁　锐、王 海 涛、全　为、刘 庆 海、齐 燕 妮、汤 才 坤、许 立 山、

宋 利 坡、罗 能 钧、赵 良 治、贺 小 军、秦 继 红、聂 美 清、党 宏 伟、

逯 庆 杰、曾 运 平、谭 志 刚（按姓氏笔画为顺序）

主　　　　编 ： 王 伍 仁

副　主　编 ： 罗 能 钧、汤 才 坤、丁　锐

编　　　辑 ： 马　楠、王　坤、石 润 苏、田 贺 维、邢 其 龙、吕　邈、刘　航、

齐　锐、杨　军、吴 祖 教、邹　骥、汪 志 生、张　磊、罗 振 宇、

郑 蓉 军、孟 禹 江、赵　臻、栗　庆、席 亚 煜、常　颖（按姓氏笔画为顺序）

序言 I
PREFACE

常振明
Chang Zhenming

2018 年 12 月 28 日，我代表中信集团郑重地向中信大厦的总承包单位中国建筑集团颁发了中信大厦初步接收证书，这意味着中信大厦的开发建设整体完成。

至此，经历 2597 个日夜，中信大厦在北京 CBD 核心地段，在林立高楼中拔地而起，以地面高度 528m 成为北京第一高楼，一张中国新的建筑名片矗立在北京东三环。

我从 1983 年进入中信工作，历经 35 年，见证了中信集团成长壮大的历程。中信主导和参与了众多带有时代特点的大型项目，不论是金融还是实业，中信始终在党的领导下，与国家的发展同频呼吸，践行国家战略，融入当地经济发展主线，协力民生发展。在中信集团 40 年发展中，建设地标性建筑，也是值得载入企业发展历史的一页。

从成立至今，中信集团为北京这座古都打造了四座具有地标性意义的建筑。

1982 年，国际大厦动工，成功建成北京首座国际水准的涉外办公大楼，它位于长安街沿线，在一片尚未开发的北京市三环以内，它以独特的巧克力色，沉稳又现代的建筑样式，与国际接轨的写字楼风格，集中体现出中国企业对外开放的气质，北京人都亲切地称它"巧克力大厦"；1991 年，京城大厦落成，这栋外观独特、集写字楼和酒店式公寓于一体的高层建筑，也是当时北京最高楼，它雄踞风光秀丽的亮马河畔，与 90 多座驻华使馆为邻，壮观雄伟、气象非凡；2008 年，北京奥运会主场馆国家体育馆（鸟巢）建成并投入使用，它伴随北京奥运会震撼世界的华丽开闭幕式享誉国际。被奥委会时任主席罗格称赞为"世界上最好的体育场之一"。

中信大厦是中信集团向北京提交的第四座地标建筑，从 2010 年底拿地，到 2018 年如期实现大厦的初步接收，历经八年，其建设难度之巨、建设周期之长、协同跨度之大，远超同期项目。

2010 年底拍得 Z15 地块之后，中信集团立即调动资源，集中力量，启动中信大厦的开发建设，力邀国内超高层建筑领域的专家王伍仁先生加入项目管理团队。中信集团多次对项目的管理体系进行调整和优化，成立了中信大厦开发全过程跟踪审计小组，在项目开发建设阶段 13 次进驻现场，帮助中信和业防范各类风险、改进工作方式，提升管理能力和效率。中信和业投资有限公司的每一任董事长都由中信集团副总经理级别以上人员担任，为项目的执行给予最大的资源调度力，为大厦顺利交付保驾护航。

中信和业团队废寝忘食、全心投入，每年都能实现集团下达的年度工程开发进度目标，更如期兑现了 2012 年向集团许下的承诺：2018 年底实现"超高速度、极高品质、造价受控"。中信大厦是中信集团对单体重大项目植入新管理模式下，团结 5 家设计公司、113 家顾问单位、2 大总承包商、400 多家参建单位的数千名工程师和数万名工人夜以继日，付出了约 816 万工日辛劳铸就的建筑成果。

建设地标性建筑，是对中信这样的企业管理单体复杂项目能力的全面呈现，也是对最新科技、建筑文化、建筑技术、科学组织、时代审美等理解和实现能力的艰巨考验。企业需要经得起这样的考试，经得起时间的检验，以一张诚恳的答卷，经受专业人士和人民群众的评价，不断完善自身建设，进化企业创新基因。

新高度，新起点。中信人会继续不忘初心，牢记使命，开启新征程，展现新作为，树立新形象，在新的时代浪潮里，再创新可能。

On Dec. 28, 2018, I, on behalf of CITIC Group, solemnly issued the preliminary acceptance certification of CITIC Tower to the general contractor - CSCEC, which indicates the completion of development and construction of CITIC Tower.

So far, after 2597 days, CITIC Tower has risen straight from the ground in the CBD of Beijing, which becomes the highest building in Beijing with the height of 528m and stands in Beijing east third Ring Road as a new landmark.

I began to work in CITIC in 1983 and witnessed the development of CITIC Group through 35 years. CITIC Group has led and participated in many large-scale projects with the characteristics of the era. Either finance or industry, CITIC Group has always been under the leadership of the Party and in step with the development of the country, fulfilling the national strategy, integrating into the principal line of local economic development, and coordinating with the development of people's livelihood. During the 40 years of development of CITIC Group, the construction of landmark building is also a page worthy of being recorded in the history of enterprise development.

Since its founding, CITIC Group has built four landmark buildings for the ancient city of Beijing.

International Tower started the construction in 1982 and successfully became the first international-level foreign-related office building in Beijing. It is located in Chang An Avenue, within the undeveloped third ring road of Beijing. Due to its unique chocolate color, calm and modern architectural style, and office style in line with international standards, it intensively reflects the opening to the outside world of Chinese enterprises. Chinese affectionately call it "chocolate building". In 1991, the Capital Mansion was completed. It was the tallest building in Beijing with a unique appearance, a collection of office buildings and hotel-style apartments and extraordinary style, located in Liangma River bank and adjacent to more than 90 embassies in China. The National Stadium (Bird Nest), the main stadium of the Beijing Olympic Games, was built and put into use in 2008. Jacques Rogge, the then President of IOC, praised it as "one of the best stadiums in the world".

CITIC Tower is the fourth landmark building submitted by CITIC Group to Beijing government. From the land disposal at the end of 2010 to the initial acceptance of the building in 2018 as scheduled, it has gone through eight years, with its construction difficulty, construction period and coordination span far greater than those in the same period.

After the land Z15 was granted at the end of 2010, CITIC Group immediately mobilized resources and concentrated its efforts to commence the development and construction of CITIC Tower, and invited Mr. Wang Wuren, an expert in the field of super high-rise buildings in China, to join the project management team. CITIC Group had adjusted and optimized the management system of the project for many times. It had set up a tracking and auditing team for the whole development process of CITIC Tower, which entered the site for 13 times in the development and construction phase of the project, helping CITIC Heye prevent various risks, improving working methods, and enhancing the management ability and efficiency. Each chairman of CITIC Heye Investment Co., Ltd. was held by a person above the position of deputy general manager of CITIC Group, providing maximum resource mobilization for the implementation of the project and escorting the smooth delivery of the building.

CITIC Heye team was dedicated to achieving the goal of annual project development schedule set by the Group every year, and fulfilling the promise made to the Group in 2012 as scheduled: achieving "ultra-high speed, super high quality and controlled cost" by the end of 2018. CITIC Tower is the architectural achievement realized through the hard work of thousands of engineers and tens of thousands of workers from five design units, 113 consulting units, two general contractors and more than 400 participating units after 8.16 million working days.

The construction of landmark buildings is not only a comprehensive presentation of CITIC's ability to manage individual and complex project, but also an arduous test of its ability to understand and realize the latest science and technology, architectural culture, architectural technology, scientific organization and era aesthetics. Enterprises need to stand up to such examination and the test of time, and then be subject to the evaluation of professionals and the people with a sincere achievement, so as to constantly improve their own construction and enterprise innovation.

New height, new start. CITIC will continue to stay true to their original aspiration, keep their mission in mind, start a new journey, present new achievements, establish a new image, and create new possibilities in the new era.

八年追梦

写在中信大厦初步接收时

王伍仁
Wang Wuren

早春的北京，太阳挣脱静谧的晨雾，沿着城市错落的建筑之间的缝隙，吐露着缕缕金光。晨光中，中信大厦（中标时称 Z15 地块项目，我更喜欢称它"中国尊"）银灰色的幕墙披着朝霞在天际间熠熠生辉，像是一支巨大的日晷上闪光的晷针，映着五彩的光，装点着整个城市，让古老而现代的北京在光影中灿烂开来。

2018 年 12 月 28 日，中信集团董事长常振明从中建集团总经理王祥明手里接过中信大厦的金钥匙，标志着中信和业投资有限公司如期兑现了 2012 年 3 月向集团党委做出的"确保 2018 年底 Z15 地块项目实现竣工"的郑重承诺。

2018 年 12 月 14 日，中信大厦取得了北京市规划和国土资源管理委员会下发的《建设工程档案预验收意见书》，中信大厦建设工程档案预验收正式通过。2018 年 12 月 19 日，中信大厦通过了消防验收，取得了消防验收合格意见书（国内第一个在竣工前取得消防验收合格证的超高层建筑）。望着眼前高耸入云的中信大厦，我不由得回想起 2011 年 10 月 20 日从香港九龙仓建设集团接盘中信大厦开发重任后的 2709 天历程。2010 年 12 月集团中标 Z15 地块后，中信大厦的开发管理模式几经变迁，在中信集团党委和中信大厦指挥部的领导下，尤其是集团副总经理李庆萍担任指挥长后大刀阔斧地调整了公司管理架构，建立了清晰的责任体系，使公司步入规范运行轨道。2016 年初集团副董事长、总经理王炯担任中信和业董事长，带领我们克服重重困难，渡过了中信大厦开发中最为紧张、最为艰难的时段。

一幢超高层建筑的建造周期普遍都需要十多年，我经常把中信大厦的开发建设比喻成一场"马拉松比赛"，因为中信大厦的开发像马拉松比赛一样，不仅考验我们前期的筹划能力（Z15 地块项目施工许可证分段申请取证，大幅压缩了整体开发建设周期）、长时距统筹管理能力（中信大厦 EPCO "规划 / 设计、采购、建造、运营一体化工程管理模式"的成功实践，为安全、高品质、造价受控，实现总开发目标奠定了坚实的基础），同时也考验着我们科学的体能分配（我们创新了"设计联合体"模式，填平了概念设计与初步设计之间的鸿沟；创新了施工总承包与机电总承包"双总包"管理体系，尽管我们明知"双总包"体系会加大业主方的责任和压力，是一种"自讨苦吃"的管理模式，但是这种模式可有效地避免大型工程建设中结构施工占压机电 / 装修施工工期，导致总工期一而再再而三地延迟的通病发生）。马拉松需要匀速、不间断的奔跑能力，我们创新了"工期进度节点奖"机制，促使施工方严控每个单项工程进度节点的用时，进而确保了合同总工期目标的实现。

中信大厦开发建设的实际速度是国内同类超高层建筑的 1.4 倍（即仅用同类工程 70% 的时间）。我们从这场漫长的马拉松赛跑的准备阶段就瞄准了安全、高品质、造价受控，如期实现开发总工期目标的"冠军杯"，但是，我们深知要实现这一目标，将要面临多大的挑战与考验，将要付出多大的努力和拼搏，要承受多大的压力、质疑甚至是非议。值此《中国建造 中信大厦建造纪实》出版之际，特撰此文，以期答谢那些信任并支持、帮助我的领导、专家、同事和家人。

中信集团常振明董事长为了中信大厦的开发多次带队拜访或致函北京市领导，推动解决了许多中信大厦建造中遇到的重大问题，例如大厦外形设计方案的选择、大厦购地红线的南移、大厦坐向调整至与长安街平行、工程提前打桩、大厦顶部的优化、大厦正式供电等重大问题。中信集团副董事长、总经理王炯自 2016 年 2 月担任中信和业公司董事长以来，先后召开了 24 次董事长办公会，对项目推进中的 36 项重大事项及时进行决策，15 次亲临施工现场。还亲自带队拜会新设立的国家应急管理部领导，亲力亲为推动中信大厦的消防验收工作。

2019 年中信大厦将如期交付使用、亮相京城，成为中信集团向中华人民共和国成立 70 周年献礼的经典之作，亦成为中信集团成立 40 周年集团发展壮大的代表作。

回顾本人加盟中信并主持中信大厦开发的八年历程，我和公司管理团队承受了各种压力，因为拥有集团党委的坚强支持，我们传承了中信的文化基因，我们坚定地勇往直前并不断创新。工程建造点滴成绩的取得，离不开所有中信大厦建造者的心血与汗水。中信大厦就是一个容量惊人的载体，承载着参建者的拼搏与奉献，纪录着参加者的辛勤与耕耘。在近三千个日日夜夜中，我们并肩作战、众志成城，铸造了中信大厦的节节高升，也完成了自我的升华。中信大厦的建造经历，已经成为我们终身受益的财富，为今后的职业发展、人生历程画上浓墨重彩的一笔。经过八年艰苦卓绝的奋斗，我们终于实现了初心，向集团、向北京、向祖国、向世界递交了一份广受业界高度赞誉的答卷。

借此书出版之际，我代表中信和业投资有限公司管理团队向北京市、朝阳区、CBD 管委会相关负责人、中信集团领导，设计五方联合体及顾问公司、施工总承包商、机电总承包商、监理公司、各专业分包商、设备和材料供应商，对所有关心、支持中信大厦建设的各方专家，对所有为如期建成中信大厦付出智慧、心血、辛劳的同仁及家人、设计师、工程师、顾问团队和工人兄弟姐妹们致以由衷的敬意与谢意！是你们的精诚协作、拼搏奉献，为北京留下了一座难以超越的时代精品，是你们推动了中国建筑工程管理与建造技术的创新，是你们铸就了中国超高层建筑开发领域难以企及的新高度。

2011 年在我们出征前常振明董事长交给我三项任务：一是建成一幢具有国际领先水准的超高层建筑；二是带出一支政治觉悟高，技术功底深厚的管理团队；三是提交一套超高层建造完整的管理、技术资料。编写本书一是为了给超高层建造、开发管理的后来者提供借鉴，也是为了完成常总嘱托的第三项任务。因为我了解到此前出版的有关超高层建筑建造书籍

的编者不是来自业主方，就是来自工程施工方甚至是专业分包方，因责任或视角不同，都无法完整地描述出超高层建筑建造的完整过程。本次编辑这本《中国建造 中信大厦建造纪实》从构思开始，我就想要整合所有参建方的智慧和贡献，所以采用建设方与两大施工总承包方联合编制的方式，以期尽可能全面地介绍超高层建筑建造攻克难关的对策和措施。这也是一次新的尝试，我要感谢所有为此书提供资料的单位和人员，没有他们的支持和帮助我的设想再好也无法落地。

本书的编撰得到了中建三局、中建安装集团项目管理团队的积极参与，北京建筑设计研究院、KPF、奥雅纳、江河幕墙、中建钢构、通力电梯、格力电器、Azbil、金螳螂装饰、亚厦装饰、北京华美装饰等单位也提供了部分资料，在此一并表示感谢。本书的资料收集工作由许立山、贺小军、石润苏、罗振宇、刘航、齐锐、赵臻、常颖、王坤等承担，他们对稿件的严格要求与慷慨的时间付出，才让这本书更加丰富和立体，在此向你们道一声"你们辛苦了"。感谢上海渡影文化传播有限公司胡文杰团队北京、上海双轮驱动，为本书提供了出色的视觉策划和摄影设计。感谢中国建筑工业出版社，尤其是该社徐纺女士为本书提出了许多很好的建议，并一直督促我们工作进度，确保本书能够赶在中信集团成立四十周年之际得以出版。

因编撰水平有限，审稿时间仓促，本书中难免有所纰漏，欢迎各位读者不吝提出批评、指正意见。

EIGHT YEARS OF PURSUING A DREAM

Written at the time of the preliminary acceptance in CITIC Tower

Wang Wuren

In early spring of Beijing, the sunlight breaks through the morning mist, and spreads golden color along the cracks between the sprawling buildings in city. In the morning light, the silver-gray curtain wall of CITIC Tower (it was called the Z15 plot project when it won the bid, I prefer to call it "China Zun") shines brightly in the sky, like a glittering needle on a giant sundial, reflecting a colorful light, decorating the whole city and making the ancient and modern Beijing glorious in the light and shade.

On December 28, 2018, CITIC Group Chairman Chang Zhenming took over the golden key of CITIC Tower from CITIC Group General Manager Wang Xiangming, marking that CITIC Heye Investment Co., Ltd. fulfilled the solemn promise made to the Party Committee of CITIC Group in March 2012 to "ensure the completion of the Z15 plot project by the end of 2018".

On December 14, 2018, CITIC Tower obtained the Opinion on Pre-acceptance of Construction Project Archives issued by Beijing Municipal Planning and Land and Resources Management Committee, meaning that the pre-acceptance of construction project archives of CITIC Tower was formally approved. On December 19, 2018, CITIC Tower passed the fire acceptance and obtained the fire acceptance qualification opinion (the first super high-rise building in China to obtain the fire acceptance certificate before completion). Looking at the towering CITIC Tower, I cannot help but recall the 2709-day development of CITIC Tower, which I took over after leaving Wharf (Holdings) Ltd. in Kowloon, Hong Kong, on October 20, 2011. After winning the bid for Z15 plot in December, 2010, the development and management mode of CITIC Tower changed several times. Under the leadership of the Party Committee of CITIC Group and the Z15 project headquarters, especially after Li Qingping, the deputy general manager of CITIC Group, took the post of commander, the management structure of CITIC Group was drastically adjusted, a clear responsibility system was established, and CITIC Group stepped into the standard operation track. In early 2016, Wang Jiong, Vice Chairman and General Manager of CITIC Group, served as Chairman of CITIC Heye, led us overcoming mountains of difficulties and through the most intense and difficult period in the development of CITIC Tower.

The construction life cycle of a super high-rise building generally takes more than 10 years. I have often likened the development of CITIC Tower to a "marathon", since its development not only tests our planning ability in the early stage (applying for the construction permit of the Z15 plot project in phases dramatically shortened the overall development and construction period), and long-term overall management capability. (The successful practice of "integrated project management mode of planning/design, purchasing, construction and operation" of EPCO laid a solid foundation for safety, high quality, cost control, and fulfilling the overall time limit), but also tests our scientific physical distribution (we innovate the "design consortium" model and bridge the gap between conceptual design and preliminary design; "Double General Contractor" management system of general construction contractor and electromechanical general contractor is innovated. Although the "Double General Contractor" system will increase the responsibility and pressure of the owner, which is a "troublesome" management mode, but this can effectively avoid the electromechanics/construction period being occupied by structural construction in large-scale engineering construction, and prevent the total construction period from being repeatedly delayed, which is a common defect in the past). Besides, it also tests our uniform and uninterrupted running ability (we innovated the "schedule node award" mechanism to urge the constructor to strictly control the completion time of each individual project schedule node, thus ensuring the realization of the total contract duration target).

The actual speed of development and construction of CITIC Tower is 1.4 times faster than that of similar super high-rise buildings in China (only 70% of the time is spent on similar projects). We've been aiming for the "Champions" of safety, quality, cost control from the start of this long marathon, and for the target of the total development period is to be achieved on schedule. However, we are fully aware of how difficult challenges and tests will be faced, how much effort and attention will be needed, and how much pressure will be put on us to achieve this goal. Queries and even criticisms will be levelled at us. On the occasion of the publication of the CITIC Tower Construction Documentation, this article is intended to thank my leaders, experts, colleagues and family for their trust and tacit support.

Chairman Chang Zhenming of CITIC Group has led many visits or sent letters to the leaders of Beijing in the development of CITIC Tower, which has promoted the solution of many major problems encountered in the construction of CITIC Tower, such as the choice of design scheme for its appearance; southward movement of the red line of land purchase of the building; sitting direction of the building adjusted to be parallel to Chang An Avenue; piling ahead of schedule; optimization of the top of the building, the official power supply of the building and other major issues. Wang Jiong, Vice Chairman and General Manager of CITIC Group, has held 24 executive meetings of the Chairman of CITIC Heye since he began to serve as the Chairman of CITIC Heye in February 2016, has made timely decisions on 36 major items in the project advancement, and visited the construction site 15 times in person. Wang jiong led a team to call on the leaders of the newly established state emergency management department for the purpose of promoting the fire acceptance of the CITIC Tower.

In 2019, CITIC Tower will be put into use as scheduled and unveiled in Beijing, which also becomes a classic work of CITIC Group saluting to the 70th anniversary of the founding of the People's Republic of China and represents CITIC Group's development and prosperity on the 40th anniversary of its founding.

In the past eight years since joining CITIC and taking charge of the development of CITIC Tower, our management team and I have been in various pressures. Because of the strong support of the Party Committee of the Group, we have carried forward the CITIC culture, and steadfastly and courageously forged ahead and constantly achieved innovation. Our achievements cannot be made without the efforts and hard work of all the builders of the CITIC Tower. CITIC Tower has an astonishing capacity that bears everyone's hard work and dedication, and records everyone's toil. In nearly 3,000 days and nights, we fought side by side with one heart and one mind, forging the CITIC Tower's construction

and completing the self-improvement. The participation in the construction of the CITIC Tower has become our lifelong wealth, adding color to our future career development and life course. After eight years of arduous journey, we have finally realized our original commitment. For the Group, Beijing, our motherland and the world, it is a satisfactory result and highly praised by the industry to the Group.

On the occasion of publication of this book, I, on behalf of CITIC Heye management team, would like to express my gratitude to the relevant responsible persons of Beijing, Chaoyang District and CBD management committee, the leaders of CITIC Group, five-party design consortium, consulting company, construction general contractor (China State Construction Engrg. Corp. Ltd./Consortium of the China Construction Third Engineering Bureau Co., Ltd.), electromechanical general contractor (China Construction Industrial & Energy Engineering Group), supervision unit, equipment and material suppliers. Thanks to all the experts for their care about and support to the construction of the CITIC Tower, and all the colleagues, designers, engineers, consultative groups and workers for their hard work and wisdom, which have led to the scheduled completion of the CITIC Tower! It is your sincere cooperation, hard work and dedication that have created a fine and unparalleled product of the times in Beijing. You have promoted the reform and innovation of China's building management and construction technology, and you have forged a new height that is unattainable in the field of China's super high-rise buildings.

In 2011 when we set out, Chairman Chang Zhenming entrusted me with three tasks: to build a super high-rise building with international leading standards; to bring out a management team with high political consciousness and profound technical skills; and to submit a set of complete management and technical data for super high-rise building. This book was prepared to accomplish the third task. To my knowledge, editors of the previous books on the construction of super high-rise buildings come from either the owner, or the project builders or even professional subcontractors. Because of their different responsibilities or perspectives, it is unable to completely describe the construction process of super high-rise buildings. It is also my hope to integrate the wisdom and contribution of all parties involved in the construction of the CITIC Tower from the beginning of the conception in the compilation of the Documentary on the Construction of CITIC Tower. Thus, the method of joint preparation of the construction party and the two general construction contractors is adopted in order to introduce the difficulties in the construction of super high-rise buildings and the measures to overcome them as comprehensively as possible. Therefore, I would like to thank all the units and persons who have provided data for this book, whose support and assistance are essential to put into practice my vision.

The book has been compiled with the active participation of the general contractor project management team from the China Construction Third Engineering Bureau Co., Ltd. In addition, thanks to Beijing Institute of Architectural Design, KPF, Arup, Jianghe Curtain Wall, China Construction Steel Structure Corp. Ltd., Kone Elevator, Gree Electric Appliances, Azibl, Gold Mantis and other units have also provided some data and pictures. The materials and data for this book are collected by Xu Lishan, He Xiaojun, Shi Runsu, Luo Zhenyu, Qi Rui, Zhao Zhen, Chang Ying, Wang Kun, etc. Thanks to their strict requirements and generous time for manuscripts that enrich this book. Thanks to the Hu Wenjie's Team from the Shanghai Pdoing Vision & Culture Communication Co., Ltd., and Beijing and Shanghai Two-wheel Drive for Excellent visual planning and photographic design for the book. Thanks to China Architecture & Building Press, especially Ms. Xu Fang, for her good suggestions and urge for the book, so that the book can be published in time for the 40th anniversary of CITIC Group.

Due to the limited editing level and review time, some mistakes may be found in the book. You are welcome to give your comments.

Wang Wuren

目录
CONTENTS

目录
CONTENTS

开篇 CITIC TOWER | A NEW ICON FOR BEIJING

OPENING

1420 故宫建成

2018 中信大厦建成

BY CHINA   CONSTRUCTION RECORD OF CITIC TOWER

项目概述
Project Overview

简介
Brief Introduction

中信大厦（中国尊）位于北京市商务中心核心区，距离首都国际机场约 25km，距离故宫约 6.6km。作为北京新地标，造型蕴含了古代尊形、城门等中国历史文化元素，形态挺拔秀美，既与首都北京庄重典雅的城市风貌相融合，又彰显了新时代面向未来追求创新之势。中信大厦占地 11,478 ㎡，总建筑面积 43.7 万㎡（地上 35 万㎡，地下 8.7 万㎡），建筑层数 121 层（地上 109 层 +4 层夹层，地下 7 层 +1 层夹层），建筑高度 528m，是首都的新地标，与国贸建筑群、中央电视台和银泰中心等构成了新的北京天际线。

中信大厦于 2013 年 7 月 29 日开工，2017 年 8 月 18 日主体结构封顶；2018 年 12 月 19 日通过消防验收；2018 年 12 月 27 日单位工程竣工验收；2018 年 12 月 28 日颁发初步接收证书。建造历时 65 个月，共计 1,978 个日历天。

中信大厦自 2010 年 12 月购地起，开发历时 8 年，共计 2,597 个日历天。项目自启动之日，始终贯彻"打造超高层建筑精品"的目标，做到了质量与效率并行。开发建设期间，经过反复优化设计，共出具 9 版设计方案和 5 版施工图；在审查初步设计的过程中，中信和业投资有限公司组织设计及顾问团队从多角度进行审核和优化，保障大楼品质；创新性地使用了"分段实施、并行推进"的开发措施与创新设计联合体、"双总包"施工管理模式，开发速度是中国已建成同类超高层建筑平均开发速度的 1.4 倍。

CITIC Tower (China Zun) is located at the core area of CBD, 25 km away from Beijing Capital International Airport and 6.6 km away from The Palace Museum. As the new landmark building of Beijing, CITIC Tower contains some Chinese historical and cultural elements like "Zun" and City Gate, making it tall, straight and beautiful, which is not only integrated with the solemn and classic Capital features but also attests to the motion of future future-oriented innovation pursuit in the new era. CITIC Tower covers an area of 11,478 m² with total structure area of 437,000 m² (above ground 350,000 m² and below ground 87,000 m²), total 121 floors (above ground 109 floors+4 mezzanines and below ground 7 floors+1 mezzanine) and height of 528 m. It is the new landmark of Beijing and consists of the new Beijing skyline together with China Beijing World Towers, CCTV Headquarters and Yintai Centre.

Construction of CITIC Tower was started on July 29, 2013 and roof of major structure was completed on August 18, 2017. Fire-fighting acceptance was achieved on December 19, 2018; initial acceptance was achieved on December 28, 2018. CITIC Tower was built over 65 months and 1,978 calendar days.

CITIC Tower began to purchase the land from December 2010, lasting for 8 years, totally 2,597 days. Since the start of this project, the target of "constructing exquisite super high-rise buildings" has been implemented all along with quality and efficiency in parallel. During the construction, nine design plans and five construction drawings were issued by continuously optimizing. In reviewing the initial design, designers and consultative groups of CITIC Heye Investment Co., Ltd. were asked to examine and optimize from different angles to ensure the building quality. The innovative adoption of the method of "phase-based application and parallel pushing forward" and the construction management mode of "Double General Contractor" make the Speed of development is 1.4 times higher than the average develop speed of similar super high-rise buildings already built in China.

中信大厦 北京新地标 CITIC TOWER A NEW ICON FOR BEIJING

ROOFTOP R=18

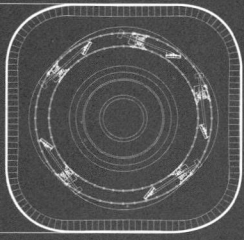

69m

L82 R=14

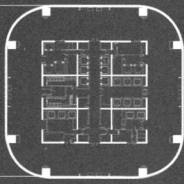

54m

528m

L8 R=14

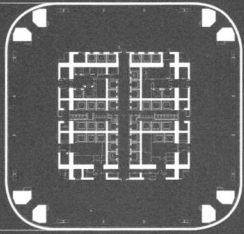

69.6m

L1 R=14

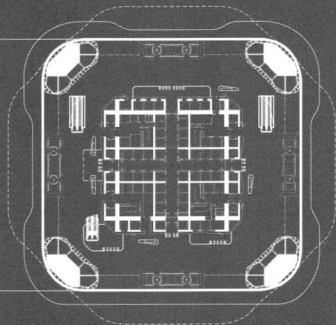

78m

±0.00

38m

46m

136.1m

商务中心
BUSINESS CENTER
103F-108F
Z8

办公区
OFFICE AREA
87F-102F
Z7

Z6

办公区
OFFICE AREA
43F-86F
Z5

Z4

Z3

办公区
OFFICE AREA
5F-42F
Z2

Z1

大堂、会议中心
LOBBY&MEETING CENTER
B1MF-4F
Z0

地下区域
UNDERGROUND AREA
B1-B7
ZB

地下桩基础
UNDERGROUND PILE
FOUNDATION

项目概述
Project Overview

设计理念
Design Philosophy

5轮
设计研究
Rounds of design research

78m
塔底几何轮廓宽度
The geometric outline width
of the tower base

54m
塔腰部几何轮廓宽度
The geometric outline width
of the tower waist

69m
塔顶部几何轮廓宽度
The geometric outline
width of the tower top

385m
建筑腰线高度
Waistline height

128个
每层幕墙竖向分格构成数量
The curtain wall in each floor is
vertically divided into 128 pieces

中信大厦历时5轮设计研究，从中国历史文化中汲取尊形塔身、孔明灯顶冠和城门入口等元素，经过抽象处理和比例优化，其形体自下至中上部逐渐缩小，同时顶部逐渐放大，形成中部略有收分的双曲线建筑造型，整体设计效果对称庄重又舒缓渐变。

塔楼剖面由4段相切的圆弧构成，平面为简洁实用的圆角房型。塔底几何轮廓宽度为78m，腰部为54m，顶部为69m，经城市天际线效果的推敲，腰线位于385m处，超过周边塔楼，整个建筑高耸直入云端，表现出顶天立地之势，展现出隽秀挺拔的恢弘气势。大厦底部为正方形，顶部内尊为圆形，塔冠用圆切构成花瓣状曲线，彰显了"外圆内方、天圆地方"的人文内涵。

塔底四角对巨柱和底部斜撑进行包合，形成庞大厚重的支座锚扎入大地，同时外幕墙壳体在四边逐渐扬起 并向外延伸成一体化的裙边形雨篷，形成富有动感的落客区。巨柱间由超白拉索玻璃幕墙连接，大堂空间通透开敞。

塔楼外壳体借鉴了古代尊与瓢的肌理构成技巧，以竖向金属条的造型和主次层次化处理，强调弧形体量的变化效果，并增强表皮肌理的丰富性。塔楼每层幕墙由128个竖向分格构成，主肋条的宽窄沿竖向渐变，形成微微收分和拉伸的外壳肌理，与塔楼形体相呼应。

首层大堂在底部沿用了尊的造型技巧，把外壳体延展拉起的同时，大堂内部的核心筒墙面和吊顶通过转角圆弧过度处理，形成向上展开的方尊造型。内尊体沿用外壳体的竖向分格和肋条处理，强调造型特征，并沿着吊顶向外延伸在雨篷边缘转折处交汇凝聚，形成内和外一体化、和谐统一的独特效果，与大堂和落客区空间在视觉上的内外贯通相呼应，塑造了开敞透明和大气通畅的塔底效果。

腰部高过周围建筑
WAIST FREE   OF SURROUNDING BUILDINGS
PREVIOUS   CENTRAL   WAIST   标地阶段腰部

After 5 rounds of design research, designers got inspired by such China historical and cultural elements as the tower shape of Zun, cap of Kongming Lanterns and entry of city gates. With abstract processing and proportion optimization of these elements, CITIC Tower shapes in gradually dwindling from the bottom to the middle-upper but gradually extending on the top, thus creating the hyperbola design with slight shrinkage in the middle. The overall design effect is symmetrical, solemn but soothing and varying.

The tower section is composed of four tangential arcs and the plan is simple and practical with the round cornered design. Geometric outline width of the bottom is 78m, waist 54 m and top 69 m. The waistline height is in 385m with consideration of Beijing skyline effect. Higher than the surroundings, the whole building towers into the clouds, manifesting its indomitable spirit and grandness of being graceful and tall. The tower bottom is square and the top inner is round. Tangential arcs constitute pedal curve on the tower crown. These designs manifest the humanistic connotation of "outwardly gentle but inwardly firm" and "spherical sky and square earth".

With wrapping of the giant columns and inclined struts from bottom by the four tower corners, the strong base anchors into the ground. The curtain wall shells raise and extend outwards gradually to form a skirt hem shaped awning and a dynamic porte cochere. Columns are connected by ultra-white cable-net supported glass curtain walls, making the lobby spacious and transparent. The outer shell of the tower draws on the technique of texture composition of ancient Zun and gourd ladle, and uses the modeling of vertical metal strips and the treatment of primary and secondary layers to emphasize the change effect of arc volume and enhance the richness of skin texture. The curtain wall of each floor of the tower is composed of 128 vertical compartments. The width of the main ribs changes gradually along the vertical direction, and the texture of the outer shell in gradually dwindling and gradually extending corresponds to the shape of the tower.

At the bottom of the lobby on the first floor, Zun modeling technique is used. When the outer shell is extended and pulled up, the core tube wall and the ceiling inside the lobby are coupled through the corner arc, forming the square Zun shape that unfolds upward. The inner Zun body is treated with vertical compartments and ribs of the outer shell, to emphasize modeling characteristics, and extends outward along the ceiling and converges and agglomerates at the turning of the edge of the awning, thus forming a unique effect of integration of inner and outer, harmony and uniformity, which echoes the visual connection between the inside and outside of the lobby and the porte cochere, and shaping an open and transparent tower bottom with a clear atmosphere.

1.3 主要参建单位
Major Participation Units

1. 投资方	
中信集团	
2. 开发建设方	
中信和业投资有限公司	
3. 设计方	
标地阶段概念设计	
TFP Farrells 建筑事务所＼北京市建筑设计研究院有限公司	
概念调整和方案设计	
Kohn Pedersen Fox Associates（设计牵头方）	
北京市建筑设计研究院有限公司（设计总负责方）	
中信建筑设计研究总院（结构试桩施工图）	
初步设计	
Kohn Pedersen Fox Associates（外壳体和主要公共空间内装设计的牵头方）	
北京市建筑设计研究院有限公司（其他部分设计的牵头方）	
施工图设计	
北京市建筑设计研究院有限公司（牵头方）	
Kohn Pedersen Fox Associates（与幕墙顾问合作幕墙招标图和文件）	
中信建筑设计研究总院（协助方）	
施工阶段	
北京市建筑设计研究院有限公司（牵头方）	
Kohn Pedersen Fox Associates（从设计意图方面审核幕墙施工图纸和样板）	
中信建筑设计研究总院（业主配合方）	
结构设计	
奥雅纳工程咨询（上海）有限公司	
北京市建筑设计研究院有限公司	
机电设计	
柏诚亚洲有限公司	
北京市建筑设计研究院有限公司	
4. 勘察设计	
北京市勘察设计研究院有限公司	
5. 主要顾问公司	
幕墙	
艾勒泰建筑工程咨询（上海）有限公司	
照明设计	
上海碧甫照明工程有限公司	
风环境	
安邸建筑环境工程咨询（上海）有限公司	
交通设计	
弘达交通咨询（深圳）有限公司	
可持续设计和 LEED 认证	
清华大学、君凯环境管理咨询（上海）有限公司	
消防安全	
奥雅纳工程咨询（上海）有限公司	
中国建筑研究院防火所	
施工文件审核	
森大厦（上海）有限公司	
华东建筑设计研究院有限公司	
标识设计	
株式会社黎设计综合计划研究所	
BIM 设计	
悉地国际设计顾问（深圳）有限公司	
物业顾问	
北京仲量联行物业管理服务有限公司、北京中际北视物业管理有限公司	
6. 监理单位	
北京远达国际工程管理咨询有限公司	
7. 总承包单位	
施工总承包：中国建筑股份有限公司／中建三局集团有限公司　联合体	
机电总承包：中建安装集团有限公司	

1. INVESTOR

CITIC Group

2. DEVELOPMENT AND CONSTRUCTION UNIT

CITIC Heye Investment Co., Ltd.

3. DESIGN UNIT

Conceptual design in biding phase

TFP Farrells construction office / Beijing Institute of Architectural Design Co., Ltd.

Conceptual adjustment and schematic design

Kohn Pederson Fox Associates (design leading unit)

Beijing Institute of Architectural Design Co., Ltd. (design general responsible unit)

CITIC General Institute of Architectural Design and Research Co., Ltd. (construction drawing of structural test piles)

Preliminary design

Kohn Pederson Fox Associates (leading unit for design of outer shell and main public space)

Beijing Institute of Architectural Design Co., Ltd. (leading unit of other designs)

Design of construction drawing

Beijing Institute of Architectural Design Co., Ltd. (leading unit)

Kohn Pederson Fox Associates (cooperate with curtain wall consultant for curtain wall bidding drawing and document)

CITIC General Institute of Architectural Design and Research Co., Ltd. (assisting unit)

Construction phase

Beijing Institute of Architectural Design Co., Ltd. (leading unit)

Kohn Pederson Fox Associates (review curtain wall drawing and template in terms of design intent)

CITIC General Institute of Architectural Design and Research Co., Ltd. (cooperating as owner)

Structural design

ARUP International Consultants (Shanghai) Co., Ltd.

Beijing Institute of Architectural Design Co., Ltd.

Electromechanical design

Parsons Brinckerhoff Asia Ltd.

Beijing Institute of Architectural Design Co., Ltd.

4. INVESTIGATION DESIGN

BGI Engineering Consultants Ltd.

5. MAIN CONSULTING UNITS

Curtain wall

ALT Engineering and Consulting (Shanghai) Co., Ltd.

Lighting design

Shanghai Brandston Partnership Inc.

Wind environment

RMDI International China Inc.

Traffic design

MVA (Shenzhen) Inc.

Sustainable design and LEED certification

Tsinghua University / Environmental Market Solutions (Shanghai) Inc.

Fire safety

ARUP International Consultants (Shanghai) Co., Ltd.

Institute of fire, China Academy of Architecture Sciences

Review on construction document

MORI Building China Co., Ltd.

ECADI Co., Ltd.

LOGO design

General Plan Institute of Li Design Co., Ltd.

BIM design

CCDI (Shenzhen) Co., Ltd.

Property consultant

JLL Beijing Property Services Co., Ltd. / Beijing Zhongjibeishi Property Management Co., Ltd.

6. SUPERVISING UNIT

Beijing Yuanda International Engineering Management Consult Co., Ltd.

7. GENERAL CONTRACTOR

Construction general contractor: CSCEC / China Construction Third Engineering Bureau Co., Ltd.

Electromechanical general contractor: China Construction Industrial & Energy Engineering Group Co., Ltd.

1.4 周边环境
Surrounding Environment

大数据 *Data*

5 个
周边地铁车站
Surrounding subway stations

500,000 m²
CBD 公共空间基础设施
The total scale of CBD area

规划有轨电车
Tram

现状公交站
Local Bus + Stops

现状地铁
Metro Existing

在建地铁
Metro Planned

地下公共停车
Public Parking Underground

5分钟步行范围
5min Walking Distance

中信大厦所在的北京市 CBD 核心区，正在实施一个巨大的都市发展计划。为了支持城市这样高强度的集中开发，由政府先行启动了 CBD 核心区城市基础设施工程，利用核心区的城市道路与公共开放空间进行规划建设，在城市道路和公共绿地之下，建设了包括两层地下市政管廊、连通周边 5 个地铁车站的地下人行交通通道、连通每个地块的机动车物流通道、服务空间和机动车停车空间。

CBD 核心区的公共空间和基础设施的工程建设总规模达 50 万 m²，将形成一个功能完善、性能优越的城市"主板"，将和其上的各个摩天大厦"插块"共同组成北京 CBD 发展的强大主板，此主板也成就了中信大厦这个世界一流水平的超高层摩天楼。

中信大厦与 CBD 核心区地下公共交通人行和货运车行系统直接相连；通过地下人行交通系统可到达周边多个轨道交通车站；通过地下公共车行环形隧道实现机动车和物资远距离进入，避免地面交通拥堵；通过地下公共管廊实现市政管线的无缝衔接，提高大厦运营效率。

The CBD area where CITIC Tower is located is implementing a huge urban development plan. In order to support the urban high-intensity and centralized development, the government has started the urban infrastructure project of the CBD area first, made use of urban roads and public open spaces in the core area for planning and construction, and has built two floors of underground municipal pipe corridor, underground pedestrian traffic channels connecting 5 surrounding subway stations, vehicle logistics channels connecting each plot, service space and vehicle parking space under the urban road and public green space.

The public space and infrastructure has a total construction scale of 500,000m², which will form an urban "main board" with complete functions and superior performance, and form a powerful "main board" for the development of Beijing CBD together with the "inserting blocks" of various skyscrapers on it. This main board also facilitates CITIC Tower, a world-class super high-rise skyscraper.

CITIC Tower is directly connected to the underground public transport pedestrian and freight traffic system in the CBD area: multiple rail transit stations can be reached through the underground pedestrian traffic system; through the underground circular tunnel of public vehicles, long-distance entry of motor vehicles and materials can be realized so as to avoid traffic congestion on the ground; through the underground public pipe corridor, the seamless connection of municipal pipelines can be realized, to improve the safety of building operation.

世界摩天大厦
WORLD SKYSCRAPER

(2)

大数据 Data

8 位
2018 年 CTBUH 排名
Ranked 8th in CTBUH in 2018

528 m
建筑高度
Building height

2013.7.29
中信大厦开工时间
CITIC Tower started

2018.12.28
中信大厦竣工时间
CITIC Tower completed

纽约 New York

北京 Beijing　　首尔 Seoul

上海 Shanghai

Guangzhou 广州　　台北 Taipei

麦加 Mecca　　迪拜 Dubai　　Shenzhen 深圳

828m 2004-2010 哈利法塔 Burj Khalifa 迪拜 Dubai **1**	541m 2006-2014 世界贸易中心一号大楼 One world Trade Center 纽约 New York **6**
632m 2008-2015 上海中心大厦 Shanghai Tower 上海 Shanghai **2**	530m 2010-2017 广州东塔 Guangzhou CTF Finance Center 广州 Guangzhou **7**
601m 2004-2012 麦加皇家钟塔酒店 Makkah Royal Clock Tower Hotel 麦加 Mecca **3**	528m 2013-2018 中信大厦 CITIC Tower 北京 Beijing **8**
599m 2007-2016 平安金融中心 Ping An Finance Center 深圳 Shenzhen **4**	508m 1999-2004 台北 101 Taipei 101 台北 Taipei **9**
555m 2011-2015 乐天世界大厦 Lotte World Tower 首尔 Seoul **5**	492m 2004-2008 上海环球金融中心 SWFC 上海 Shanghai **10**

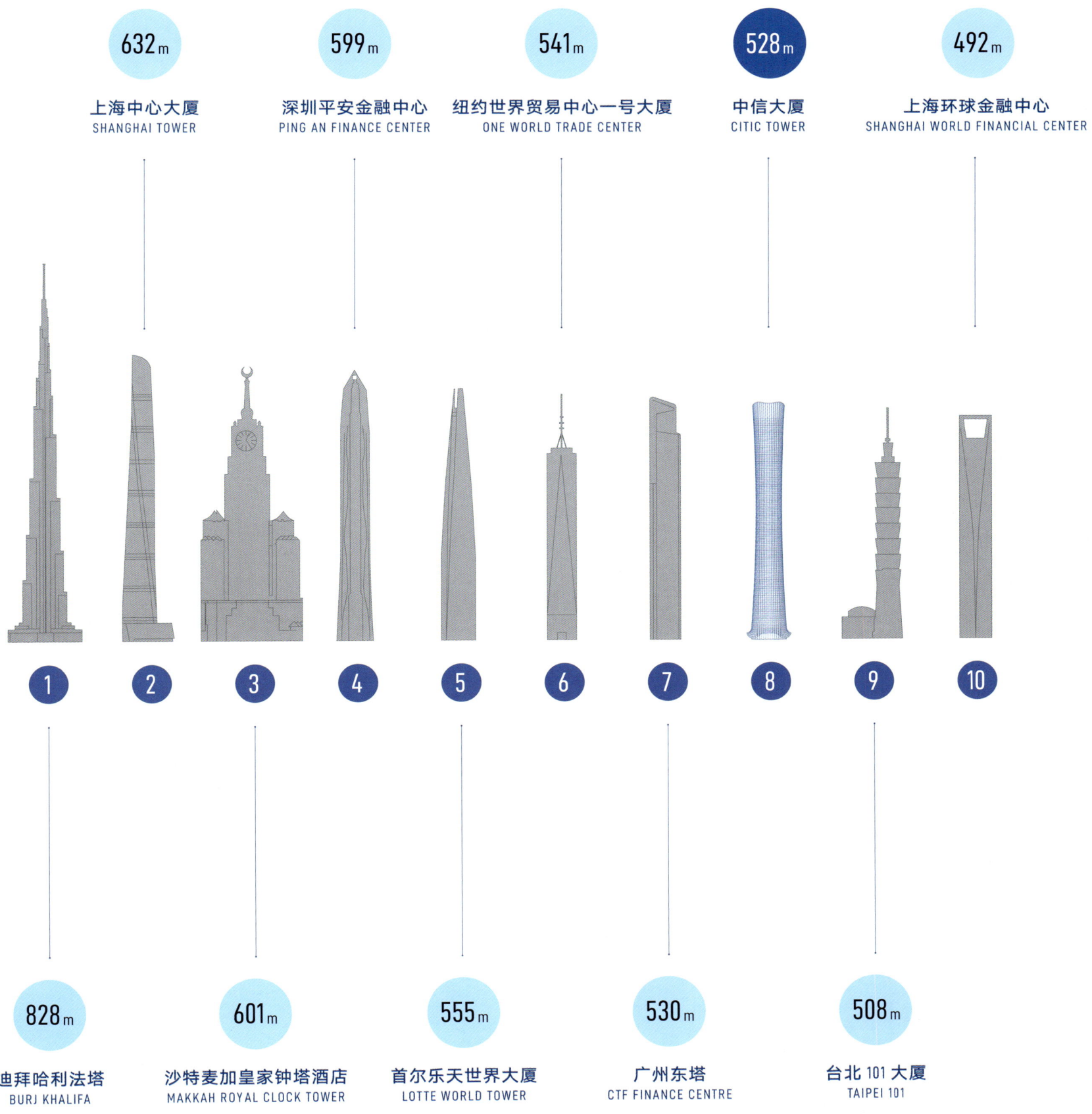

632m

上海中心大厦
SHANGHAI TOWER

599m

深圳平安金融中心
PING AN FINANCE CENTER

541m

纽约世界贸易中心一号大厦
ONE WORLD TRADE CENTER

528m

中信大厦
CITIC TOWER

492m

上海环球金融中心
SHANGHAI WORLD FINANCIAL CENTER

1   2   3   4   5   6   7   8   9   10

828m

迪拜哈利法塔
BURJ KHALIFA

601m

沙特麦加皇家钟塔酒店
MAKKAH ROYAL CLOCK TOWER

555m

首尔乐天世界大厦
LOTTE WORLD TOWER

530m

广州东塔
CTF FINANCE CENTRE

508m

台北 101 大厦
TAIPEI 101

建造 CITIC TOWER ‖ A NEW ICON FOR BEIJING
CONSTRUCTION

10 天 2138 根锚栓 165 个锚栓套架安装，总计起吊次数 253 吊次，最大误差为 8 mm

2014.06.18

结构总用钢量约 **140,000**t
The total steel consumption of CITIC Tower is about 140,000t

高强钢 **Q390GJ** 用量约 **50,000**t
High-strength steelQ390GJ is about 50,000t

智能化施工装备集成平台安装
INSTALLATION OF INTELLIGENT CONSTRUCTION EQUIPMENT INTEGRATED PLATFORM SYSTEM

2015.02.26

中国建筑 中国建筑

FAVELLE
FAVCO

BUILT BY CHINA   CONSTRUCTION RECORD OF CITIC TOWER

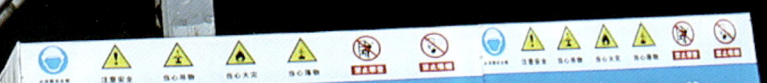

BUILT BY CHINA · CONSTRUCTION RECORD OF CITIC TOWER

2017.06.23

FAVCO

MAIN
50T 1-
100T
M1280D S

提质量　创品牌
加快建设质量强国

运营第一

4.1 超高层建筑的 EPCO 管理体系
EPCO Management System For Super High-Rise Buildings

大数据 Data

E
规划、设计
Engineering

P
采购
Procurement

C
建造
Construction

O
运维
Operation

延伸阅读 Links

P310

在中信大厦建设之初，中信集团领导便确立了"规划、开发、建造、运维一体化"的全生命周期管理理念。借鉴国际上较为成熟的 EPC 管理模式，中信和业开创性地在中信大厦开发中采用 EPCO（Engineering Procurement Construction Operation）模式，即："规划、设计、采购、建造、运维一体化管理模式"。中信和业投资有限公司在中信大厦开发建设中构建了设计总包联合体、施工总承包和机电总承包等单一责任主体管理模式，解决了国内超高层建筑普遍存在的"设计鸿沟"和"机电滞后"问题。从自持物业保值增值视角推进中信大厦开发建设，在设计和施工阶段就考虑大厦建成后的运维需求，针对大厦的安全、舒适、效率、节能及智能方面采用了大量新技术和新设备，在中信大厦全生命周期价值层面追求投入和产出的均衡点。

规划、设计 Engineering：

- 建设工程的总体策划（功能、业态、品质、性能等）
- 工程实施建设周期的测算和谋划（开发周期、现金流、参建方责任划分等预测）
- 工程设计工作（建筑外形、结构体系、流线、垂直运力、消防体系、机电设备等）
- 工程造价的测算（概算、开发期财务成本预测）及管控途径等

采购 Procurement：

- 合同体系构建
- 专业责任划分
- 满足本工程所需的设备、设施、材料的寻找、对比、筛选
- 针对特殊需求产品的研发，如超薄窗边空调机组的研制推动、加湿功能革新
- 采购体系的建立
- 预制、供应、运输计划
- 施工用时、安装用工测算
- 材料、设备的仓储及倒运库房等

建造 Construction：

- 工程施工管理体系构建、共享资源体系建立及调配
- 政府审批要件的准备及申办
- 利用监理对品质的旁站式监督
- 针对工程特性的辅助性措施
- 针对超高层垂直运力构建的思考
- 超高层施工降效超常的对策
- 为工程安全、高品质、造价受控的快速推进扫清障碍

运维 Operation：

- 充分利用规划、设计、建设阶段奠定的基础条件
- 建立一套安全、稳定的、低成本的运维体系
- 使工程设备、设施处于优良状态的技术和操控者能力的保障
- 不降低安全指数、不牺牲舒适度、快捷性地降低成本的运维体系
- 优质、高满意度的客户服务体系构建

At the beginning of CITIC Tower, leaders of CITIC Group had established the full life cycle management concept of "integration of planning, development, construction and operation and maintenance". Drawing lessons from a mature international EPC management mode, CITIC Heye pioneered the EPCO (Engineering, Procurement, Construction, Operation) mode in CITIC Tower development, namely: "The mode of integrated management of planning, design, procurement, construction and operation and maintenance". In the development and construction of CITIC Tower, CITIC Heye Investment Co., Ltd. has established the single responsibility theme management mode, such as design general contractor consortium, construction general contractor and electromechanical general contractor, etc., and solved the problems of "design gap" and "electromechanical lag" that commonly exist in China's super high-rise buildings. It promotes the development and construction of CITIC Tower from the perspective of self-maintained property preservation and appreciation, considers the operation and maintenance requirements of the building after it is completed during the design and construction phase, adopts a lot of new technologies and equipment for the safety, comfort, efficiency, energy conservation and intelligence of the building, and pursues the balance point of input and output in the full life cycle value of CITIC Tower.

Engineering:

- Overall planning of construction project (function, type of business, quality, performance, etc.)
- Calculation and planning of construction period of project implementation (prediction of development period, cash flow, division of responsibilities of participants, etc.)
- Engineering design (architectural form, structural system, streamline, vertical transportation capacity, fire-fighting system, electromechanical equipment, etc.)
- Calculation of project cost (budget estimate, financial cost forecast during the development period) and control methods.

Procurement:

- Contract system construction
- Division of professional responsibilities
- Finding, comparing and screening the equipment, facilities and materials needed for the project
- Research and development of products for special needs, such as the development of ultra-thin window-side air-conditioning units, humidification function innovation
- Establishment of purchasing system
- Prefabrication, supply, transportation plan
- Estimation of construction time and installation labor
- Warehousing and backhauling of materials and equipment, etc.

Construction:

- Building the construction management system and constructing and allocating the resource sharing system
- Preparation and application of documents for government examination and approval
- The supervisor's on-spot supervision of the quality
- Auxiliary measures for engineering characteristics
- Reflection on the construction of vertical transportation capacity of super high-rise buildings
- Countermeasures for extremely low efficiency of the super high-rise building construction
- Clearing obstacles for the rapid progress of engineering safety, high quality and cost control

Operation:

- Taking full advantage of the basic conditions laid during the planning, design and construction phases
- Establishing a safe, stable and low-cost operation and maintenance system
- Guarantee of the technology and operator's ability to keep the engineering equipment and facilities in good condition
- Cost-saving operation and maintenance system without reducing safety index, comfort and rapidity
- Construction of customer service system with high quality and high satisfaction

传统业主管理与中信大厦管理对比
Comparison between traditional owner management and CITIC Tower management

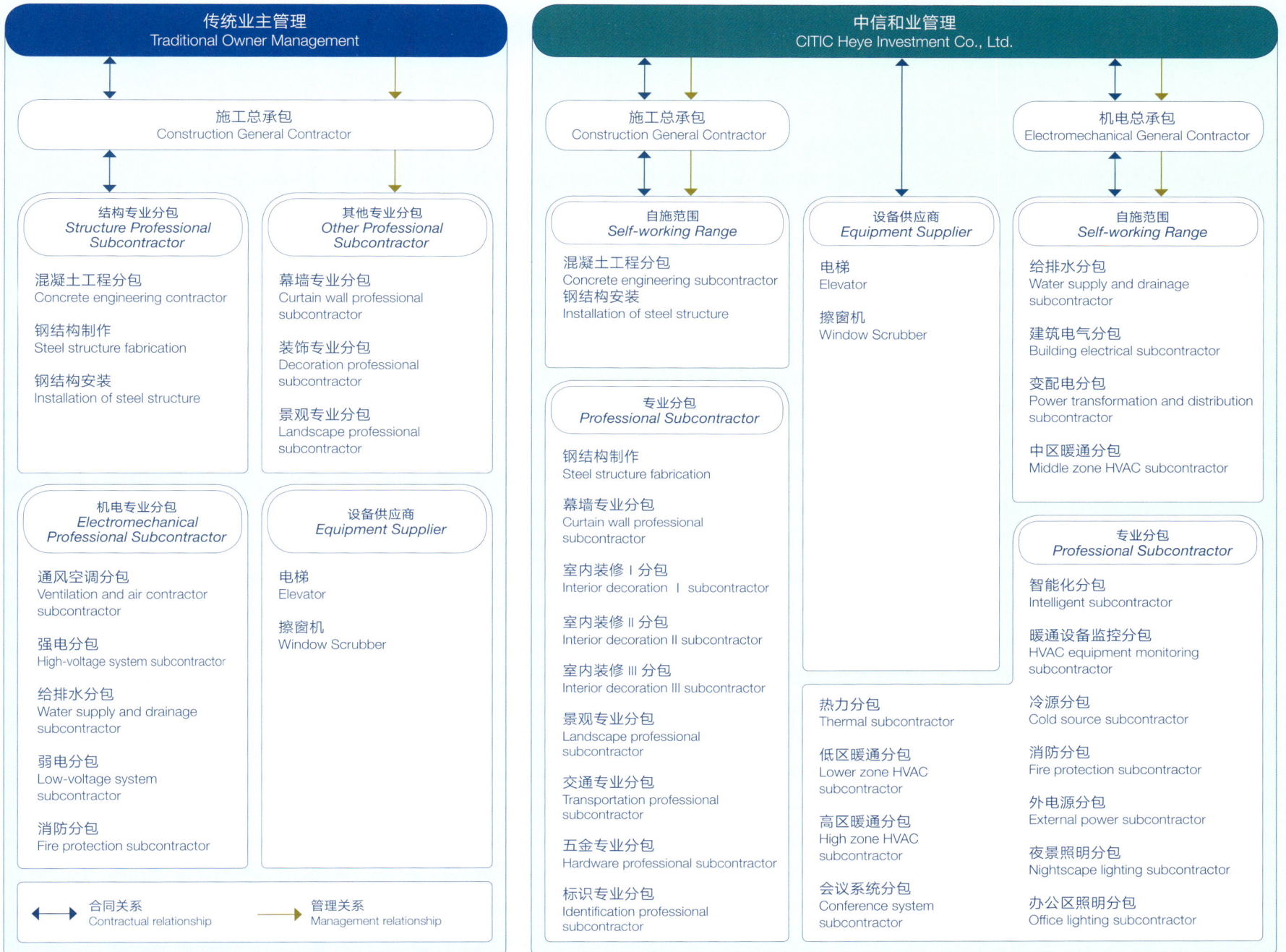

传统业主管理
Traditional Owner Management

施工总承包
Construction General Contractor

结构专业分包
Structure Professional Subcontractor

混凝土工程分包
Concrete engineering contractor

钢结构制作
Steel structure fabrication

钢结构安装
Installation of steel structure

其他专业分包
Other Professional Subcontractor

幕墙专业分包
Curtain wall professional subcontractor

装饰专业分包
Decoration professional subcontractor

景观专业分包
Landscape professional subcontractor

机电专业分包
Electromechanical Professional Subcontractor

通风空调分包
Ventilation and air contractor subcontractor

强电分包
High-voltage system subcontractor

给排水分包
Water supply and drainage subcontractor

弱电分包
Low-voltage system subcontractor

消防分包
Fire protection subcontractor

设备供应商
Equipment Supplier

电梯
Elevator

擦窗机
Window Scrubber

合同关系
Contractual relationship

管理关系
Management relationship

中信和业管理
CITIC Heye Investment Co., Ltd.

施工总承包
Construction General Contractor

自施范围
Self-working Range

混凝土工程分包
Concrete engineering subcontractor
钢结构安装
Installation of steel structure

专业分包
Professional Subcontractor

钢结构制作
Steel structure fabrication

幕墙专业分包
Curtain wall professional subcontractor

室内装修Ⅰ分包
Interior decoration Ⅰ subcontractor

室内装修Ⅱ分包
Interior decoration Ⅱ subcontractor

室内装修Ⅲ分包
Interior decoration Ⅲ subcontractor

景观专业分包
Landscape professional subcontractor

交通专业分包
Transportation professional subcontractor

五金专业分包
Hardware professional subcontractor

标识专业分包
Identification professional subcontractor

设备供应商
Equipment Supplier

电梯
Elevator

擦窗机
Window Scrubber

热力分包
Thermal subcontractor

低区暖通分包
Lower zone HVAC subcontractor

高区暖通分包
High zone HVAC subcontractor

会议系统分包
Conference system subcontractor

机电总承包
Electromechanical General Contractor

自施范围
Self-working Range

给排水分包
Water supply and drainage subcontractor

建筑电气分包
Building electrical subcontractor

变配电分包
Power transformation and distribution subcontractor

中区暖通分包
Middle zone HVAC subcontractor

专业分包
Professional Subcontractor

智能化分包
Intelligent subcontractor

暖通设备监控分包
HVAC equipment monitoring subcontractor

冷源分包
Cold source subcontractor

消防分包
Fire protection subcontractor

外电源分包
External power subcontractor

夜景照明分包
Nightscape lighting subcontractor

办公区照明分包
Office lighting subcontractor

传统设计管理模式与中信大厦设计管理模式对比
Comparison between traditional design management mode and CITIC Tower design management mode

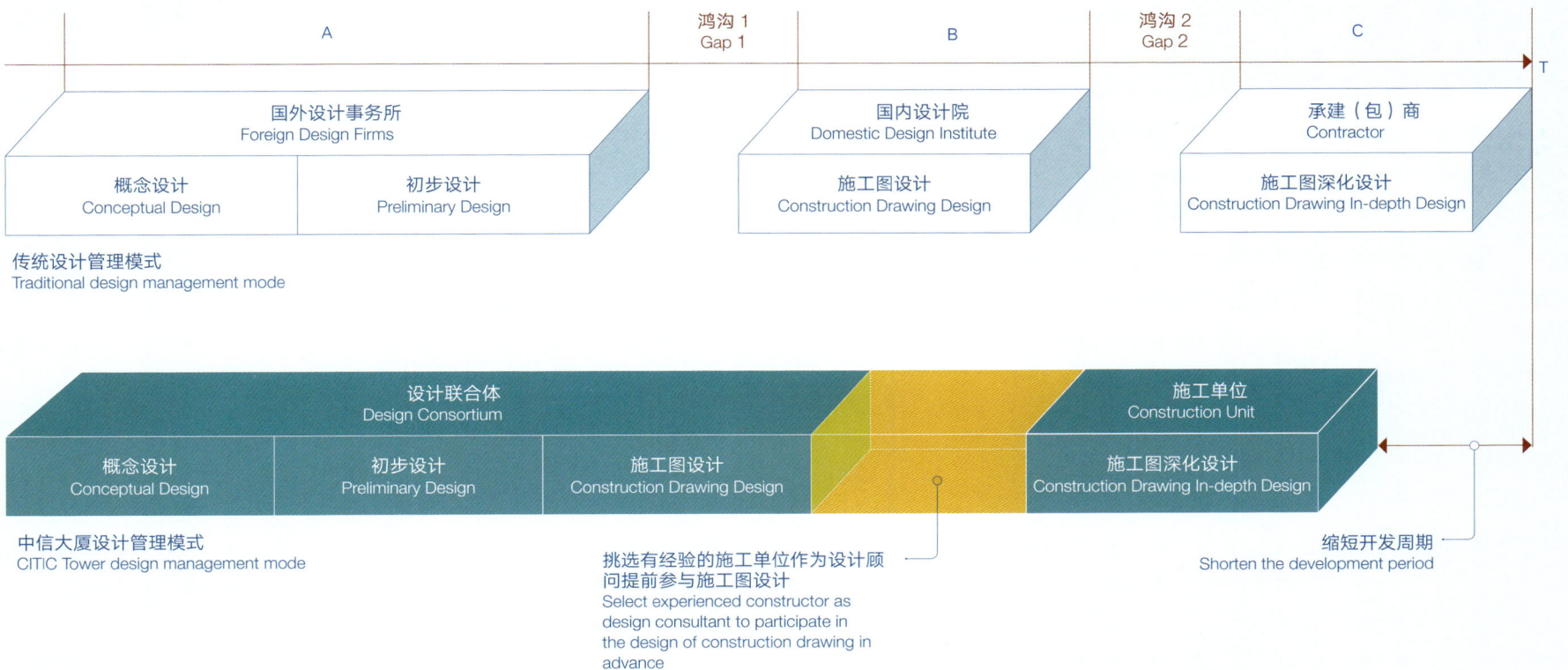

A

鸿沟1
Gap 1

B

鸿沟2
Gap 2

C

T

国外设计事务所
Foreign Design Firms

概念设计
Conceptual Design

初步设计
Preliminary Design

国内设计院
Domestic Design Institute

施工图设计
Construction Drawing Design

承建（包）商
Contractor

施工图深化设计
Construction Drawing In-depth Design

传统设计管理模式
Traditional design management mode

设计联合体
Design Consortium

施工单位
Construction Unit

概念设计
Conceptual Design

初步设计
Preliminary Design

施工图设计
Construction Drawing Design

施工图深化设计
Construction Drawing In-depth Design

中信大厦设计管理模式
CITIC Tower design management mode

挑选有经验的施工单位作为设计顾问提前参与施工图设计
Select experienced constructor as design consultant to participate in the design of construction drawing in advance

缩短开发周期
Shorten the development period

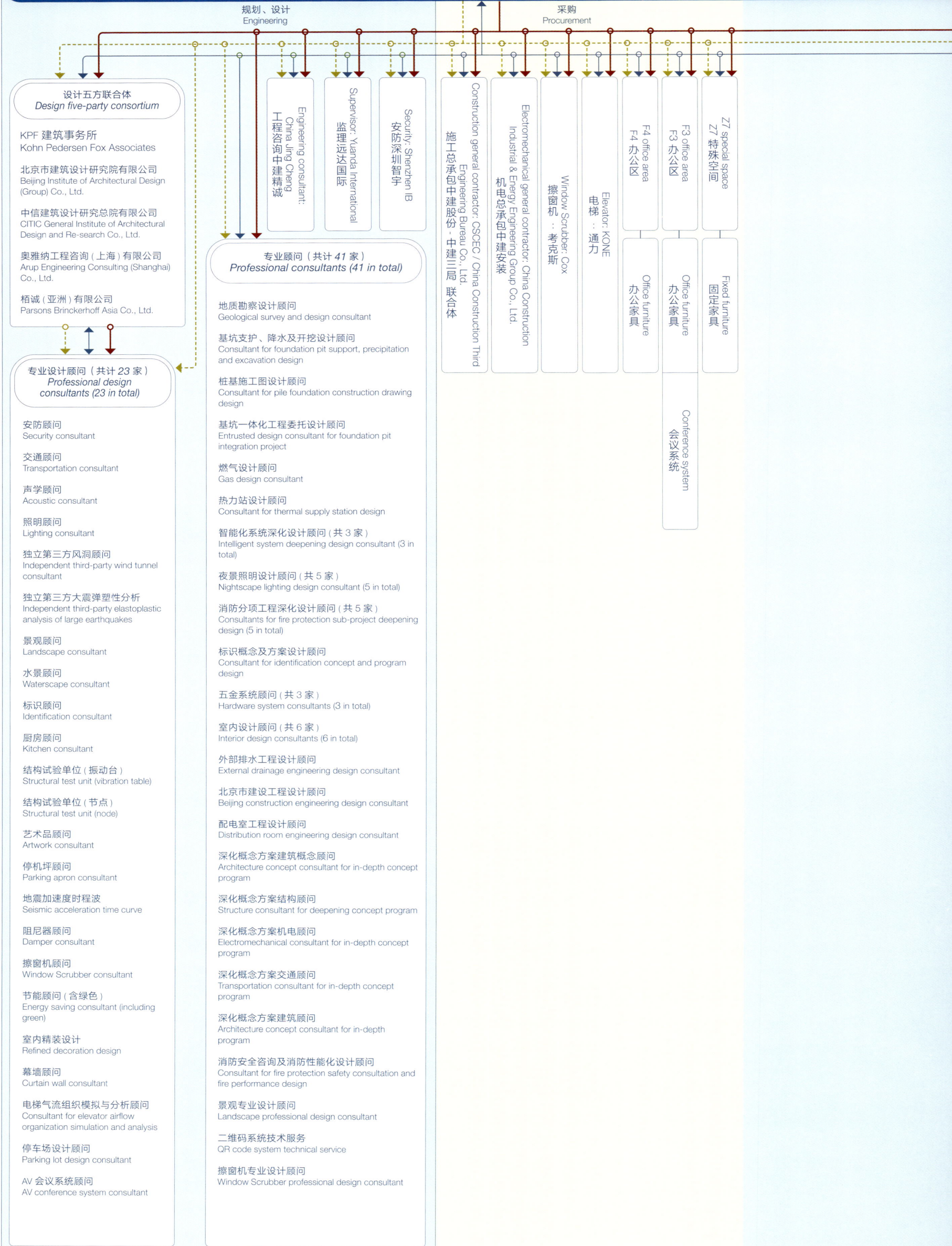

中信和业投资有限公司
CITIC Heye Investment Co., Ltd.

规划、设计
Engineering

采购
Procurement

设计五方联合体
Design five-party consortium

KPF 建筑事务所
Kohn Pedersen Fox Associates

北京市建筑设计研究院有限公司
Beijing Institute of Architectural Design (Group) Co., Ltd.

中信建筑设计研究总院有限公司
CITIC General Institute of Architectural Design and Re-search Co., Ltd.

奥雅纳工程咨询（上海）有限公司
Arup Engineering Consulting (Shanghai) Co., Ltd.

梧诚（亚洲）有限公司
Parsons Brinckerhoff Asia Co., Ltd.

工程咨询中建精诚
Engineering consultant: China Jing Cheng

监理远达国际
Supervisor: Yuanda International

安防深圳智宇
Security: Shenzhen IB

施工总承包中建股份-中建三局 联合体
Construction general contractor: CSCEC / China Construction Engineering Bureau Co., Ltd. / China Construction Third

机电总承包中建安装
Electromechanical general contractor: China Construction Industrial & Energy Enginnering Group Co., Ltd.

擦窗机：考克斯
Window Scrubber: Cox

电梯：通力
Elevator: KONE

F4 办公区
F4 office area

办公家具
Office furniture

F3 办公区
F3 office area

办公家具
Office furniture

会议系统
Conference system

Z7 特殊空间
Z7 special space

固定家具
Fixed furniture

专业设计顾问（共计 23 家）
Professional design consultants (23 in total)

安防顾问
Security consultant

交通顾问
Transportation consultant

声学顾问
Acoustic consultant

照明顾问
Lighting consultant

独立第三方风洞顾问
Independent third-party wind tunnel consultant

独立第三方大震弹塑性分析
Independent third-party elastoplastic analysis of large earthquakes

景观顾问
Landscape consultant

水景顾问
Waterscape consultant

标识顾问
Identification consultant

厨房顾问
Kitchen consultant

结构试验单位（振动台）
Structural test unit (vibration table)

结构试验单位（节点）
Structural test unit (node)

艺术品顾问
Artwork consultant

停机坪顾问
Parking apron consultant

地震加速度时程波
Seismic acceleration time curve

阻尼器顾问
Damper consultant

擦窗机顾问
Window Scrubber consultant

节能顾问（含绿色）
Energy saving consultant (including green)

室内精装设计
Refined decoration design

幕墙顾问
Curtain wall consultant

电梯气流组织模拟与分析顾问
Consultant for elevator airflow organization simulation and analysis

停车场设计顾问
Parking lot design consultant

AV 会议系统顾问
AV conference system consultant

专业顾问（共计 41 家）
Professional consultants (41 in total)

地质勘察设计顾问
Geological survey and design consultant

基坑支护、降水及开挖设计顾问
Consultant for foundation pit support, precipitation and excavation design

桩基施工图设计顾问
Consultant for pile foundation construction drawing design

基坑一体化工程委托设计顾问
Entrusted design consultant for foundation pit integration project

燃气设计顾问
Gas design consultant

热力站设计顾问
Consultant for thermal supply station design

智能化系统深化设计顾问（共 3 家）
Intelligent system deepening design consultant (3 in total)

夜景照明设计顾问（共 5 家）
Nightscape lighting design consultant (5 in total)

消防分项工程深化设计顾问（共 5 家）
Consultants for fire protection sub-project deepening design (5 in total)

标识概念及方案设计顾问
Consultant for identification concept and program design

五金系统顾问（共 3 家）
Hardware system consultants (3 in total)

室内设计顾问（共 6 家）
Interior design consultants (6 in total)

外部排水工程设计顾问
External drainage engineering design consultant

北京市建设工程设计顾问
Beijing construction engineering design consultant

配电室工程设计顾问
Distribution room engineering design consultant

深化概念方案建筑概念顾问
Architecture concept consultant for in-depth concept program

深化概念方案结构顾问
Structure consultant for deepening concept program

深化概念方案机电顾问
Electromechanical consultant for in-depth concept program

深化概念方案交通顾问
Transportation consultant for in-depth concept program

深化概念方案建筑顾问
Architecture concept consultant for in-depth program

消防安全咨询及消防性能化设计顾问
Consultant for fire protection safety consultation and fire performance design

景观专业设计顾问
Landscape professional design consultant

二维码系统技术服务
QR code system technical service

擦窗机专业设计顾问
Window Scrubber professional design consultant

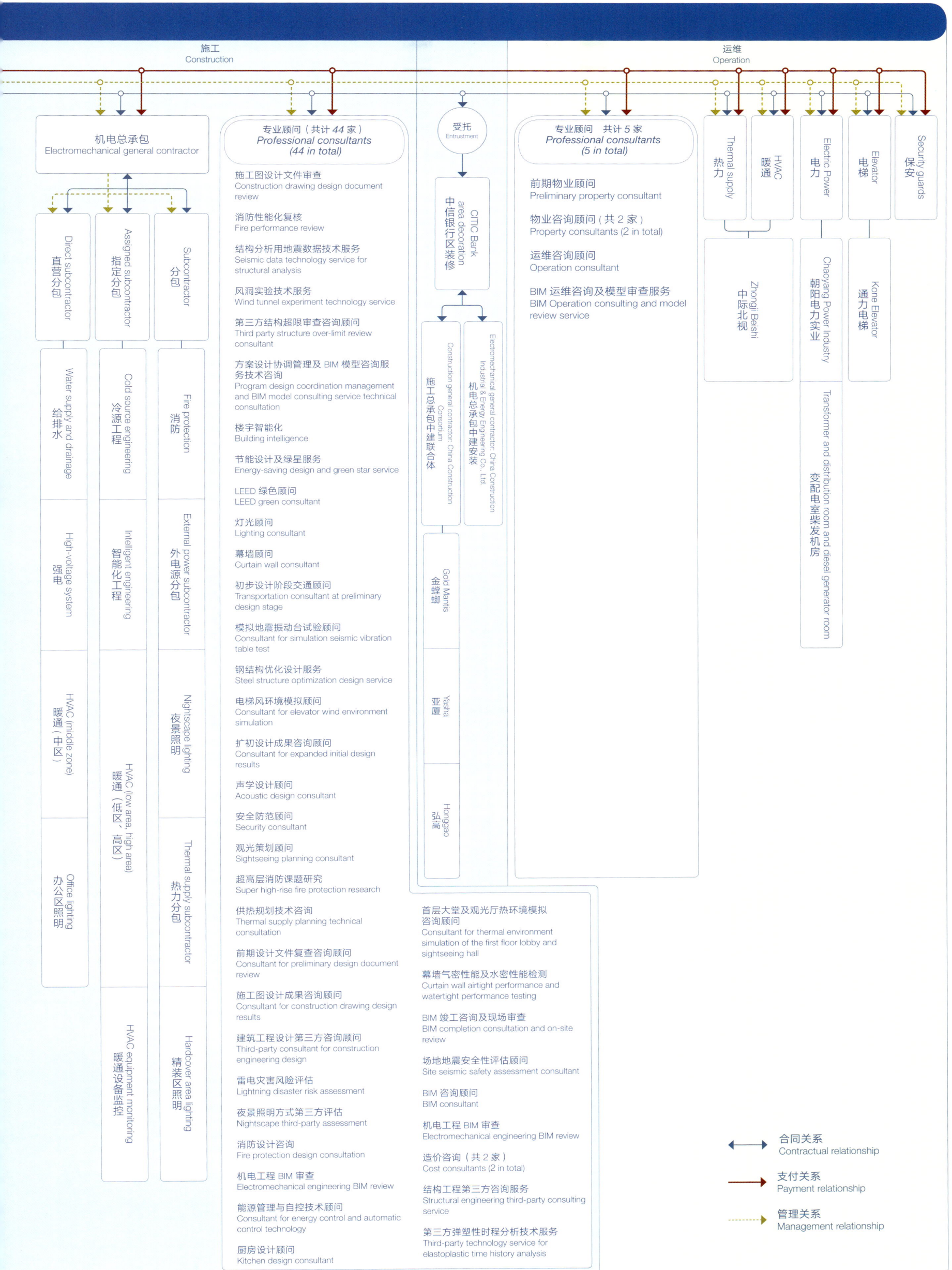

施工
Construction

运维
Operation

机电总承包
Electromechanical general contractor

专业顾问（共计 44 家）
Professional consultants
(44 in total)

受托
Entrustment

专业顾问 共计 5 家
Professional consultants
(5 in total)

热力
Thermal supply

暖通
HVAC

电力
Electric Power

电梯
Elevator

保安
Security guards

直营分包
Direct subcontractor

指定分包
Assigned subcontractor

分包
Subcontractor

CITIC Bank area decoration
中信银行行区装修

施工图设计文件审查
Construction drawing design document review

消防性能化复核
Fire performance review

结构分析用地震数据技术服务
Seismic data technology service for structural analysis

风洞实验技术服务
Wind tunnel experiment technology service

第三方结构超限审查咨询顾问
Third party structure over-limit review consultant

方案设计协调管理及 BIM 模型咨询服务技术咨询
Program design coordination management and BIM model consulting service technical consultation

楼宇智能化
Building intelligence

节能设计及绿星服务
Energy-saving design and green star service

LEED 绿色顾问
LEED green consultant

灯光顾问
Lighting consultant

幕墙顾问
Curtain wall consultant

初步设计阶段交通顾问
Transportation consultant at preliminary design stage

模拟地震振动台试验顾问
Consultant for simulation seismic vibration table test

钢结构优化设计服务
Steel structure optimization design service

电梯风环境模拟顾问
Consultant for elevator wind environment simulation

扩初设计成果咨询顾问
Consultant for expanded initial design results

声学设计顾问
Acoustic design consultant

安全防范顾问
Security consultant

观光策划顾问
Sightseeing planning consultant

超高层消防课题研究
Super high-rise fire protection research

供热规划技术咨询
Thermal supply planning technical consultation

前期设计文件复查咨询顾问
Consultant for preliminary design document review

施工图设计成果咨询顾问
Consultant for construction drawing design results

建筑工程设计第三方咨询顾问
Third-party consultant for construction engineering design

雷电灾害风险评估
Lightning disaster risk assessment

夜景照明方式第三方评估
Nightscape third-party assessment

消防设计咨询
Fire protection design consultation

机电工程 BIM 审查
Electromechanical engineering BIM review

能源管理与自控技术顾问
Consultant for energy control and automatic control technology

厨房设计顾问
Kitchen design consultant

首层大堂及观光厅热环境模拟咨询顾问
Consultant for thermal environment simulation of the first floor lobby and sightseeing hall

幕墙气密性能及水密性能检测
Curtain wall airtight performance and watertight performance testing

BIM 竣工咨询及现场审查
BIM completion consultation and on-site review

场地地震安全性评估顾问
Site seismic safety assessment consultant

BIM 咨询顾问
BIM consultant

机电工程 BIM 审查
Electromechanical engineering BIM review

造价咨询（共 2 家）
Cost consultants (2 in total)

结构工程第三方咨询服务
Structural engineering third-party consulting service

第三方弹塑性时程分析技术服务
Third-party technology service for elastoplastic time history analysis

Consortium 施工总承包中建联合体
Consortium Construction general contractor: China Construction

机电总承包中建安装
Electromechanical general contractor: China Construction Industrial & Energy Engineering Co., Ltd

金螳螂
Gold Mantis

亚厦
Yasha

弘高
Honggao

前期物业顾问
Preliminary property consultant

物业咨询顾问（共 2 家）
Property consultants (2 in total)

运维咨询顾问
Operation consultant

BIM 运维咨询及模型审查服务
BIM Operation consulting and model review service

中际北视
Zhongji Beishi

朝阳电力实业
Chaoyang Power Industry

通力电梯
Kone Elevator

变配电室及柴发机房
Transformer and distribution room and diesel generator room

给排水
Water supply and drainage

强电
High-voltage system

暖通（中区）
HVAC (middle zone)

办公区照明
Office lighting

冷源工程
Cold source engineering

智能化工程
Intelligent engineering

暖通（低区、高区）
HVAC (low area, high area)

暖通设备监控
HVAC equipment monitoring

消防
Fire protection

外电源分包
External power subcontractor

夜景照明
Nightscape lighting

热力分包
Thermal supply subcontractor

精装区照明
Hardcover area lighting

合同关系
Contractual relationship

支付关系
Payment relationship

管理关系
Management relationship

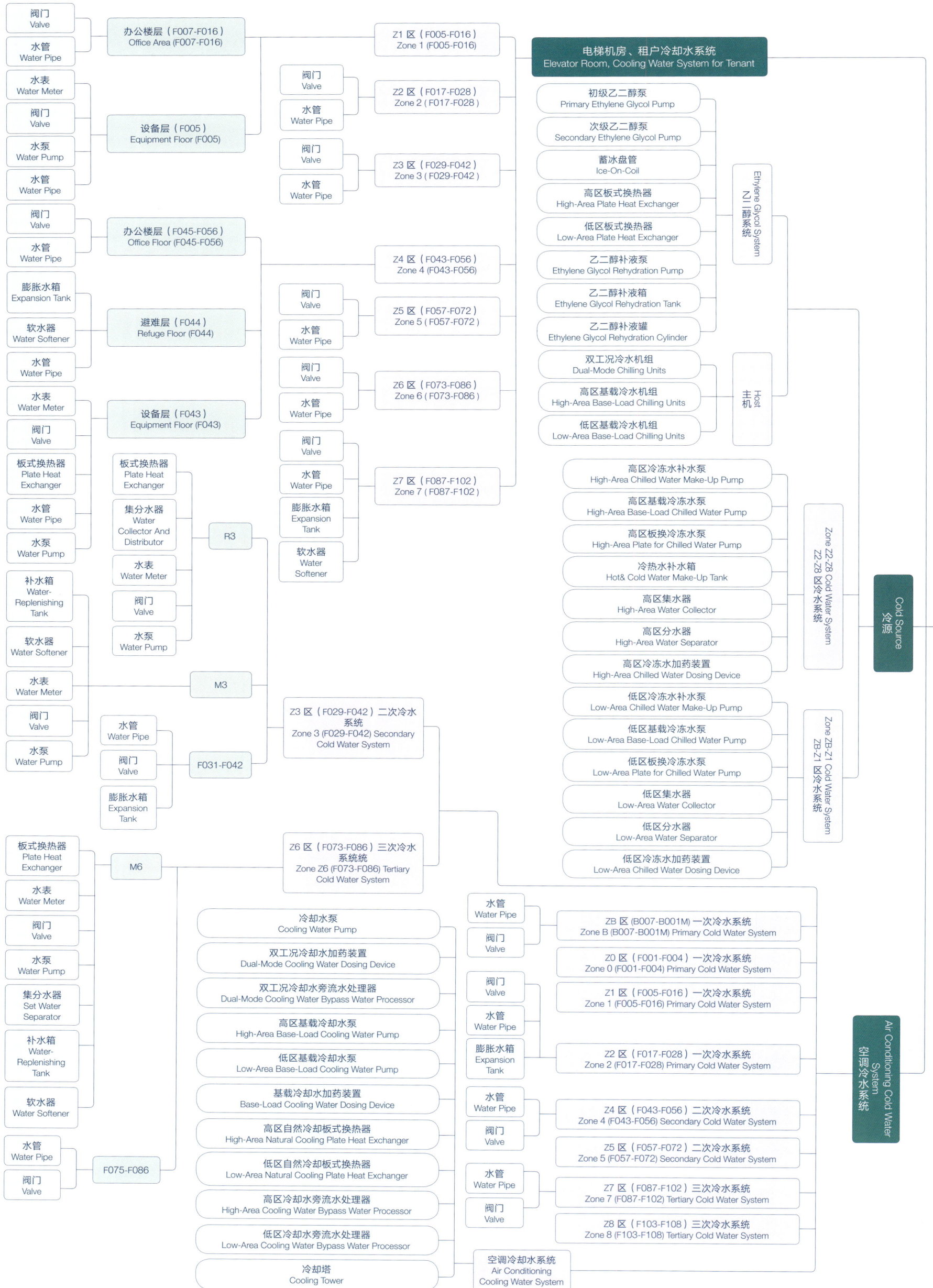

阀门
Valve

水管
Water Pipe

办公楼层（F007-F016）
Office Area (F007-F016)

Z1 区（F005-F016）
Zone 1 (F005-F016)

电梯机房、租户冷却水系统
Elevator Room, Cooling Water System for Tenant

水表
Water Meter

阀门
Valve

阀门
Valve

水管
Water Pipe

Z2 区（F017-F028）
Zone 2 (F017-F028)

初级乙二醇泵
Primary Ethylene Glycol Pump

水泵
Water Pump

次级乙二醇泵
Secondary Ethylene Glycol Pump

水管
Water Pipe

设备层（F005）
Equipment Floor (F005)

蓄冰盘管
Ice-On-Coil

阀门
Valve

水管
Water Pipe

Z3 区（F029-F042）
Zone 3 (F029-F042)

高区板式换热器
High-Area Plate Heat Exchanger

阀门
Valve

低区板式换热器
Low-Area Plate Heat Exchanger

水管
Water Pipe

办公楼层（F045-F056）
Office Floor (F045-F056)

乙二醇补液泵
Ethylene Glycol Rehydration Pump

乙二醇补液箱
Ethylene Glycol Rehydration Tank

Z4 区（F043-F056）
Zone 4 (F043-F056)

乙二醇补液罐
Ethylene Glycol Rehydration Cylinder

Ethylene Glycol System
乙二醇系统

膨胀水箱
Expansion Tank

阀门
Valve

Z5 区（F057-F072）
Zone 5 (F057-F072)

双工况冷水机组
Dual-Mode Chilling Units

软水器
Water Softener

水管
Water Pipe

避难层（F044）
Refuge Floor (F044)

高区基载冷水机组
High-Area Base-Load Chilling Units

水管
Water Pipe

阀门
Valve

低区基载冷水机组
Low-Area Base-Load Chilling Units

Host
主机

水表
Water Meter

水管
Water Pipe

Z6 区（F073-F086）
Zone 6 (F073-F086)

阀门
Valve

设备层（F043）
Equipment Floor (F043)

水管
Water Pipe

高区冷冻水补水泵
High-Area Chilled Water Make-Up Pump

板式换热器
Plate Heat Exchanger

板式换热器
Plate Heat Exchanger

阀门
Valve

高区基载冷冻水泵
High-Area Base-Load Chilled Water Pump

水管
Water Pipe

集分水器
Water Collector And Distributor

水管
Water Pipe

Z7 区（F087-F102）
Zone 7 (F087-F102)

高区板换冷冻水泵
High-Area Plate for Chilled Water Pump

水泵
Water Pump

R3

水表
Water Meter

膨胀水箱
Expansion Tank

冷热水补水箱
Hot& Cold Water Make-Up Tank

补水箱
Water-Replenishing Tank

阀门
Valve

软水器
Water Softener

高区集水器
High-Area Water Collector

软水器
Water Softener

水泵
Water Pump

高区分水器
High-Area Water Separator

水表
Water Meter

M3

高区冷冻水加药装置
High-Area Chilled Water Dosing Device

阀门
Valve

低区冷冻水补水泵
Low-Area Chilled Water Make-Up Pump

水管
Water Pipe

Z3 区（F029-F042）二次冷水系统
Zone 3 (F029-F042) Secondary Cold Water System

低区基载冷冻水泵
Low-Area Base-Load Chilled Water Pump

水泵
Water Pump

阀门
Valve

F031-F042

低区板换冷冻水泵
Low-Area Plate for Chilled Water Pump

膨胀水箱
Expansion Tank

低区集水器
Low-Area Water Collector

Zone Z2-Z8 Cold Water System
Z2-Z8区冷水系统

低区分水器
Low-Area Water Separator

板式换热器
Plate Heat Exchanger

M6

Z6 区（F073-F086）三次冷水系统
Zone Z6 (F073-F086) Tertiary Cold Water System

低区冷冻水加药装置
Low-Area Chilled Water Dosing Device

Zone ZB-Z1 Cold Water System
ZB-Z1区冷水系统

水表
Water Meter

阀门
Valve

冷却水泵
Cooling Water Pump

水管
Water Pipe

水泵
Water Pump

双工况冷却水加药装置
Dual-Mode Cooling Water Dosing Device

阀门
Valve

ZB 区（B007-B001M）一次冷水系统
Zone B (B007-B001M) Primary Cold Water System

集分水器
Set Water Separator

双工况冷却水旁流水处理器
Dual-Mode Cooling Water Bypass Water Processor

Z0 区（F001-F004）一次冷水系统
Zone 0 (F001-F004) Primary Cold Water System

补水箱
Water-Replenishing Tank

高区基载冷却水泵
High-Area Base-Load Cooling Water Pump

阀门
Valve

Z1 区（F005-F016）一次冷水系统
Zone 1 (F005-F016) Primary Cold Water System

软水器
Water Softener

低区基载冷却水泵
Low-Area Base-Load Cooling Water Pump

水管
Water Pipe

基载冷却水加药装置
Base-Load Cooling Water Dosing Device

膨胀水箱
Expansion Tank

Z2 区（F017-F028）一次冷水系统
Zone 2 (F017-F028) Primary Cold Water System

水管
Water Pipe

高区自然冷却板式换热器
High-Area Natural Cooling Plate Heat Exchanger

阀门
Valve

F075-F086

低区自然冷却板式换热器
Low-Area Natural Cooling Plate Heat Exchanger

水管
Water Pipe

Z4 区（F043-F056）二次冷水系统
Zone 4 (F043-F056) Secondary Cold Water System

高区冷却水旁流水处理器
High-Area Cooling Water Bypass Water Processor

阀门
Valve

Z5 区（F057-F072）二次冷水系统
Zone 5 (F057-F072) Secondary Cold Water System

低区冷却水旁流水处理器
Low-Area Cooling Water Bypass Water Processor

水管
Water Pipe

Z7 区（F087-F102）三次冷水系统
Zone 7 (F087-F102) Tertiary Cold Water System

阀门
Valve

冷却塔
Cooling Tower

空调冷却水系统
Air Conditioning Cooling Water System

Z8 区（F103-F108）三次冷水系统
Zone 8 (F103-F108) Tertiary Cold Water System

Cold Source
冷源

Air Conditioning Cold Water System
空调冷水系统

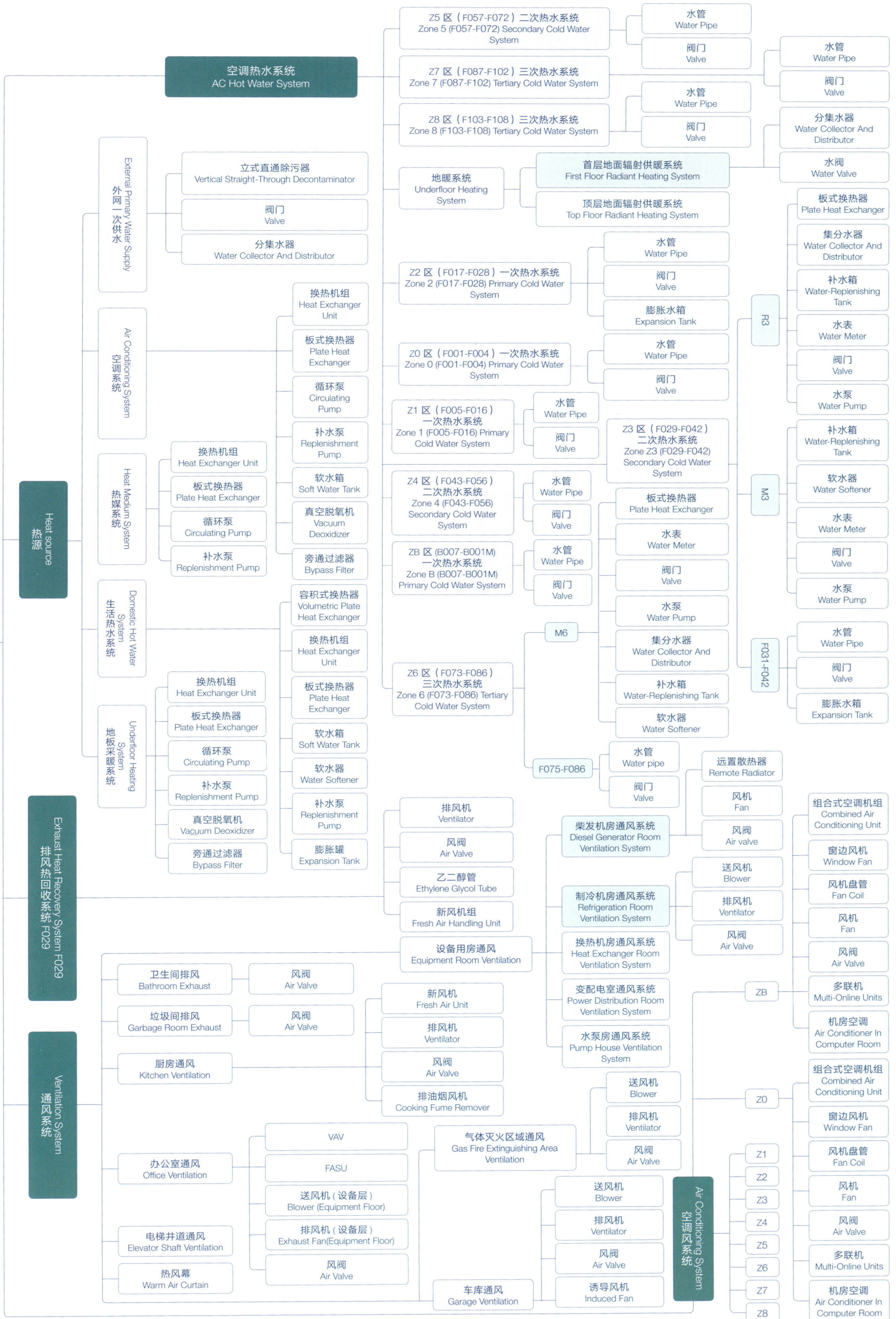

中信大厦暖通系统 CITIC Tower HVAC System

热源 Heat source

外网一次供水 External Primary Water Supply
- 立式直通除污器 Vertical Straight-Through Decontaminator
- 阀门 Valve
- 分集水器 Water Collector And Distributor

空调系统 Air Conditioning System
- 换热机组 Heat Exchanger Unit
- 板式换热器 Plate Heat Exchanger
- 循环泵 Circulating Pump
- 补水泵 Replenishment Pump
- 软水箱 Soft Water Tank
- 真空脱氧机 Vacuum Deoxidizer
- 旁通过滤器 Bypass Filter

热媒系统 Heat Medium System
- 换热机组 Heat Exchanger Unit
- 板式换热器 Plate Heat Exchanger
- 循环泵 Circulating Pump
- 补水泵 Replenishment Pump

生活热水系统 Domestic Hot Water System
- 容积式换热器 Volumetric Plate Heat Exchanger
- 换热机组 Heat Exchanger Unit
- 板式换热器 Plate Heat Exchanger
- 软水箱 Soft Water Tank
- 软水器 Water Softener
- 补水泵 Replenishment Pump
- 膨胀罐 Expansion Tank

地板采暖系统 Underfloor Heating System
- 换热机组 Heat Exchanger Unit
- 板式换热器 Plate Heat Exchanger
- 循环泵 Circulating Pump
- 补水泵 Replenishment Pump
- 真空脱氧机 Vacuum Deoxidizer
- 旁通过滤器 Bypass Filter

空调热水系统 AC Hot Water System
- Z5区（F057-F072）二次热水系统 Zone 5 (F057-F072) Secondary Cold Water System
  - 水管 Water Pipe
  - 阀门 Valve
    - 水管 Water Pipe
    - 阀门 Valve
- Z7区（F087-F102）三次热水系统 Zone 7 (F087-F102) Tertiary Cold Water System
  - 水管 Water Pipe
  - 阀门 Valve
    - 分集水器 Water Collector And Distributor
    - 水阀 Water Valve
- Z8区（F103-F108）三次热水系统 Zone 8 (F103-F108) Tertiary Cold Water System

地暖系统 Underfloor Heating System
- 首层地面辐射供暖系统 First Floor Radiant Heating System
- 顶层地面辐射供暖系统 Top Floor Radiant Heating System

- Z2区（F017-F028）一次热水系统 Zone 2 (F017-F028) Primary Cold Water System
  - 水管 Water Pipe
  - 阀门 Valve
  - 膨胀水箱 Expansion Tank
- Z0区（F001-F004）一次热水系统 Zone 0 (F001-F004) Primary Cold Water System
  - 水管 Water Pipe
  - 阀门 Valve
- Z1区（F005-F016）一次热水系统 Zone 1 (F005-F016) Primary Cold Water System
  - 水管 Water Pipe
  - 阀门 Valve
- Z4区（F043-F056）二次热水系统 Zone 4 (F043-F056) Secondary Cold Water System
  - 水管 Water Pipe
  - 阀门 Valve
- ZB区（B007-B001M）一次热水系统 Zone B (B007-B001M) Primary Cold Water System
  - 水管 Water Pipe
  - 阀门 Valve

R3
- 板式换热器 Plate Heat Exchanger
- 集分水器 Water Collector And Distributor
- 补水箱 Water-Replenishing Tank
- 水表 Water Meter
- 阀门 Valve
- 水泵 Water Pump

- Z3区（F029-F042）二次热水系统 Zone Z3 (F029-F042) Secondary Cold Water System
  - 板式换热器 Plate Heat Exchanger
  - 水表 Water Meter
  - 阀门 Valve
  - 水泵 Water Pump
  - 集分水器 Water Collector And Distributor
  - 补水箱 Water-Replenishing Tank
  - 软水器 Water Softener

M3
- 补水箱 Water-Replenishing Tank
- 软水器 Water Softener
- 水表 Water Meter
- 阀门 Valve
- 水泵 Water Pump

F031-F042
- 水管 Water Pipe
- 阀门 Valve
- 膨胀水箱 Expansion Tank

M6
- Z6区（F073-F086）三次热水系统 Zone 6 (F073-F086) Tertiary Cold Water System

F075-F086
- 水管 Water pipe
- 阀门 Valve
  - 远置散热器 Remote Radiator

排风热回收系统 F029 Exhaust Heat Recovery System F029
- 排风机 Ventilator
- 风阀 Air Valve
- 乙二醇管 Ethylene Glycol Tube
- 新风机组 Fresh Air Handling Unit

通风系统 Ventilation System

卫生间排风 Bathroom Exhaust
- 风阀 Air Valve

垃圾间排风 Garbage Room Exhaust
- 风阀 Air Valve

厨房通风 Kitchen Ventilation
- 新风机 Fresh Air Unit
- 排风机 Ventilator
- 风阀 Air Valve
- 排油烟风机 Cooking Fume Remover

办公室通风 Office Ventilation
- VAV
- FASU
- 送风机（设备层）Blower (Equipment Floor)
- 排风机（设备层）Exhaust Fan (Equipment Floor)
- 风阀 Air Valve

电梯井道通风 Elevator Shaft Ventilation

热风幕 Warm Air Curtain

设备用房通风 Equipment Room Ventilation
- 柴发机房通风系统 Diesel Generator Room Ventilation System
  - 风机 Fan
  - 风阀 Air valve
- 制冷机房通风系统 Refrigeration Room Ventilation System
  - 送风机 Blower
  - 排风机 Ventilator
  - 风阀 Air Valve
- 换热机房通风系统 Heat Exchanger Room Ventilation System
- 变配电室通风系统 Power Distribution Room Ventilation System
- 水泵房通风系统 Pump House Ventilation System

气体灭火区域通风 Gas Fire Extinguishing Area Ventilation

车库通风 Garage Ventilation
- 送风机 Blower
- 排风机 Ventilator
- 风阀 Air Valve
- 诱导风机 Induced Fan

空调风系统 Air Conditioning System
- ZB
  - 组合式空调机组 Combined Air Conditioning Unit
  - 窗边风机 Window Fan
  - 风机盘管 Fan Coil
  - 风机 Fan
  - 风阀 Air Valve
  - 多联机 Multi-Online Units
  - 机房空调 Air Conditioner In Computer Room
- Z0
  - 组合式空调机组 Combined Air Conditioning Unit
  - 窗边风机 Window Fan
- Z1
- Z2
- Z3
- Z4
- Z5
- Z6
- Z7
- Z8
  - 风机盘管 Fan Coil
  - 风机 Fan
  - 风阀 Air Valve
  - 多联机 Multi-Online Units
  - 机房空调 Air Conditioner In Computer Room

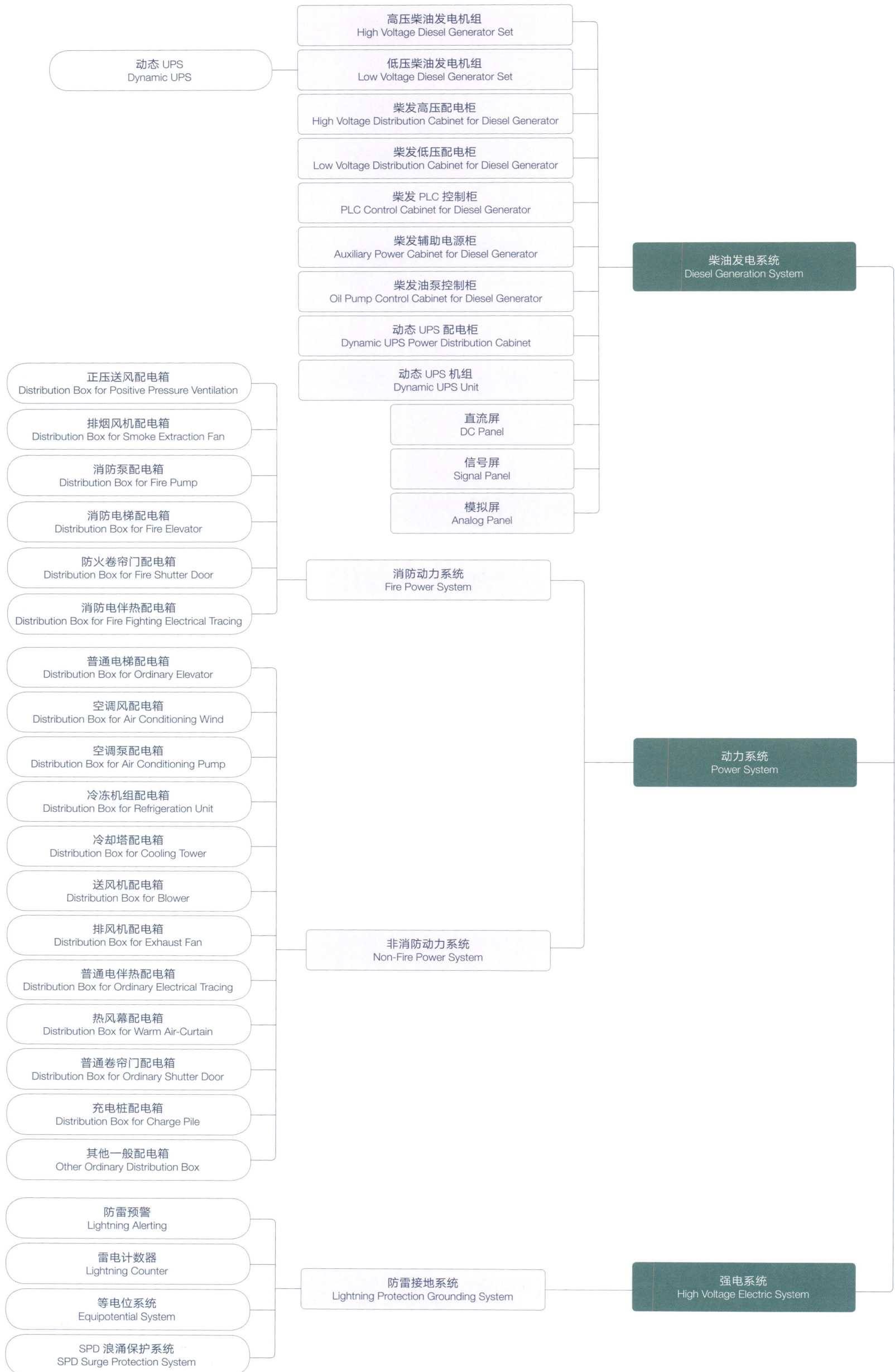

动态 UPS
Dynamic UPS

高压柴油发电机组
High Voltage Diesel Generator Set

低压柴油发电机组
Low Voltage Diesel Generator Set

柴发高压配电柜
High Voltage Distribution Cabinet for Diesel Generator

柴发低压配电柜
Low Voltage Distribution Cabinet for Diesel Generator

柴发 PLC 控制柜
PLC Control Cabinet for Diesel Generator

柴发辅助电源柜
Auxiliary Power Cabinet for Diesel Generator

柴发油泵控制柜
Oil Pump Control Cabinet for Diesel Generator

动态 UPS 配电柜
Dynamic UPS Power Distribution Cabinet

动态 UPS 机组
Dynamic UPS Unit

直流屏
DC Panel

信号屏
Signal Panel

模拟屏
Analog Panel

柴油发电系统
Diesel Generation System

正压送风配电箱
Distribution Box for Positive Pressure Ventilation

排烟风机配电箱
Distribution Box for Smoke Extraction Fan

消防泵配电箱
Distribution Box for Fire Pump

消防电梯配电箱
Distribution Box for Fire Elevator

防火卷帘门配电箱
Distribution Box for Fire Shutter Door

消防电伴热配电箱
Distribution Box for Fire Fighting Electrical Tracing

消防动力系统
Fire Power System

普通电梯配电箱
Distribution Box for Ordinary Elevator

空调风配电箱
Distribution Box for Air Conditioning Wind

空调泵配电箱
Distribution Box for Air Conditioning Pump

冷冻机组配电箱
Distribution Box for Refrigeration Unit

冷却塔配电箱
Distribution Box for Cooling Tower

送风机配电箱
Distribution Box for Blower

排风机配电箱
Distribution Box for Exhaust Fan

普通电伴热配电箱
Distribution Box for Ordinary Electrical Tracing

热风幕配电箱
Distribution Box for Warm Air-Curtain

普通卷帘门配电箱
Distribution Box for Ordinary Shutter Door

充电桩配电箱
Distribution Box for Charge Pile

其他一般配电箱
Other Ordinary Distribution Box

非消防动力系统
Non-Fire Power System

动力系统
Power System

防雷预警
Lightning Alerting

雷电计数器
Lightning Counter

等电位系统
Equipotential System

SPD 浪涌保护系统
SPD Surge Protection System

防雷接地系统
Lightning Protection Grounding System

强电系统
High Voltage Electric System

CITIC Tower HV Electric System
中信大厦强电系统

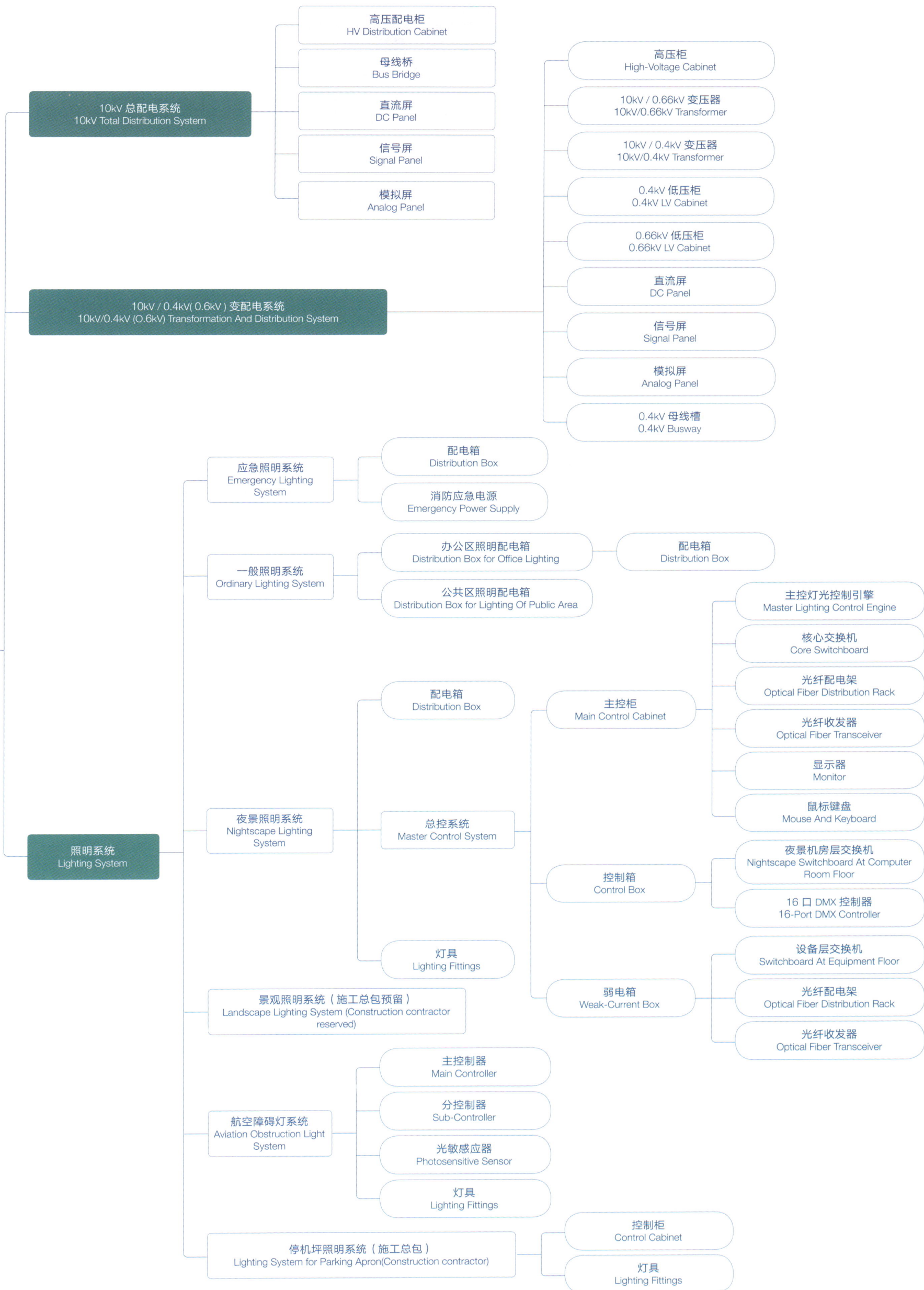

10kV 总配电系统 / 10kV Total Distribution System

- 高压配电柜 / HV Distribution Cabinet
- 母线桥 / Bus Bridge
- 直流屏 / DC Panel
- 信号屏 / Signal Panel
- 模拟屏 / Analog Panel

10kV / 0.4kV(0.6kV) 变配电系统 / 10kV/0.4kV (O.6kV) Transformation And Distribution System

- 高压柜 / High-Voltage Cabinet
- 10kV / 0.66kV 变压器 / 10kV/0.66kV Transformer
- 10kV / 0.4kV 变压器 / 10kV/0.4kV Transformer
- 0.4kV 低压柜 / 0.4kV LV Cabinet
- 0.66kV 低压柜 / 0.66kV LV Cabinet
- 直流屏 / DC Panel
- 信号屏 / Signal Panel
- 模拟屏 / Analog Panel
- 0.4kV 母线槽 / 0.4kV Busway

照明系统 / Lighting System

应急照明系统 / Emergency Lighting System
- 配电箱 / Distribution Box
- 消防应急电源 / Emergency Power Supply

一般照明系统 / Ordinary Lighting System
- 办公区照明配电箱 / Distribution Box for Office Lighting — 配电箱 / Distribution Box
- 公共区照明配电箱 / Distribution Box for Lighting Of Public Area

夜景照明系统 / Nightscape Lighting System
- 配电箱 / Distribution Box
- 总控系统 / Master Control System
  - 主控柜 / Main Control Cabinet
    - 主控灯光控制引擎 / Master Lighting Control Engine
    - 核心交换机 / Core Switchboard
    - 光纤配电架 / Optical Fiber Distribution Rack
    - 光纤收发器 / Optical Fiber Transceiver
    - 显示器 / Monitor
    - 鼠标键盘 / Mouse And Keyboard
  - 控制箱 / Control Box
    - 夜景机房层交换机 / Nightscape Switchboard At Computer Room Floor
    - 16 口 DMX 控制器 / 16-Port DMX Controller
  - 弱电箱 / Weak-Current Box
    - 设备层交换机 / Switchboard At Equipment Floor
    - 光纤配电架 / Optical Fiber Distribution Rack
    - 光纤收发器 / Optical Fiber Transceiver
- 灯具 / Lighting Fittings

景观照明系统（施工总包预留）/ Landscape Lighting System (Construction contractor reserved)

航空障碍灯系统 / Aviation Obstruction Light System
- 主控制器 / Main Controller
- 分控制器 / Sub-Controller
- 光敏感应器 / Photosensitive Sensor
- 灯具 / Lighting Fittings

停机坪照明系统（施工总包）/ Lighting System for Parking Apron(Construction contractor)
- 控制柜 / Control Cabinet
- 灯具 / Lighting Fittings

室内温度传感器
Indoor Temperature Sensor

风管温湿度传感器
Temperature And Humidity Sensor In Air Duct

风管静压传感器
Static Pressure Sensor In Air Duct

风压差开关
Wind Pressure Difference Switch

风管二氧化碳浓度传感器
Carbon Dioxide Concentration Sensor In Air Duct

水管温度传感器
Temperature Sensor In Water Pipe

水管压力传感器
Pressure Sensor In Water Pipe

防冻开关
Antifreeze Switch

风管温度传感器
Temperature Sensor In Air Duct

水流开关
Flow Switch

水压差传感器
Water Pressure Difference Sensor

一氧化碳浓度传感器
Carbon Monoxide Concentration Sensor

传感器
Sensor

投影机
Projector

投影屏幕
Projection Screen

会议表决
Meeting Voting

会议总控制器
Conference General Controller

同声传译
Simultaneous Interpretation

会议系统
Conference System

物业设施管理系统
Property& Facility Management System

系统服务器
System Server

网络控制器
Network Controller

PLC 现场控制器
PLC Field Controller

控制面板
Control Panel

风阀执行器
Damper Actuator

水阀执行器
Water Valve Actuator

智慧阀
Intelligent Valve

VAV 控制器
VAV controller

FCU 控制器
FCU controller

暖通监控
HVAC Monitoring

系统服务器
System Server

PLC 现场控制器
PLC Field Controller

网络控制器
Network Controller

传感器
Sensor

水阀执行器
Water Valve Actuator

给排水监控
Water Supply And Drainage Monitoring

冷源监控
Cold Source Monitoring

热源监控
Hot Source Monitoring

建筑设备监控系统
Construction Equipment Monitoring System

水管温度传感器
Temperature Sensor In Water Pipe

水管压力传感器
Pressure Sensor In Water Pipe

水流开关
Flow Switch

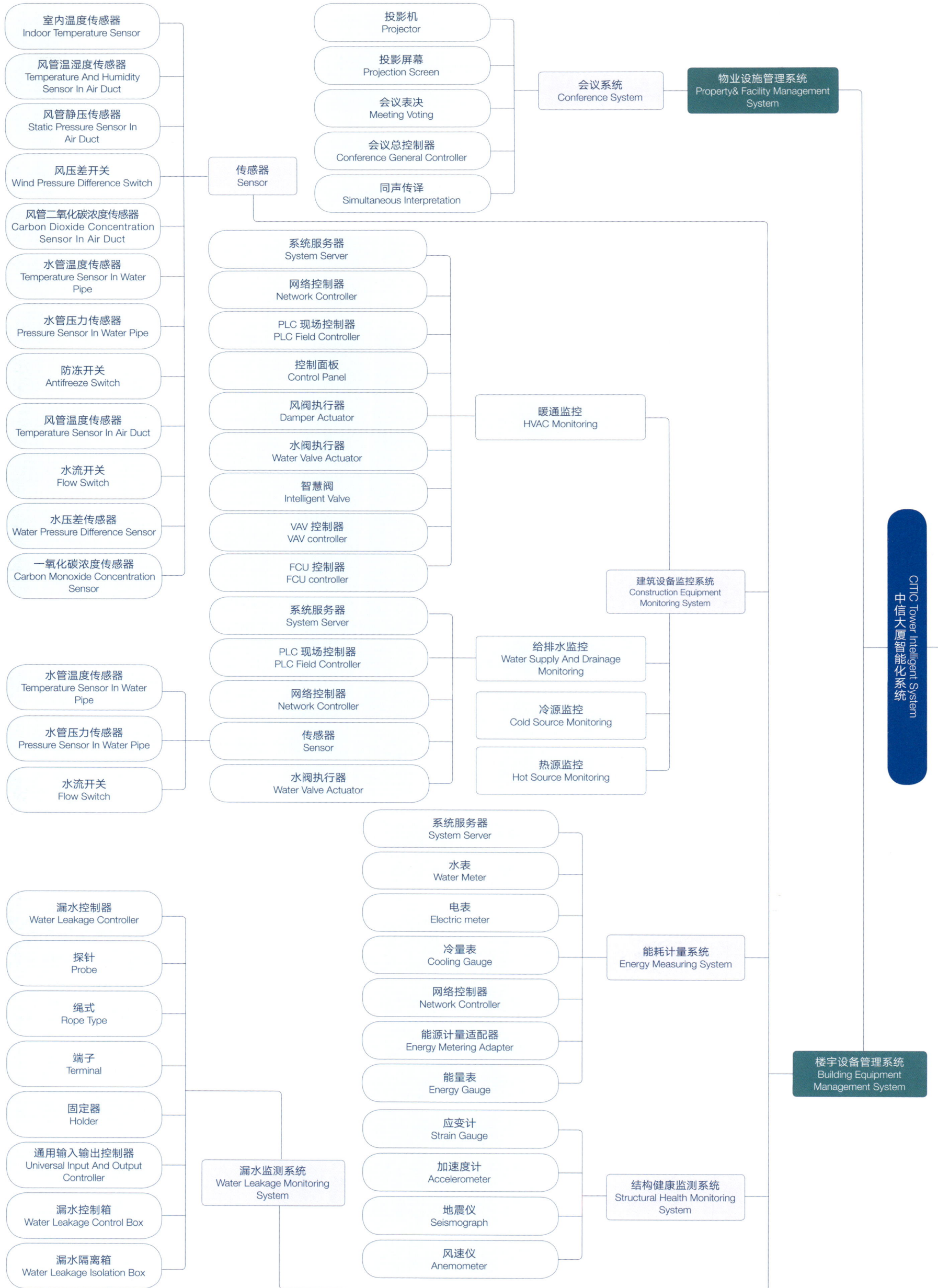

CITIC Tower Intelligent System
中信大厦智能化系统

漏水控制器
Water Leakage Controller

探针
Probe

绳式
Rope Type

端子
Terminal

固定器
Holder

通用输入输出控制器
Universal Input And Output Controller

漏水控制箱
Water Leakage Control Box

漏水隔离箱
Water Leakage Isolation Box

漏水监测系统
Water Leakage Monitoring System

系统服务器
System Server

水表
Water Meter

电表
Electric meter

冷量表
Cooling Gauge

网络控制器
Network Controller

能源计量适配器
Energy Metering Adapter

能量表
Energy Gauge

能耗计量系统
Energy Measuring System

应变计
Strain Gauge

加速度计
Accelerometer

地震仪
Seismograph

风速仪
Anemometer

结构健康监测系统
Structural Health Monitoring System

楼宇设备管理系统
Building Equipment Management System

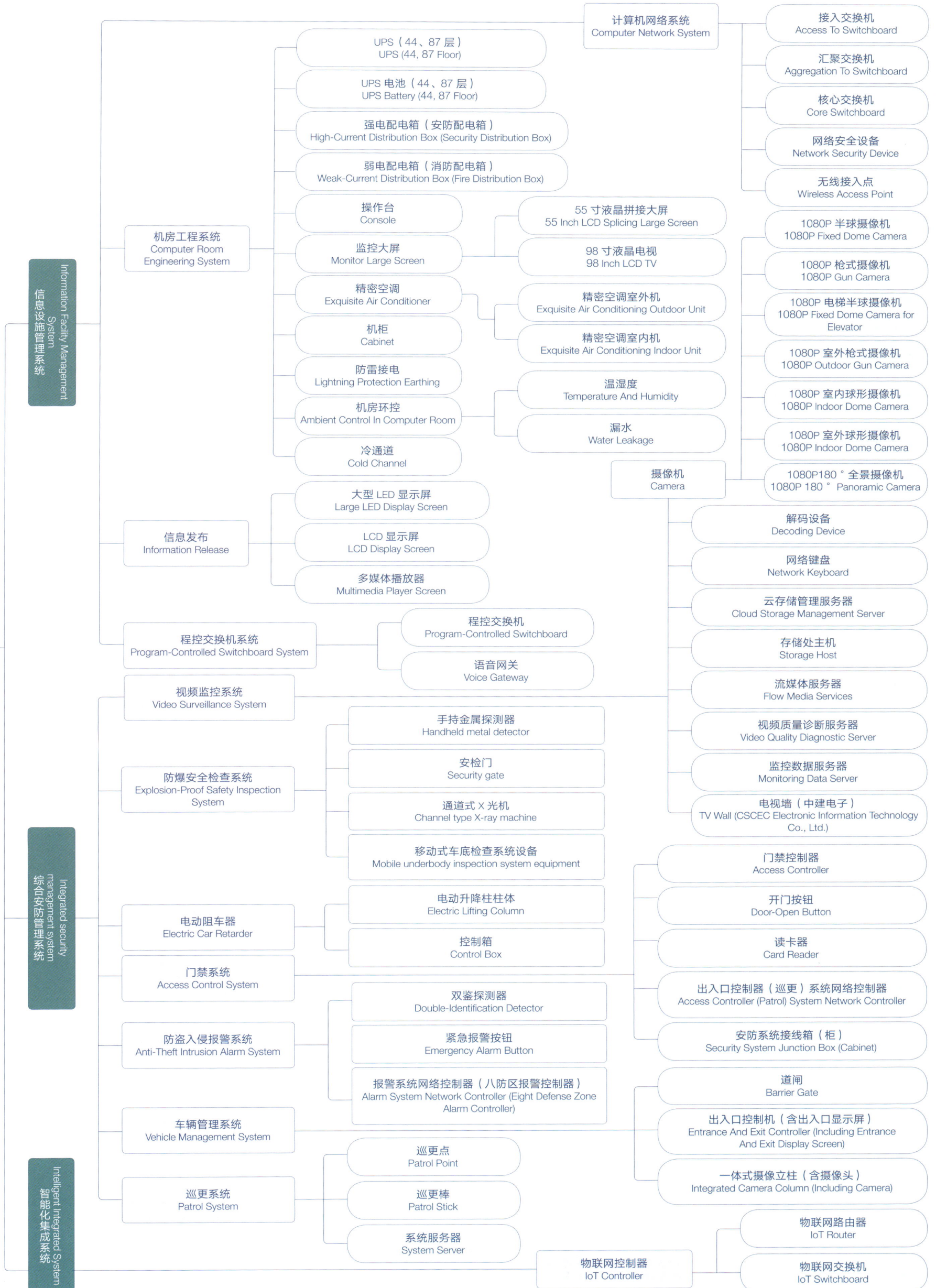

信息设施管理系统 Information Facility Management System

机房工程系统 Computer Room Engineering System
- UPS（44、87 层） UPS (44, 87 Floor)
- UPS 电池（44、87 层） UPS Battery (44, 87 Floor)
- 强电配电箱（安防配电箱） High-Current Distribution Box (Security Distribution Box)
- 弱电配电箱（消防配电箱） Weak-Current Distribution Box (Fire Distribution Box)
- 操作台 Console
- 监控大屏 Monitor Large Screen
  - 55 寸液晶拼接大屏 55 Inch LCD Splicing Large Screen
  - 98 寸液晶电视 98 Inch LCD TV
- 精密空调 Exquisite Air Conditioner
  - 精密空调室外机 Exquisite Air Conditioning Outdoor Unit
  - 精密空调室内机 Exquisite Air Conditioning Indoor Unit
- 机柜 Cabinet
- 防雷接电 Lightning Protection Earthing
- 机房环控 Ambient Control In Computer Room
  - 温湿度 Temperature And Humidity
  - 漏水 Water Leakage
- 冷通道 Cold Channel

信息发布 Information Release
- 大型 LED 显示屏 Large LED Display Screen
- LCD 显示屏 LCD Display Screen
- 多媒体播放器 Multimedia Player Screen

程控交换机系统 Program-Controlled Switchboard System
- 程控交换机 Program-Controlled Switchboard
- 语音网关 Voice Gateway

计算机网络系统 Computer Network System
- 接入交换机 Access To Switchboard
- 汇聚交换机 Aggregation To Switchboard
- 核心交换机 Core Switchboard
- 网络安全设备 Network Security Device
- 无线接入点 Wireless Access Point

综合安防管理系统 Integrated security management system

视频监控系统 Video Surveillance System
- 摄像机 Camera
  - 1080P 半球摄像机 1080P Fixed Dome Camera
  - 1080P 枪式摄像机 1080P Gun Camera
  - 1080P 电梯半球摄像机 1080P Fixed Dome Camera for Elevator
  - 1080P 室外枪式摄像机 1080P Outdoor Gun Camera
  - 1080P 室内球形摄像机 1080P Indoor Dome Camera
  - 1080P 室外球形摄像机 1080P Indoor Dome Camera
  - 1080P180°全景摄像机 1080P 180° Panoramic Camera
- 解码设备 Decoding Device
- 网络键盘 Network Keyboard
- 云存储管理服务器 Cloud Storage Management Server
- 存储处主机 Storage Host
- 流媒体服务器 Flow Media Services
- 视频质量诊断服务器 Video Quality Diagnostic Server
- 监控数据服务器 Monitoring Data Server
- 电视墙（中建电子） TV Wall (CSCEC Electronic Information Technology Co., Ltd.)

防爆安全检查系统 Explosion-Proof Safety Inspection System
- 手持金属探测器 Handheld metal detector
- 安检门 Security gate
- 通道式 X 光机 Channel type X-ray machine
- 移动式车底检查系统设备 Mobile underbody inspection system equipment

电动阻车器 Electric Car Retarder
- 电动升降柱柱体 Electric Lifting Column
- 控制箱 Control Box

门禁系统 Access Control System
- 门禁控制器 Access Controller
- 开门按钮 Door-Open Button
- 读卡器 Card Reader
- 出入口控制器（巡更）系统网络控制器 Access Controller (Patrol) System Network Controller
- 安防系统接线箱（柜） Security System Junction Box (Cabinet)

防盗入侵报警系统 Anti-Theft Intrusion Alarm System
- 双鉴探测器 Double-Identification Detector
- 紧急报警按钮 Emergency Alarm Button
- 报警系统网络控制器（八防区报警控制器） Alarm System Network Controller (Eight Defense Zone Alarm Controller)

车辆管理系统 Vehicle Management System
- 道闸 Barrier Gate
- 出入口控制机（含出入口显示屏） Entrance And Exit Controller (Including Entrance And Exit Display Screen)
- 一体式摄像立柱（含摄像头） Integrated Camera Column (Including Camera)

巡更系统 Patrol System
- 巡更点 Patrol Point
- 巡更棒 Patrol Stick
- 系统服务器 System Server

智能化集成系统 Intelligent Integrated System

- 物联网控制器 IoT Controller
  - 物联网路由器 IoT Router
  - 物联网交换机 IoT Switchboard

中信大厦消防系统
CITIC Tower Fire Protection System

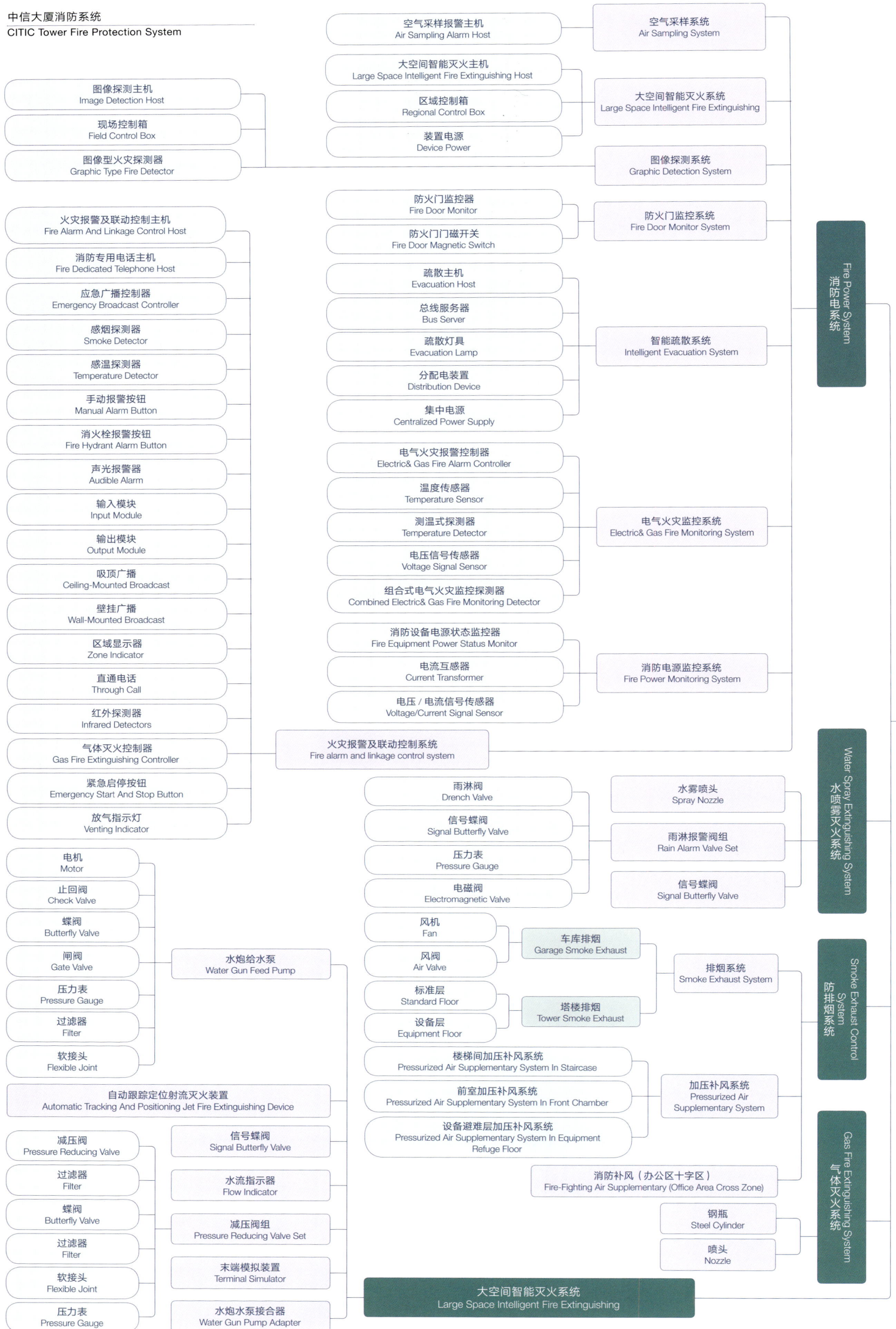

中信大厦消防系统 / CITIC Tower Fire Protection System

消防电系统 / Fire Power System

- 空气采样报警主机 / Air Sampling Alarm Host → 空气采样系统 / Air Sampling System
- 大空间智能灭火主机 / Large Space Intelligent Fire Extinguishing Host
- 区域控制箱 / Regional Control Box
- 装置电源 / Device Power
  → 大空间智能灭火系统 / Large Space Intelligent Fire Extinguishing

- 图像探测主机 / Image Detection Host
- 现场控制箱 / Field Control Box
- 图像型火灾探测器 / Graphic Type Fire Detector
  → 图像探测系统 / Graphic Detection System

- 防火门监控器 / Fire Door Monitor
- 防火门磁开关 / Fire Door Magnetic Switch
  → 防火门监控系统 / Fire Door Monitor System

- 火灾报警及联动控制主机 / Fire Alarm And Linkage Control Host
- 消防专用电话主机 / Fire Dedicated Telephone Host
- 应急广播控制器 / Emergency Broadcast Controller
- 感烟探测器 / Smoke Detector
- 感温探测器 / Temperature Detector
- 手动报警按钮 / Manual Alarm Button
- 消火栓报警按钮 / Fire Hydrant Alarm Button
- 声光报警器 / Audible Alarm
- 输入模块 / Input Module
- 输出模块 / Output Module
- 吸顶广播 / Ceiling-Mounted Broadcast
- 壁挂广播 / Wall-Mounted Broadcast
- 区域显示器 / Zone Indicator
- 直通电话 / Through Call
- 红外探测器 / Infrared Detectors
- 气体灭火控制器 / Gas Fire Extinguishing Controller
- 紧急启停按钮 / Emergency Start And Stop Button
- 放气指示灯 / Venting Indicator

- 疏散主机 / Evacuation Host
- 总线服务器 / Bus Server
- 疏散灯具 / Evacuation Lamp
- 分配电装置 / Distribution Device
- 集中电源 / Centralized Power Supply
  → 智能疏散系统 / Intelligent Evacuation System

- 电气火灾报警控制器 / Electric& Gas Fire Alarm Controller
- 温度传感器 / Temperature Sensor
- 测温式探测器 / Temperature Detector
- 电压信号传感器 / Voltage Signal Sensor
- 组合式电气火灾监控探测器 / Combined Electric& Gas Fire Monitoring Detector
  → 电气火灾监控系统 / Electric& Gas Fire Monitoring System

- 消防设备电源状态监控器 / Fire Equipment Power Status Monitor
- 电流互感器 / Current Transformer
- 电压／电流信号传感器 / Voltage/Current Signal Sensor
  → 消防电源监控系统 / Fire Power Monitoring System

火灾报警及联动控制系统 / Fire alarm and linkage control system

水喷雾灭火系统 / Water Spray Extinguishing System

- 雨淋阀 / Drench Valve
- 信号蝶阀 / Signal Butterfly Valve
- 压力表 / Pressure Gauge
- 电磁阀 / Electromagnetic Valve

- 水雾喷头 / Spray Nozzle
- 雨淋报警阀组 / Rain Alarm Valve Set
- 信号蝶阀 / Signal Butterfly Valve

防排烟系统 / Smoke Exhaust Control System

- 电机 / Motor
- 止回阀 / Check Valve
- 蝶阀 / Butterfly Valve
- 闸阀 / Gate Valve
- 压力表 / Pressure Gauge
- 过滤器 / Filter
- 软接头 / Flexible Joint
  → 水炮给水泵 / Water Gun Feed Pump

- 风机 / Fan
- 风阀 / Air Valve
  → 车库排烟 / Garage Smoke Exhaust
- 标准层 / Standard Floor
- 设备层 / Equipment Floor
  → 塔楼排烟 / Tower Smoke Exhaust
  → 排烟系统 / Smoke Exhaust System

- 楼梯间加压补风系统 / Pressurized Air Supplementary System In Staircase
- 前室加压补风系统 / Pressurized Air Supplementary System In Front Chamber
- 设备避难层加压补风系统 / Pressurized Air Supplementary System In Equipment Refuge Floor
  → 加压补风系统 / Pressurized Air Supplementary System

消防补风（办公区十字区） / Fire-Fighting Air Supplementary (Office Area Cross Zone)

气体灭火系统 / Gas Fire Extinguishing System

- 钢瓶 / Steel Cylinder
- 喷头 / Nozzle

自动跟踪定位射流灭火装置 / Automatic Tracking And Positioning Jet Fire Extinguishing Device

- 减压阀 / Pressure Reducing Valve
- 过滤器 / Filter
- 蝶阀 / Butterfly Valve
- 过滤器 / Filter
- 软接头 / Flexible Joint
- 压力表 / Pressure Gauge

- 信号蝶阀 / Signal Butterfly Valve
- 水流指示器 / Flow Indicator
- 减压阀组 / Pressure Reducing Valve Set
- 末端模拟装置 / Terminal Simulator
- 水炮水泵接合器 / Water Gun Pump Adapter

大空间智能灭火系统 / Large Space Intelligent Fire Extinguishing

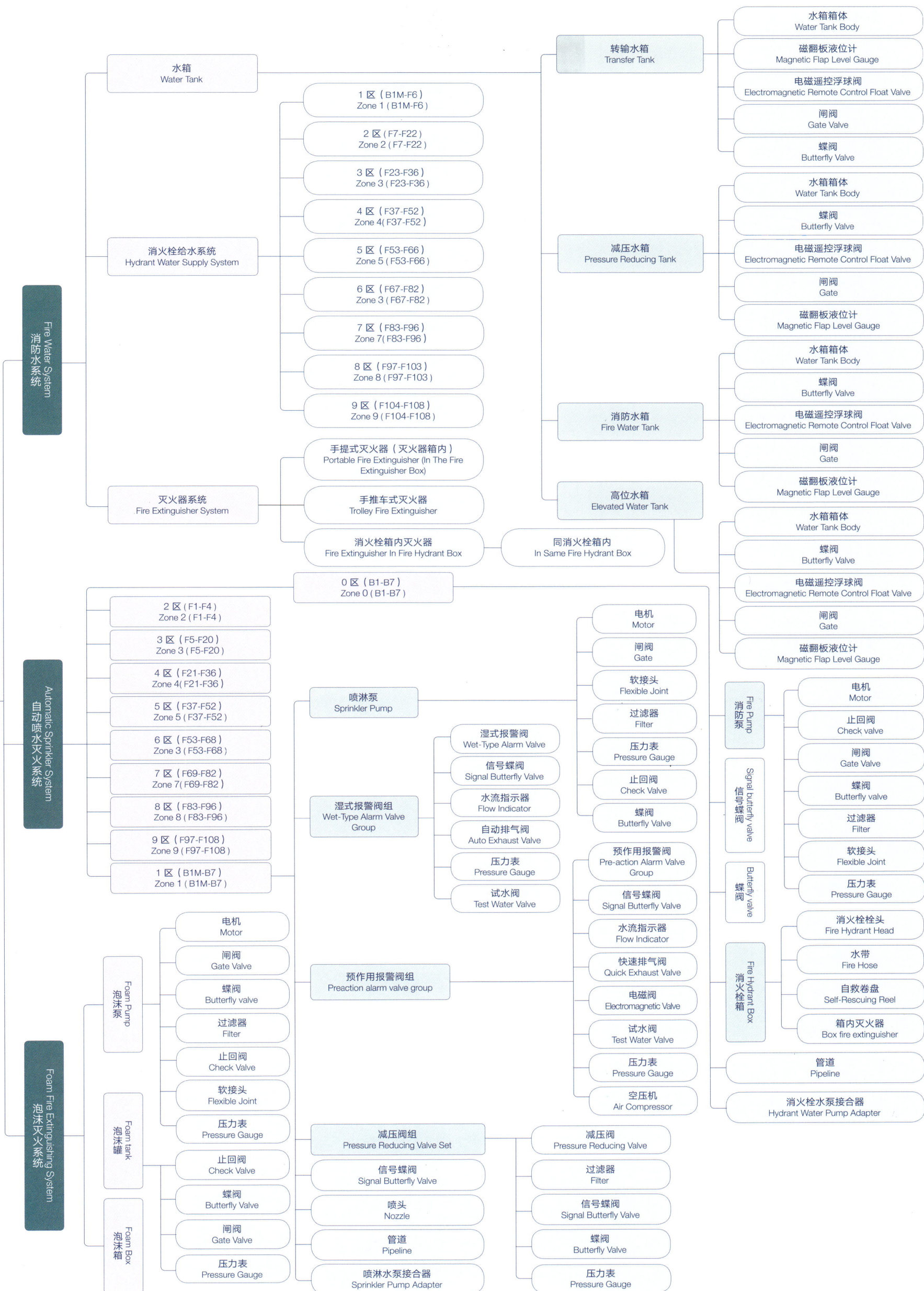

消防水系统 Fire Water System

```
消防水系统
Fire Water System
├── 水箱 Water Tank
│   └── 转输水箱 Transfer Tank
│       ├── 水箱箱体 Water Tank Body
│       ├── 磁翻板液位计 Magnetic Flap Level Gauge
│       ├── 电磁遥控浮球阀 Electromagnetic Remote Control Float Valve
│       ├── 闸阀 Gate Valve
│       └── 蝶阀 Butterfly Valve
│   └── 减压水箱 Pressure Reducing Tank
│       ├── 水箱箱体 Water Tank Body
│       ├── 蝶阀 Butterfly Valve
│       ├── 电磁遥控浮球阀 Electromagnetic Remote Control Float Valve
│       ├── 闸阀 Gate
│       └── 磁翻板液位计 Magnetic Flap Level Gauge
│   └── 消防水箱 Fire Water Tank
│       ├── 水箱箱体 Water Tank Body
│       ├── 蝶阀 Butterfly Valve
│       ├── 电磁遥控浮球阀 Electromagnetic Remote Control Float Valve
│       ├── 闸阀 Gate
│       └── 磁翻板液位计 Magnetic Flap Level Gauge
│   └── 高位水箱 Elevated Water Tank
│       ├── 水箱箱体 Water Tank Body
│       ├── 蝶阀 Butterfly Valve
│       ├── 电磁遥控浮球阀 Electromagnetic Remote Control Float Valve
│       ├── 闸阀 Gate
│       └── 磁翻板液位计 Magnetic Flap Level Gauge
├── 消火栓给水系统 Hydrant Water Supply System
│   ├── 1 区（B1M-F6）Zone 1 (B1M-F6)
│   ├── 2 区（F7-F22）Zone 2 (F7-F22)
│   ├── 3 区（F23-F36）Zone 3 (F23-F36)
│   ├── 4 区（F37-F52）Zone 4 (F37-F52)
│   ├── 5 区（F53-F66）Zone 5 (F53-F66)
│   ├── 6 区（F67-F82）Zone 3 (F67-F82)
│   ├── 7 区（F83-F96）Zone 7 (F83-F96)
│   ├── 8 区（F97-F103）Zone 8 (F97-F103)
│   └── 9 区（F104-F108）Zone 9 (F104-F108)
└── 灭火器系统 Fire Extinguisher System
    ├── 手提式灭火器（灭火器箱内）Portable Fire Extinguisher (In The Fire Extinguisher Box)
    ├── 手推车式灭火器 Trolley Fire Extinguisher
    └── 消火栓箱内灭火器 Fire Extinguisher In Fire Hydrant Box
        └── 同消火栓箱内 In Same Fire Hydrant Box
```

自动喷水灭火系统 Automatic Sprinkler System

```
自动喷水灭火系统
Automatic Sprinkler System
├── 2 区（F1-F4）Zone 2 (F1-F4)
├── 3 区（F5-F20）Zone 3 (F5-F20)
├── 4 区（F21-F36）Zone 4 (F21-F36)
├── 5 区（F37-F52）Zone 5 (F37-F52)
├── 6 区（F53-F68）Zone 3 (F53-F68)
├── 7 区（F69-F82）Zone 7 (F69-F82)
├── 8 区（F83-F96）Zone 8 (F83-F96)
├── 9 区（F97-F108）Zone 9 (F97-F108)
├── 1 区（B1M-B7）Zone 1 (B1M-B7)
└── 0 区（B1-B7）Zone 0 (B1-B7)
    ├── 喷淋泵 Sprinkler Pump
    │   ├── 电机 Motor
    │   ├── 闸阀 Gate
    │   ├── 软接头 Flexible Joint
    │   ├── 过滤器 Filter
    │   ├── 压力表 Pressure Gauge
    │   ├── 止回阀 Check Valve
    │   └── 蝶阀 Butterfly Valve
    ├── 湿式报警阀组 Wet-Type Alarm Valve Group
    │   ├── 湿式报警阀 Wet-Type Alarm Valve
    │   ├── 信号蝶阀 Signal Butterfly Valve
    │   ├── 水流指示器 Flow Indicator
    │   ├── 自动排气阀 Auto Exhaust Valve
    │   ├── 压力表 Pressure Gauge
    │   └── 试水阀 Test Water Valve
    ├── 预作用报警阀组 Preaction alarm valve group
    │   ├── 预作用报警阀 Pre-action Alarm Valve Group
    │   ├── 信号蝶阀 Signal Butterfly Valve
    │   ├── 水流指示器 Flow Indicator
    │   ├── 快速排气阀 Quick Exhaust Valve
    │   ├── 电磁阀 Electromagnetic Valve
    │   ├── 试水阀 Test Water Valve
    │   ├── 压力表 Pressure Gauge
    │   └── 空压机 Air Compressor
    └── 减压阀组 Pressure Reducing Valve Set
        ├── 减压阀 Pressure Reducing Valve
        ├── 过滤器 Filter
        ├── 信号蝶阀 Signal Butterfly Valve
        ├── 蝶阀 Butterfly Valve
        └── 压力表 Pressure Gauge
```

消防泵 Fire Pump
- 电机 Motor
- 止回阀 Check valve
- 闸阀 Gate Valve
- 蝶阀 Butterfly valve
- 过滤器 Filter
- 软接头 Flexible Joint
- 压力表 Pressure Gauge

信号蝶阀 Signal butterfly valve
蝶阀 Butterfly valve

消火栓箱 Fire Hydrant Box
- 消火栓栓头 Fire Hydrant Head
- 水带 Fire Hose
- 自救卷盘 Self-Rescuing Reel
- 箱内灭火器 Box fire extinguisher
- 管道 Pipeline
- 消火栓水泵接合器 Hydrant Water Pump Adapter

泡沫灭火系统 Foam Fire Extinguishing System

```
泡沫灭火系统
Foam Fire Extinguishing System
├── 泡沫泵 Foam Pump
├── 泡沫罐 Foam tank
└── 泡沫箱 Foam Box
    ├── 电机 Motor
    ├── 闸阀 Gate Valve
    ├── 蝶阀 Butterfly valve
    ├── 过滤器 Filter
    ├── 止回阀 Check Valve
    ├── 软接头 Flexible Joint
    ├── 压力表 Pressure Gauge
    ├── 止回阀 Check Valve
    ├── 蝶阀 Butterfly Valve
    ├── 闸阀 Gate Valve
    └── 压力表 Pressure Gauge
```

喷头 Nozzle
管道 Pipeline
喷淋水泵接合器 Sprinkler Pump Adapter

中信大厦给排水系统
CITIC Tower Water Supply and Drainage System

不锈钢软接
Stainless Steel Flexible Joint

磁翻板液位计
Magnetic Flap Level Gauge

蝶阀
Butterfly Valve

浮球阀
Float Valve

过滤器
Filter

水表
Water Meter

自动排气阀
Auto Exhaust Valve

减压阀
Pressure Reducing Valve

青铜闸阀
Bronze Gate Valve

压力表
Pressure Gauge

止回阀
Check Valve

远传水表
Remote Water Meter

薄壁不锈钢管道
Thin-Walled Stainless Steel Pipe

球墨铸铁雨水管
Ductile Iron Rain Pipe

A 型机制铸铁排水管
Type A Mechanism Cast Iron Drain Pipe

不锈钢水箱
Stainless Steel Water Tank

臭氧自洁消毒器
Ozone Self-cleaning Sterilizer

水锤吸纳器
Water Hammer Absorber

中水转输泵组
Recycled Water Transfer Pump Set

给水转输泵组
Supply Water Transfer Pump Set

紫外线消毒器
UV Sterilizer

变频中水转输泵组
Frequency Conversion Recycled Water Transfer Pump Set

新鲜油脂分离器
Fresh Grease Separator

阀部件
Valve Component

管材
Pipe

设备
Equipment

Water System
中水系统

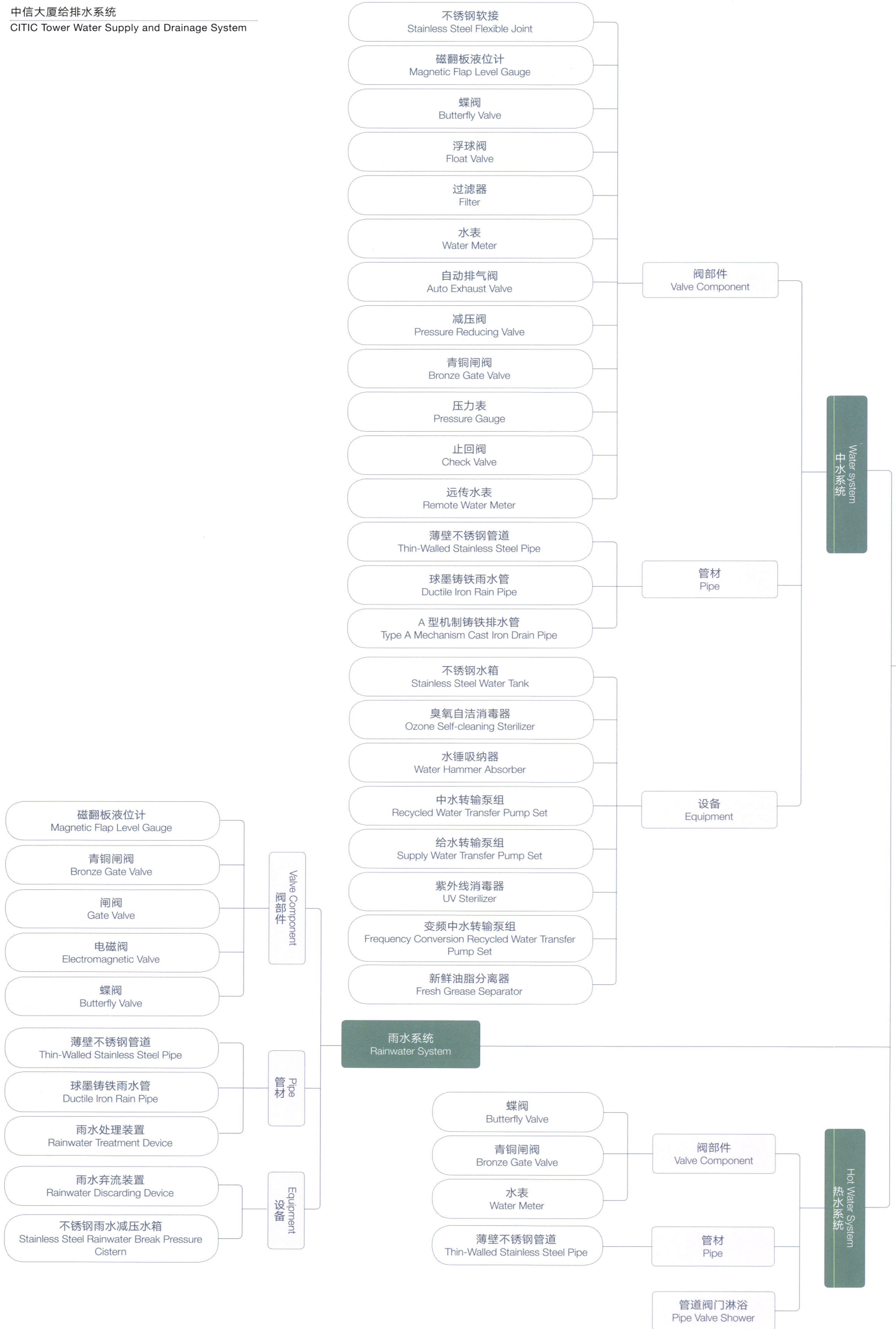

CITIC Tower Drainage System
中信大厦给排水系统

磁翻板液位计
Magnetic Flap Level Gauge

青铜闸阀
Bronze Gate Valve

闸阀
Gate Valve

电磁阀
Electromagnetic Valve

蝶阀
Butterfly Valve

薄壁不锈钢管道
Thin-Walled Stainless Steel Pipe

球墨铸铁雨水管
Ductile Iron Rain Pipe

雨水处理装置
Rainwater Treatment Device

雨水弃流装置
Rainwater Discarding Device

不锈钢雨水减压水箱
Stainless Steel Rainwater Break Pressure Cistern

Valve Component
阀部件

Pipe
管材

Equipment
设备

雨水系统
Rainwater System

蝶阀
Butterfly Valve

青铜闸阀
Bronze Gate Valve

水表
Water Meter

薄壁不锈钢管道
Thin-Walled Stainless Steel Pipe

阀部件
Valve Component

管材
Pipe

Hot Water System
热水系统

管道阀门淋浴
Pipe Valve Shower

4.2 设计管理体系及初步设计、施工图设计、施工图深化管理

Design Management System And Preliminary Design, Construction Drawing Design And Construction Drawing Deepening Management

大数据 Data

97,805 张
双总包合计深化及设计图纸
Total deepening and design drawings of double general contractor

2,773 个
双总包合计 BIM 设计模型
Total BIM design models of double general contractor

11,981 项
地下部分审核意见
Underground section review opinions

8,825 项
解决设计问题
Design problems solved

延伸阅读 Links

P162

设计管理贯穿于项目建设全过程，设计管理质量影响着整个工程项目的成本、进度和质量目标以及交付使用后的运营效果。中信大厦工程体量大、造型独特、结构复杂、系统繁多、科技含量高，项目采用"设计联合体＋双总包（施工总承包和机电总承包）"的工程总承包管理模式，设计联合体由北京市建筑设计研究院作为设计总负责单位联合多家设计顾问公司组成。

在业主提供的 4,723 张图纸基础上，经过施工总承包和机电总承包设计完善和优化提升后，最终完成施工图纸 97,805 张。施工总承包 BIM 设计模型 2,150 个，模型总量共计 86.53GB；机电总承包 BIM 设计模型 623 个，机电专用族库模型 543 个，模型总量共 131.6 GB，出具 BIM 审核报告共 557 份，双总包合计 BIM 设计模型 2,773 个。

在施工图设计阶段，业主协调两个总承包单位以及专业分包商与设计院沟通配合，联合各方提前介入深化设计，弥补设计图纸存在的不足或缺陷，由设计院进行整合，形成有效的施工图设计成果文件。除钢结构外，其余专业工程的深化设计比对应的施工计划日期要提前一年，即各专业工程开始实施时已经完成该专业的深化设计及专业间的协调工作。中信和业主还制定了建设全生命周期 BIM 技术运用的目标，要求各参建单位在施工图设计阶段利用 BIM 模型初步进行专业协同，审核施工图纸，从而在未施工前，进行图纸升版，大幅减少施工过程中因碰撞、拆改及设备未选型而造成的浪费、工期延误、造价增大等问题发生的概率。根据统计，项目设计阶段审核初步设计、施工图设计 BIM 模型共 56 批次，优化和调整施工图设计问题 6,526 项；施工阶段，施工总承包牵头组织 52 次全专业 BIM 综合协调，共解决问题共计 2,129 项，提升了工程的整体品质。

同时，以《中信大厦工程界面划分表》作为深化设计接口管理的指导依据。组织专业单位相互审核接口专业深化设计图纸的关联性和相容性，在深化图纸中理顺各类接口关系，统筹各专业交叉点接口概念的表达，对接口信息审核无误后进行专业会签，充分减少专业冲突、提升设计成果质量、助推设计管理水平，促进后续施工工序合理衔接及顺利推进。

Design management runs through the whole process of project construction, and its quality affects the cost, schedule and quality objectives of the whole project and the operation effect after delivery. CITIC Tower has a large volume, unique shape, complex structure, various systems and high technology content. The general contractor management mode of "Design Consortium + Double Main Contractor (general construction contractor and electromechanical general contractor)". The design consortium is composed of Beijing Institute of Architectural Design as the design unit in charge and multiple design consulting companies.

Based on the 4,723 drawings provided by CITIC Heye, 97,805 construction drawings are completed at last after the optimization and perfection of general construction contractor and electromechanical general contractor. The total number of

BIM design models for the construction general contractor is 2,150, with a total of 86.53GB; There are 623 BIM design models and 543 special family library models for electromechanical general contractor, with a total of 131.6GB of models, and 557 BIM audit reports have been issued in total. A total of 2,773 BIM design models of double general contractor are made.

In the stage of construction drawing design, CITIC Heye actively coordinates the communication and cooperation between the two general contractors, the Professional subcontractor and the design institute, joins all parties in advance to deepen the design, makes up for the deficiencies or shortcomings of the design drawings, and integrates the design institute to form an effective construction drawing design outcome document. Except for steel structure, the detailed design of other specialized projects is one year ahead of the corresponding construction plan, that is, the detailed design and coordination among the specialized projects have been completed when the implementation of each specialized project begins. The project also formulated the goal of BIM technology application in the full life cycle of construction, requiring all participating units to use BIM model to initially carry out professional collaboration and audit construction drawings on the stage of construction drawings design, so as to upgrade the drawings before construction, and greatly reduce the problems caused by collision, dismantling and modification and unselected equipment in the construction process. Probability of problems such as cost, time delay and cost increase. According to statistics, 56 batches of preliminary design and construction drawing design BIM models were reviewed in the project design stage, and 6,526 problems of construction drawing design were optimized and adjusted. During the construction stage, the general construction contractor led the organization of 52 times of professional BIM comprehensive coordination, which solved a total of 2,129 problems and improved the overall quality of the project.

At the same time, CITIC Tower Engineering Interface Division Table acts as a guide to deepen the design interface management. Organize professional units to audit interface specialty to deepen the relevance and compatibility of design drawings, straighten out the interface relations among various types of drawings, coordinate the expression of interface concepts of professional intersections, and conduct professional signature after the interface information is audited correctly, so as to reduce professional conflicts, improve the quality of design results and promote design management level, promote the reasonable convergence and smooth implementation for the follow-up construction process.

中信和业投资有限公司
CITIC Heye Investment Co., Ltd.

设计五方联合体
Design Five-party Consortium

KPF 建筑事务所
Kohn Pedersen Fox Associates

北京市建筑设计研究院有限公司
Beijing Institute of Architectural Design (Group) Co., Ltd.

中信建筑设计研究总院有限公司
CITIC General Institute of Architectural Design and Research Co., Ltd.

奥雅纳工程咨询（上海）有限公司
Arup Engineering Consulting (Shanghai) Co., Ltd.

栢诚（亚洲）有限公司
Parsons Brinckerhoff Asia Co., Ltd.

监督
Supervision

专业顾问（共计 44 家）
Professional Consultants (44 in total)

独立第三方风洞顾问
Independent third-party wind tunnel consultant

独立第三方大震弹塑性分析
Independent third-party elastoplastic analysis of large earthquakes

艺术品顾问
Artwork consultant

停机坪顾问
Parking apron consultant

地震加速度时程波
Seismic acceleration time curve

阻尼器顾问
Damper consultant

AV 会议系统顾问
AV conference system consultant

地质勘察设计顾问
Geological survey and design consultant

基坑支护、降水及开挖设计顾问
Consultant for foundation pit support, precipitation and excavation design

桩基施工图设计顾问
Consultant for pile foundation construction drawing design

基坑一体化工程委托设计顾问
Entrusted design consultant for foundation pit integration project

夜景照明设计顾问（共 5 家）
Nightscape lighting design consultants (5 in total)

标识概念及方案设计顾问
Consultant for identification concept and program design

配电室工程设计顾问
Distribution room engineering design consultant

深化概念方案建筑概念顾问
Architecture concept consultant for in-depth concept program

深化概念方案结构顾问
Structure consultant for deepening concept program

深化概念方案机电顾问
Electromechanical consultant for in-depth concept program

深化概念方案交通顾问
Transportation consultant for in-depth concept program

深化概念方案建筑顾问
Architecture concept consultant for in-depth program

消防安全咨询及消防性能化设计顾问
Consultant for fire protection safety consultation and fire performance design

五金系统顾问（共 3 家）
Hardware System Consultant (3 in total)

外部排水工程设计顾问
External drainage engineering design consultant

结构分析用地震数据技术服务
Seismic data technology service for structural analysis

风洞实验技术服务
Wind tunnel experimental Technology Service

第三方结构超限审查咨询顾问
Consultant for third-party review on out-of -codes structure

方案设计协调管理及 BIM 模型咨询服务
Program design coordination management and BIM model consulting services

楼宇智能化
Building intelligence

节能设计及绿星服务
Energy-saving design and green star service

LEED 绿色顾问
LEED green consultant

初步设计阶段交通顾问
Transportation consultant at preliminary design stage

模拟地震振动台试验顾问
Consultant for simulation seismic vibration table test

声学设计顾问
Acoustic design consultant

安全防范顾问
Security consultant

观光策划顾问
Sightseeing planning consultant

电梯气流组织模拟与分析顾问
Consultant for elevator airflow organization simulation and analysis

停车场设计顾问
Parking lot design consultant

燃气设计顾问
Gas design consultant

热力站设计顾问
Consultant for thermal supply station design

设计及施工顾问（共计 24 家）
Design and Construction Consultants (24 in total)

结构试验单位（节点）
Structural test unit (node)

智能化系统深化设计顾问（共 3 家）
Intelligent system deepening design consultant (3 in total)

消防分项工程深化设计顾问（共 5 家）
Consultants for fire protection sub-project deepening design (5 in total)

室内设计顾问（共 6 家）
Interior design consultant (6 in total)

二维码系统技术服务
QR code system technical service

擦窗机专业设计顾问
Window Scrubber professional design consultant

灯光顾问
Lighting consultant

幕墙顾问
Curtain wall consultant

钢结构优化设计服务
Steel structure optimization design service

超高层消防课题研究
Super high-rise fire protection research

厨房设计顾问（共 2 家）
Kitchen design consultant (2 in total)

BIM 咨询顾问
BIM consultant

造价及品质顾问（共计 21 家）
Cost And Quality Consultants (21 in total)

施工图设计文件审查
Construction drawing design document review

消防性能化复核
Fire performance review

扩初设计成果咨询顾问
Consultant for expanded initial design results

供热规划技术咨询前期设计文件复查咨询顾问
Review on the preliminary design document of heating plan technical consultation

施工图设计成果咨询顾问
Consultant for construction drawing design results

建筑工程设计第三方咨询顾问
Third-party consultant for construction engineering design

雷电灾害风险评估
Lightning disaster risk assessment

夜景照明方式第三方评估
Nightscape third-party assessment

消防设计咨询
Fire protection design consultation

机电工程 BIM 审查
Electromechanical engineering BIM review

首层大堂及观光厅热环境模拟咨询顾问
Consultant for thermal environment simulation of the first floor lobby and sightseeing hall

能源管理与自控技术顾问
Consultant for energy control and automatic control technology

幕墙气密性能及水密性能检测
Curtain wall airtight performance and watertight performance testing

BIM 竣工咨询及现场审查
BIM completion consultation and on-site review

场地地震安全性评估顾问
Site seismic safety assessment consultant

机电工程 BIM 审查
Electromechanical engineering BIM review

造价咨询（共 3 家）
Cost consultation (3 in total)

结构工程第三方咨询服务
Structural engineering third-party consulting services

第三方弹塑性时程分析技术服务
Third-party technology service for elastoplastic time history analysis

---

运维顾问（共计 5 家）
Operation Consultants (5 in total)

前期物业顾问
Preliminary property consultant

物业咨询顾问（共 2 家）
Property consultants (2 in total)

运维咨询顾问
operation consultants

BIM 运维咨询及模型审查服务
BIM operation, Model review service

管理关系
Management relationship

合同关系
Contractual relationship

支付关系
Payment relationship

| 4.3 | **商务管理**
Business Management |

大数据 Data

147 份
已签订业主
合同及补充协议
Owner's contracts and
supplementary agreements

237 份
专业分包及补充协议
Professional subcontractor and
supplementary agreements

214 份
物资采购合同
Material procurement contracts

27 个
专项工程
Special projects

480 个
施工界面
Construction interfaces

>100 个
累计签订目标责任状
The cumulative target
responsibility forms

图片说明 comments

1、专业分包工程量形象
进度确认流程表
Schedule of image progress
confirmation for professional
subcontractor project quantity

中信大厦业主与施工单位间的合同为固定综合单价 + 固定总价措施费合同，其中暂估价工程由施工单位在北京市建设工程二级市场独立招标确定。与业主签订合同及补充协议 147 份，专业分包及补充协议 237 份，物资采购协议 214 份，其他类合同 112 份。

中信和业投资有限公司始终追求"高性价比"及总造价最低，从投资决策、设计、施工、运维全生命周期考量阶段性造价。

1、建设前期便编制了《中信大厦项目投资计划》，确定总投资控制线，并经中信集团批复；

2、针对投资占比最大的建安成本建立"分阶段限额控制"管理体系，严格执行阶段性控制指标。

超高层建筑施工难度巨大，分包单位众多，工序交接繁杂，这些因素均造成了商务管理工作的难度及复杂性，为快速推动项目建设，确保项目履约，中信大厦建设采用了如下管理方法：

清晰的合约界面管理

超高层建筑参与单位众多，各专业间的作业容易产生纠纷，在合同执行过程中，发包方与承包方、承包方与分包方对合同界面的理解容易产生分歧，这可能导致项目在实施过程中出现漏项甚至无人施工管理情况。为解决前述项目管理难题，在项目建设前期，中信大厦项目基于价值工程的理念，参考行业内先进管理经验，结合项目实际情况，编制了工程界面划分表用于指导项目各项工作的界面划分，在宏观上将整个项目的施工工作分为 27 个专项工程，在微观上将项目的主要施工工作划分为 480 个施工界面，为项目快速推进奠定了扎实基础。

上述的管理行动为后续招采工作清晰了各标段责任范围、过程管理也明确了责任范围，规避了冲突，减少过程争议及索赔，发挥各专业资源优势，多维度管理控制项目，从而有效推进项目实施。

招采前移与过程考核相结合

中信大厦项目涉及专业广，过程中含有大量采购工作，且项目现场材料零场地堆放，采购管理水平直接影响项目进度，各专业单位进场时间、材料进厂时间与总体施工计划能否完美衔接成为影响项目进度的重要因素。针对前述项目招标采购管理工作面临的巨大压力，为了能顺利完成招标采购任务，中信大厦项目总承包方基于价值链的思想，从招采前移和过程考核两方面优化项目采购管理来保障工程的质量、成本以及进度目标。

招采前移：即强调将各专业分包介入工作前置到专业工程图纸设计阶段，结合潜在分包商在专业方面的经验优势与专业优势，提高设计与施工的契合度，降低工程总体造价。同时强调将各专业分包工程采购工作提前，在专业分包工程图纸审核完成后即开展该专业分包工程招采工作，确保现场条件

具备后即刻开展施工，通过招采与设计的有机结合，有效地缩短现场分包进场时间，推动项目建设。

过程考核：施工总承包方根据项目年施工进度计划，分解并细化项目各节点目标，以过程中各目标的实现作为衡量各专业分包单位管理水平及分包实力的重要指标，过程考核得以量化。同时在年初与各分包单位负责人签订目标责任状，通过与各分包单位风险共担、利益共享，以激励项目各目标节点的快速实现，保障履约。项目大面积开展施工时期最多半年设置百余个节点目标，累计签订目标责任状超 100 余份，通过对各个节点目标实现的严格把控，最终确保了项目按期履约。

系统联动，多维度商务管理

项目自开工以来，施工总包就树立"服务业主无大小事、服务分包无份外之事"的理念，坚持"履约就是最大节约"的成本观，严格践行契约精神，坚持契约是建立信任的基础。同时针对分包商、分供商，无论合约规模大小、实施周期长短，施工总承包方都以一视同仁的态度进行管控，相互履责，实现项目均衡履约。

施工过程中，项目部通过加强商务部与技术部、工程部及财务部的协调沟通，使得整个商务管理工作，不仅满足经济性的要求，也满足施工进度安排。商务部与技术部联动，以技术创新、方案优化为重点，结合各施工方案成本，选择经济效益及技术效益最优方案，在保证项目质量的同时降低成本；商务部与工程部联动，根据项目现场施工工序及计划，对潜在影响现场施工的商务问题优先解决，为材料及设备进场提供支撑；加强与财务部门联动，严格执行合同关于进度付款的约定，建立专门的资金监管账户，做到专款专用，以此来监督分包对劳务及材料供应商的付款，通过对各分包单位及时付款及对其下游单位付款监督，为现场施工生产提供资金支持与保障，确保现场充足的劳动力及材料和设备，为项目履约奠定基础。

The contract between CITIC Tower and the construction unit is fixed comprehensive unit price + fixed total price measure fee, in which the temporary appraisal project is determined by the construction unit through independent bidding in the secondary market of Beijing construction project. 147 owner contracts and supplementary agreements, 237 professional subcontractor and supplementary agreements, 214 material procurement agreements and 112 other contracts have been signed.

With "High Cost Performance" and the lowest total cost as the guidelines, CITIC Heye Investment Co., Ltd. considers the stage costs of investment decision-making, design, construction, Operation in the full life cycle.

1. In the early stage of construction, the investment plan for CITIC Tower was compiled, and the total investment control line was determined, which was approved by CITIC Group.

2. In view of the construction and security cost with the largest

专业分包筛选流程
Management process for preselection of professional subcontractors

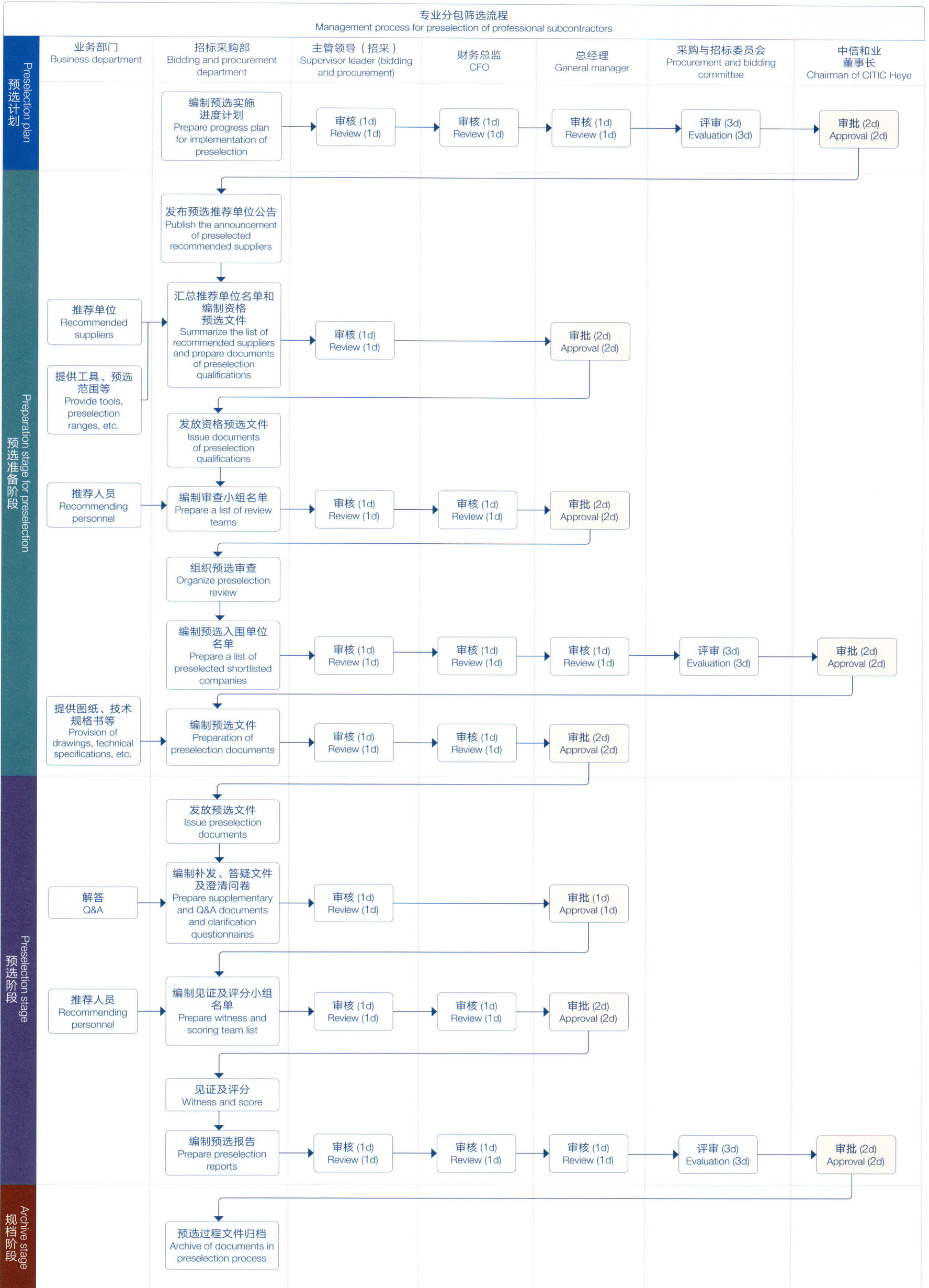

	业务部门 Business department	招标采购部 Bidding and procurement department	主管领导（招采） Supervisor leader (bidding and procurement)	财务总监 CFO	总经理 General manager	采购与招标委员会 Procurement and bidding committee	中信和业董事长 Chairman of CITIC Heye
预选计划 Preselection plan		编制预选实施进度计划 Prepare progress plan for implementation of preselection	审核 (1d) Review (1d)	审核 (1d) Review (1d)	审核 (1d) Review (1d)	评审 (3d) Evaluation (3d)	审批 (2d) Approval (2d)
预选准备阶段 Preparation stage for preselection		发布预选推荐单位公告 Publish the announcement of preselected recommended suppliers					
	推荐单位 Recommended suppliers 提供工具、预选范围等 Provide tools, preselection ranges, etc.	汇总推荐单位名单和编制资格预选文件 Summarize the list of recommended suppliers and prepare documents of preselection qualifications	审核 (1d) Review (1d)		审批 (2d) Approval (2d)		
		发放资格预选文件 Issue documents of preselection qualifications					
	推荐人员 Recommending personnel	编制审查小组名单 Prepare a list of review teams	审核 (1d) Review (1d)	审核 (1d) Review (1d)	审批 (2d) Approval (2d)		
		组织预选审查 Organize preselection review					
		编制预选入围单位名单 Prepare a list of preselected shortlisted companies	审核 (1d) Review (1d)	审核 (1d) Review (1d)	审核 (1d) Review (1d)	评审 (3d) Evaluation (3d)	审批 (2d) Approval (2d)
	提供图纸、技术规格书等 Provision of drawings, technical specifications, etc.	编制预选文件 Preparation of preselection documents	审核 (1d) Review (1d)	审核 (1d) Review (1d)	审批 (2d) Approval (2d)		
预选阶段 Preselection stage		发放预选文件 Issue preselection documents					
	解答 Q&A	编制补发、答疑文件及澄清问卷 Prepare supplementary and Q&A documents and clarification questionnaires	审核 (1d) Review (1d)		审批 (1d) Approval (1d)		
	推荐人员 Recommending personnel	编制见证及评分小组名单 Prepare witness and scoring team list	审核 (1d) Review (1d)	审核 (1d) Review (1d)	审批 (2d) Approval (2d)		
		见证及评分 Witness and score					
		编制预选报告 Prepare preselection reports	审核 (1d) Review (1d)	审核 (1d) Review (1d)	审核 (1d) Review (1d)	评审 (3d) Evaluation (3d)	审批 (2d) Approval (2d)
归档阶段 Archive stage		预选过程文件归档 Archive of documents in preselection process					

合同分类及责任部门 CONTRACT CATEGORY AND RESPONSIBLE DEPARTMENT			
合同类别 Contract category			合同履行责任部门 Contract performance de-partment
合同类别 Contract category	合同细类 Contract details		合同编码 Contract code
支出类 Expenditure category	勘察/设计/专业顾问类 Survey/design/professional consultants	勘察合同 Survey contract	结构及装饰工程协调部 Structural and decoration engineering coordination department
		设计合同 Design contract	设计统筹部 Design co-ordination department
		与设计有关的顾问合同 Design-related consulting agreement	机电工程协调部 Electromechanical engineering coordination department 结构及装饰工程协调部 Structural and decoration engineering coordination department
		与报批报建有关的合同 Contracts related to submitting applications for approval and construction	行政综合部 Comprehensive administration department
		与BIM有关的顾问合同 BIM-related consulting agreement	运营管理部 Operation management department
		与造价有关的顾问合同 Cost-related consulting agreement	成本造价部 Cost department 财务预算部 Financial budget department 成本造价部 Cost department 财务预算部 Financial budget department
		与采购有关的顾问合同 Procurement-related consulting agreement	招标采购部 Bidding and procurement department
	施工类及工程服务类 Construction and engineering services	施工合同 Construction contract	结构及装饰工程协调部 Structural and decoration engineering coordination department 机电工程协调部 Electromechanical engineering coordination department 设计统筹部 Design co-ordination department
		监理合同 Supervisor agreement	
		工程服务类合同 Engineering service contract	
	材料/设备类 Material/equipment category	/	
	其他服务类 Other service cate-gory	行政类合同 Administrative contract	行政综合部 Comprehensive administration department
		人力类合同 HR contract	人力资源部 HR department
		财务类合同 Financial contract	财务预算部 Financial budget department
		法律服务类合同 Legal service contract	法律合约部 Legal contracts department
收入类 Incomes	出售类 Sales	/	法律合约部 Legal contracts department
	租赁类 Lease	/	
	咨询类等等 Consulting, etc.	/	
资金类 Funds	贷款类 Loans	/	财务预算部 Financial budget department
	其他 Others	/	

proportion of investment, a "phase-based quota control" management system is established, and phase-based control indicators are strictly implemented.

The construction of super high-rise project is very difficult, there are many subcontractors, and the process handover is complicated. All these factors have brought the difficulty and complexity of commercial management. In order to promote the project construction quickly and ensure the project performance, the following management methods have been adopted for CITIC Tower:

I. Clear contract interface management
There are many participants in super high-rise projects, and disputes easily arise among different specialties. In the process of contract execution, the understanding of contract interface between the developer and the contractor, as well as the contractor and the subcontractor is easily divergent, which may lead to omissions or even no construction management in the process of project implementation. In order to solve the above-mentioned project management problems, in the early stage of

project construction, based on the concept of value engineering, referring to advanced management experience in the industry, CITIC Tower was combined with the actual situation of the project, and a division table of the project interface was compiled to guide the division of the work of the project, and in macro terms, the construction work of the whole project is is divided into 27 special projects. The main construction work is divided into 480 construction interfaces, which lays a solid foundation for the rapid progress of the project.

The above-mentioned management actions clarify the scope of responsibility for the subsequent recruitment work, clarify the scope of responsibility for the process management, avoid conflicts, reduce process disputes and claims, give full play to the advantages of professional resources, multi-dimensional management and control projects, so as to effectively promote project implementation.

II. Combining forward bidding and procurement with process assessment
CITIC Tower involves a wide range of specialties, and contains

测算
Calculation

估算
Estimation

概算
Budget estimate

预算
Budget

工程量审核及结算
Review and settlement of engineering quantity

决算
Final accounting

方案设计
Scheme design

初步设计
Preliminary design

施工图
Construction drawing

招投标
（合同额≤预算额）
Bidding (Contract amount ≤ Budget amount)

施工（结算额与合同额的偏差度≤15%）
Construction (Deviation between settlement amount and contract amount ≤ 15%)

竣工
（决算值≤测算值）
Completion (Final accounting value ≤ calculation value)

修改
Modification

N   Y

修改
Modification

N   Y

修改
Modification

N   Y

估算额≤测算额×80%
Estimated amount ≤ Calculated amount x80%

概算额＝估算额×（1±5%）
Budget estimate amount = Estimation amount x (1 ± 5 %)

概算额＝概算额×（1±5%）
Estimate budget amount = Estimate budget amount x (1 ± 5%)

75% 和 100% 两次循环
75% and 100% cycles

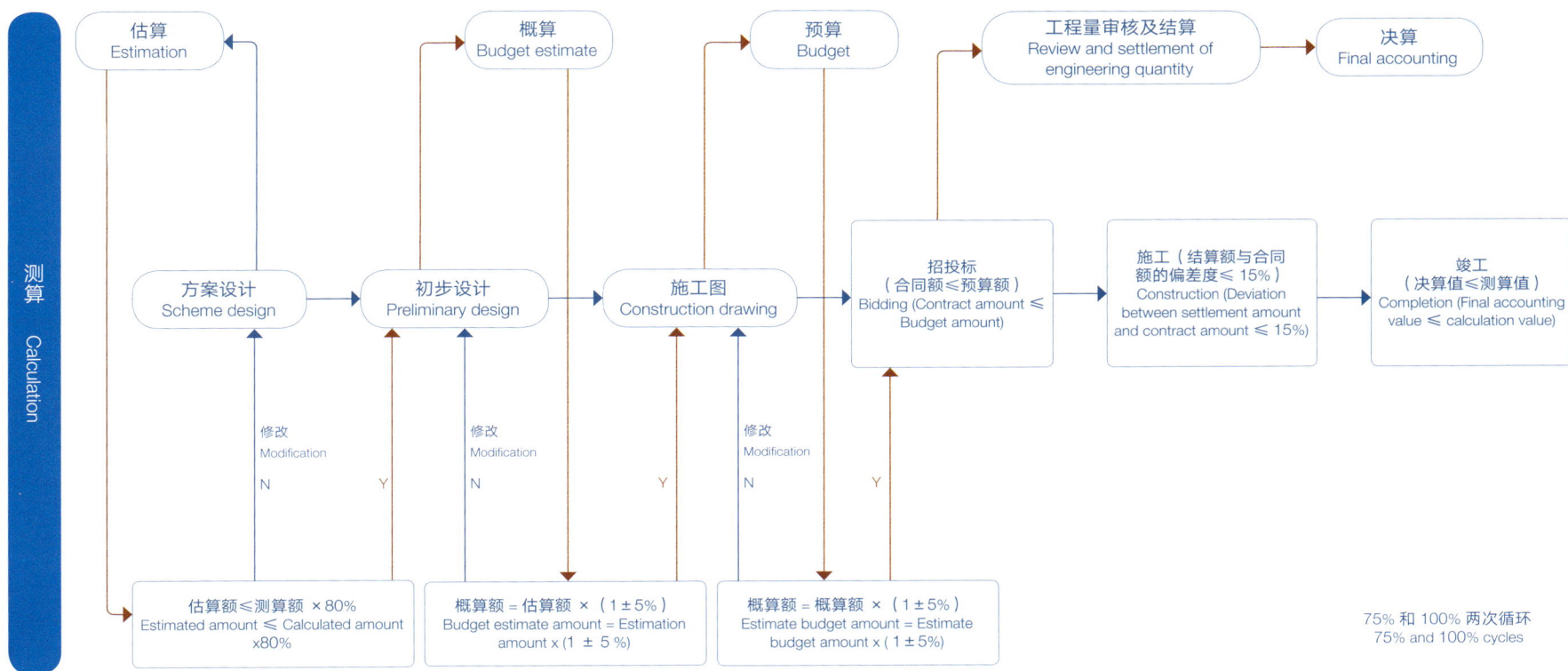

a lot of procurement work in the process, and the zero site stacking of project site material, so procurement management level directly affects the progress of the project. Whether the professional unit access time, material access time and the overall construction plan can be perfectly linked becomes an important factor affecting the progress of the project. In view of the tremendous pressure faced by the aforementioned project bidding and procurement management and in order to successfully complete the bidding and procurement task, the general contractor of CITIC Tower, based on the idea of value chain, optimizes the project procurement management from two aspects of recruitment forward and process assessment to ensure the quality, cost and progress objectives of the project.

Forward bidding and procurement: That is to say, it is emphasized to put the professional subcontractor into the professional engineering drawing design stage, combine the professional experience and professional advantages of potential subcontractors, improve the fit between design and construction, and reduce the overall cost of the project. At the same time, it is emphasized that the procurement work of subcontractor projects can be advanced, and the recruitment work of subcontractor projects can be carried out immediately after the examination and verification of the drawings of the subcontractor projects, so as to ensure the construction of the subcontractor projects immediately after the site conditions are met. Through the organic combination of recruitment and design, the entry time of subcontractor projects on the site can be effectively shortened and the project construction can be promoted.

Process assessment: according to the annual construction progress plan of the project, the general construction contractor decomposes and refines the objectives of each node of the project, and takes the realization of the objectives in the process as an important index to measure the management level and subcontractor strength of each professional subcontractor unit, so that the process assessment can be quantified. Meanwhile, at the beginning of the year, the responsible person of each subcontractor signed the target responsibility form, sharing risks and benefits with each subcontractor, in order to stimulate the rapid realization of each target node of the project and ensure the performance of the contract. During the construction period of large-scale project, more than 100 node targets were set up for a maximum of half a year, and over 100 target responsibility forms were signed. Through strict control over the realization of each node target, the performance on schedule was finally achieved for the project.

III. System linkage and multi-dimensional business management

Since the start of the project, The construction contractor has established the concept of "All matters are important to serve owner, and each matter is necessary to serve subcontractor", adhered to the cost concept of "performance is the greatest savings", strictly practiced the spirit of contract, and adhered to the contract as the basis for building trust. At the same time, regardless of the size of the contract and the length of the implementation cycle, the general construction contractor of the construction works manages and controls to subcontractors and sub-suppliers with a non-discriminatory attitude, making mutual performance of responsibilities to achieve the balanced performance of the project.

During the construction process, the project department strengthens the coordination and communication between the commerce department, technology department, engineering department and finance department, so that the whole business management work not only meets the economic requirements, but also meets the construction schedule. The commerce department and the technology department work together, focusing on technological innovation and scheme optimization, and combining the cost of various construction schemes, to select the best scheme of economic and technological benefits, so as to ensure the quality of the project and reduce the cost at the same time. Commerce department and Engineering department, in accordance with the project site construction procedures and plans, give priority to solving the business problems that potentially affect the site construction, and provide support for materials and equipment entering the site. They strengthen linkages with financial departments, implement the contract agreement on progress payment strictly, establish a special fund supervision account and make special funds to supervise the payment of labor services and materials suppliers by subcontractor, and provide on-site construction production by timely payment of each subcontractor unit and payment supervision of its downstream units. Financial support and guarantee are provided to ensure sufficient labor force, materials and equipment on site, lay the foundation for project implementation.

北京市朝阳区CBD核心区中信大厦工程——施工总承包界面划分表（摘录）

符号注释：● 实施　○ 配合　■ 总承包管理　◆ 机电总承包协调

		施工总承包/土建结构	钢结构制作	钢结构安装	幕墙	室内装修I	室内装修II	室内装修III	景观	电梯	擦窗机	停机坪	标识	交通	机电总承包	暖通	给排水	强电	智能化	消防水	电气防火	燃气	热力	变配电	照明工程1	照明工程2	冰蓄冷
1	**施工总承包**																										
1.1	**临时设施及其他**																										
(1)	平面布置的整体规划和协调管理	■/●																									
(2)	现场临时办公室	■/●	●	●	●	●	●	●	●	●	●	●	●	●	●	●	●	●	●	●	●	●	●	●	●	●	●
(3)	场外工人宿舍	■/●	●	●	●	●	●	●	●	●	●	●	●	●	●	●	●	●	●	●	●	●	●	●	●	●	●
(4)	现场临时库房	■/●	○	○	○	○	○	○	○	○	○	○	○	○	○	○	○	○	○	○	○	○	○	○	○	○	○
(5)	场外临时材料存储和周转场地	■/●	○	○	○	○	○	○	○	○	○	○	○	○	○	○	○	○	○	○	○	○	○	○	○	○	○
(6)	楼层指定堆放点的施工废料和垃圾的运输	■/●	○	○	○	○	○	○	○	○	○	○	○	○	○	○	○	○	○	○	○	○	○	○	○	○	○
(7)	自施范围施工废料和垃圾的清理，堆放至楼层指定地点	●	●	●	●	●	●	●	●	●	●	●	●	●	●	●	●	●	●	●	●	●	●	●	●	●	●
(8)	现场临时道路、围墙、工地出入口及门禁系统	■/●																									
(9)	现场保卫	■/●																									
(10)	视频监控系统	■/●																									
(11)	自施范围作业面安全防护措施	■/●	●	●	●	●	●	●	●																		
(12)	硬隔离防护的搭设、维护和拆除	■/●	●	●	●	●	●	●	●																		
(13)	防飘洒措施	■/●																									
(14)	塔吊的采购、安装、维护、爬升和拆除，提供相关单位使用	●	○																								
(15)	专用卸货和起重设备的安装、维护和使用	●	○	○	○	○	○	○	○	○	○	○	○	○	○	○	○	○	○	○	○	○	○	○	○	○	○
(16)	卸料平台的搭设、维护和拆除（每层不少于2个安放位置）	■/●	○	○	○	○	○	○		○																	●
(17)	临时施工电梯的采购、安装、维护和拆除（专业分包无偿使用）	■/●	○	○	○	○	○	○	○	○	○	○	○	○	○	○	○	○	○	○	○	○	○	○	○	○	○
(18)	正式电梯的临时使用	■								●																	
(19)	正式电梯临时使用时的厅门、门套及呼叫按钮的保护	■/●								●																	
(20)	临时供电设备的维护、使用和安全管理（包括变压器移位）	■/●																									
1.2	**土建结构工程**																										
1.2.1	**二次混凝土结构**																										
(1)	设备基础施工	●													◆	○	○	○	○	○				○			
(2)	屋顶消防水池的混凝土结构施工	●													◆		○			○							
(3)	蓄冰槽、变电站夹层等混凝土结构施工	●													◆												○
1.2.2	**砌筑**																										
(1)	砌筑隔墙、过梁、圈梁及构造柱等施工	●													◆	○	○	○	○	○							○
(2)	加气混凝土条板施工及洞口开孔	●													◆	○	○	○	○	○							○
1.2.3	**防水保温**																										
(1)	地下室顶板及坡道、楼梯等出入口的防水及保温施工	●													◆												
(2)	屋面保温及防水施工	●													◆												
(3)	设备层风百叶范围防水及保温施工	●													◆												
(4)	排水沟、集水井的防水处理	●													◆												
(5)	房间内防水施工（各自装修范围）（除蓄冰槽以外）	●			●	●	●								◆												
(6)	地下室房间内保温施工（除蓄冰槽以外）	●													◆												
1.2.4	**隔音减震**																										
(1)	浮筑楼面隔振混凝土结构施工	●													◆	○	○	○	○	○							
(2)	设备机房墙面及吊顶施工（除热力机房和智能化中控机房外）	■/●			●										◆	○	○	○	○	○							
(3)	设备机房隔声门的供应和安装	■			●										◆	○	○	○	○	○							
1.2.5	**预留预埋**																										
(1)	边长或直径≥300mm的机电洞口预留（定位由相关专业分包复核）	●								○	○	○			◆	○	○	○	○	○				○			
(2)	边长或直径＜300mm的洞口预留	●													◆	●	●	●	●	●	●						●
(3)	电梯专业洞口和预埋件的预留预埋（按照施工预留预埋，埋件由总承包提供，定位由电梯专业复核）	●									○				◆												
(4)	所有与钢结构连接的套管和预埋件的供应和焊接（定位由相关专业分包复核）	●									○				◆												
1.3	**钢结构安装工程**																										
(1)	地上钢结构的安装（包括主体构件、钢楼梯、雨篷、首层出入口、塔冠等钢结构）	●		●	○																						
(2)	电梯层间分隔梁、厅门立柱及横梁等钢构件的安装	■/●		●						○																	
(3)	地上钢结构预埋件的安装	■/●		●																							
(4)	地上钢结构防腐涂料的现场修补	■/●		●																							
(5)	钢结构防火涂料的供应、施工和修补	■/●		●																							
(6)	压型钢板及钢筋桁架板的供应和安装	■/○		●																							
1.4	**钢结构制作工程**																										
(1)	钢结构制作详图设计	■/○	●	○	○										◆												
(2)	钢构件制作加工、预拼装和运输（包括主体构件、钢楼梯、雨篷、首层出入口、塔冠等钢结构）	■	●	○																							
(3)	地上钢结构预埋件的制作及供应	■/○	●																								
(4)	地上钢结构防腐涂料的施工	■	●																								
(5)	与钢结构相连的幕墙埋件的供应和焊接（埋件图纸由幕墙施工单位提供）	■	●		●																						
(6)	BIM深化设计		●																								
1.5	**幕墙工程**																										
(1)	幕墙系统的深化设计	■/○	○	○	●	○	○	○		○					◆		○	○									
(2)	幕墙与层间楼板衔接处的铝背板、收边板、盖板、端板、水槽以及其他与幕墙相关的组件的供应和安装	■		●	●	○				○																	
(3)	幕墙系统所需的托架、扣件、边缘挡板的设计、供应和安装	■/○		●	●																						
(4)	其他幕墙埋件的设计、供应和安装	■		●	●																						
(5)	对幕墙的穿透及防水密封（为满足景观照明灯具、标识、航空障碍灯、擦窗机及其他设备的安装需求）	■			●					○	○															○	
1.6	**室内装修工程**																										
1.6.1	**室内装修工程Ⅰ（后勤区、机房（除智能化系统集成机房、消防及安防总控/分控室、各区公共网络设备机房、各区UPS机房外）、避难区、车库及库房的普通装修）**																										
(1)	建筑地面施工（包括防水层、基层和混凝土耐磨地面）	●		●																							
(2)	建筑地面装饰面层（地砖、胶地板、架空地板等）施工	■		●																							
(3)	抹灰施工，包括一般抹灰和装饰抹灰	■		○																							
(4)	轻质隔墙施工（轻钢龙骨石膏板隔墙、玻璃隔断墙、卫生间隔断等）	■		●																							
(5)	墙面装饰层施工（包括面砖、涂料及油漆、防火板/钢板保护墙边，穿孔铝板/硅酸钙穿孔板隔音墙面等）	■		●											◆	○	○	○	○							○	
(6)	吊顶施工（含天花系统提供及安装、铝格栅提供及安装、开放天花等）	■		●											◆	○	○	○	○							○	
(7)	淋浴间、清洁间和卫生间的洁具及小五金件供应及安装（含与卫生间内机电管井预留、下水及通气管道接口至卫生洁具的所有管道、保温、密闭地漏和清扫口的供应、安装、接驳及试水、试压工作；卫生洁具预留孔洞、开槽及恢复工作）	■													◆		●										
(8)	给排水管井并不在卫生间、清洁间内或与相关房间距离超过1m时，负责接驳给排水专业预留管道之后的所有管道的供应、安装以及相关工作（含保温材料供应、安装及管道试水、墙面开洞及回复等工作）	■													◆		●										
(9)	墙、顶、地机电末端追位（室内装修配合定位及开孔工作）	■			○										◆	●	●	●	●	●				●			
(10)	室内装修工程Ⅰ&室内装修工程Ⅱ标准装修区域的防火门（含五金）的供应和安装	■		●											◆		○		○								
(11)	普通门窗的加工制作和安装（包括木门、钢质门等）	■		●																							
(12)	装修工程成品保护及清洁	■		●																							
1.6.2	**室内装修工程Ⅱ（Z1~Z7区核心筒外的办公区标准装修、办公区核心筒内精装修区域）**																										
(1)	建筑地面施工（包括防水层和基层处理）	■			●																						
(2)	建筑地面面层施工（石材地面、面砖地面、胶地板、OA网络架空地板等）	■			●										◆												
(3)	抹灰施工，包括一般抹灰和装饰抹灰	■			○																						
(4)	轻质隔墙施工（轻钢龙骨石膏板隔墙、玻璃隔断墙、卫生间隔断等）	■			●										◆		○									○	
(5)	墙面装饰层施工（包括石材、面砖、金属板、涂料及油漆、镜面等）	■			●										◆	○	○	○	○							○	
(6)	外框筒装饰面层施工	■		○	●																						
(7)	吊顶施工（含天花系统提供及安装、窗帘盒提供及安装、石膏板天花提供及安装等）	■			●										◆	○	○	○	○							○	
(8)	固定家具提供及安装	■			●										◆												
(9)	窗帘系统提供及安装			○	●										◆												
1.6.3	**室内装修工程Ⅲ（观光区、多功能中心、会议中心、首层及地下大堂、空中大堂）**																										
(1)	建筑地面施工（包括防水层和基层处理）	■						●																			
(2)	建筑地面面层施工（石材地面、面砖地面、地毯、胶地板、OA网络架空地板等）	■						●																			
(3)	抹灰施工，包括一般抹灰和装饰抹灰	■						○																			
(4)	吊顶施工（含天花系统提供及安装、石膏板天花提供及安装、装饰天花提供及安装等）	■						●							◆	○	○	○	○							○	
(5)	轻质隔墙施工（轻钢龙骨石膏板隔墙、玻璃隔墙、活动隔墙等）	■						●							◆		○									○	
(6)	墙面装饰层施工包括石材、面砖、金属板、涂料及油漆、裱糊和软包、镜面等	■						●							◆												
1.7	**景观工程**																										
(1)	室外道路石材供应和施工	■/●							○					○													
(2)	室外地面石材供应和施工（包括地面、台阶、坡道、残疾人通道等）	■/●		○					○																		
(3)	室外墙面石材供应和施工（包括坡道、楼梯出入口等）	■/●		○					○																		
(4)	景观建筑物、灌溉及排水系统的施工	●							●																		
(5)	室外灯具的供应及安装	■							●						◆									●			
1.8	**停机坪工程**																										
(1)	停机坪的深化设计	■	●									●															
(2)	停机坪钢结构的加工制作、安装和施工（如有）	■	●	●								●															
(3)	停机坪的地面处理和标识画线施工	■										●															
1.9	**标识工程**																										
(1)	标识系统的深化设计（含室内标识，户外标识，楼体标识）	■		○	○	○	○						●		◆									○	○		
(2)	室内标识系统的供应及安装	■											●		◆									○			
(3)	标识照明插座的供应及安装	■											●		◆									●			
1.10	**交通工程**																										
(1)	交通工程的深化设计	■												●	◆												
(2)	相关材料的供应、制作和安装（包括标线、喷号、护角、挡车器、减速垄、道钉、反光镜、交通标牌等）	■												●	◆												
(3)	配套混凝土基础的施工	■												○	◆												
(4)	提供电源接口	■												○	◆									●			

符号注释：● 实施　○ 配合　■ 总承包管理　◆ 机电总承包协调

项目	施工总承包/土建结构	钢结构制作	钢结构安装	幕墙	室内装修I	室内装修II	室内装修III	景观	电梯	擦窗机	停机坪	标识	交通	机电总承包	暖通	给排水	强电	智能化	消防水	电气防火	燃气	热力	变配电	照明工程1	照明工程2	冰蓄冷
4　机电总承包工程																										
4.1　暖通工程																										
(1) 空调膨胀水箱补水管关断阀后管道（含浮球阀）的供应、安装及调试														◆	●	○										
(2) 热媒管线接至各生活热水换热机房内0.5m处														◆	●											
(3) 防排烟系统的所有风阀、空调通风系统所有防火阀及事故通风系统所有阀的供应、安装及调试，并提供消防信号反馈线端子（包括消防及通风空调双重控制的电动风阀）														◆	●					○						
(4) VAV BOX箱的安装														◆	●			○								
(5) 暖通电动风阀、电动水阀阀体的安装														◆	●			○								
(6) 加湿及静电除尘设备的供应、安装及调试，提供数据接口														◆	●			○								
(7) 空调机组及新风机组控制柜（包括变频器、启动装置）的供应、安装及调试，提供数据接口														◆	●			○								
(8) 空调机组新风机组控制柜与控制箱之间的配管（含桥架）、配线的供应、安装及接驳														◆	●			○								
(9) 蓄冰机房内空调风系统及通风系统（含防排烟）的供应、安装及调试														◆	●											○
(10) 热力站内空调通风系统（含防排烟）的供应、安装及调试														◆	●							○				
(11) 租户冷却水系统的供应、安装及调试，并提供数据接口	○													◆	●											
4.1.1　燃气子项工程																										
(1) 与燃气市政管网接驳														◆							●					
(2) 燃气管道与末端燃气设备的接驳	■					○								◆							●					
(3) 燃气设备控制盘的供应、安装及调试														◆							●					
4.1.2　冰蓄冷子项工程																										
(1) 冷源群控系统[工作内容包括但不限于：提供网络控制器（引擎）、数据交换机、服务器及操作平台]供应、安装及调试，为建筑设备管理平台（BMS）提供数据接口														◆				○								●
(2) 高、低压制冷机组、冷却塔（包括机房及租户合用冷却塔）、水泵控制柜（包括变频器、启动装置、电容补偿装置）供应、安装及调试														◆												●
(3) 控制柜与冷却塔、水泵及高、低压制冷机组之间的电气配管（含桥架）、配线的供应、安装及接驳														◆			○									●
(4) 冷却塔供应、安装及调试（机房及租户合用冷却塔提供冷却水供回水接口法兰）	○													◆		○										●
4.1.3　热力子项工程																										
(1) 热力站内电源箱、电源箱至配套设备之间的动力系统配管、配线及接驳（电源电缆由强电专业供应、安装，并接至热力电源箱内）														◆			○					●				
(2) 热力站内照明系统实施														◆								●				
(3) 热力站内地面、设备基础及排水沟（含箅子）实施	■/●													◆		○						●				
(4) 热力站自动控制系统供应、安装及调试并提供数据接口														◆				○				●				
(5) 热媒管线出热力站0.5m														◆	○	○						●				
4.2　给排水工程																										
(1) 预留园林灌溉系统、水景给水系统接驳用水点	○							○						◆		●										
(2) 给排水各系统管道与市政接口接驳	○													◆		●										
(3) 室外埋地管道地面挖土及回填	○													◆		●										
(4) 负责整套给水、中水泵组、污水泵的控制柜之后接驳至各设备的电缆、电线、线槽的供应、安装、调试工作并提供数据接口														◆		●		○								
(5) 污水泵控制箱与液位控制器、电源箱、控制阀的供应、安装及调试														◆		●										
(6) 冷却塔补水管线接至冷却塔，并与浮球阀接驳（冷却塔浮球阀由冰蓄冷子项供应、安装及调试）														◆		●										○
(7) 给水、中水各水池（箱）的液位传感器、溢流报警设备供应、安装及提供信号接口														◆		●		○								
(8) 冷冻机房（含蓄冰）给水补水管道进机房内0.5m处														◆		●										○
(9) 室外埋地消火栓管道供应、安装及调试														◆		●										
(10) 室外埋地消火栓阀门井的建造\并盖供应安装	■/●													◆		○										
(11) 预留厨房上、下水管道接口（给水管道含阀门及水表，排水管道立管甩口）	■													◆		○										
(12) 厨房内给排水管道与机电预留管道接口的接驳、安装及调试	■						●							◆		○										
4.3　强电工程																										
4.3.1　强电																										
(1) 按各专业及图纸要求为各设备提供电源				○	○	○	○	○	○	○	○	○	○	◆	○	○	●	○	○	○	○	○	○	○	○	○
(2) 防雷和接地结构内预埋	○	○	○	○	○	○	○	○						◆	○	○	●	○					○			
(3) 提供接地装置至图纸指定位置				○	○	○	○	○	○					◆	○	○	●	○					○			
(4) 接驳已预留接地位置，完成设备、管线有关接地系统	■													◆	○	○	●	○					○			
(5) 提供除零频、转输给水及中水泵组、污水泵以外的用电设备（如水箱外置自洁消毒器、紫外线消毒器、水泵及变压机等电源接驳）的动力电源及控制接驳，并将所需的供电明装线槽、线管、插座及电源线甩至用电设备旁														◆		○	●	○								
(6) 热水循环泵的控制供应、安装及最后的接线工作														◆		○	●									
(7) 污水泵电源动力线接至污水泵控制箱内														◆		○	●									
(8) 管道电伴热动力电源接驳至控制箱内														◆		○	●						○			
(9) 幕墙融雪装置的供应、安装及调试				○										◆			●									
(10) 供应、安装及接驳从低压开关柜或配电箱至暖通专业的控制屏及就地控制屏的电源电缆及配件														◆	○		●									○
(11) 供应、安装及接驳从低压开关柜或配电箱至给排水专业的控制屏及就地控制屏的电源电缆及配件														◆		○	●									
(12) 供应、安装及接驳从低压开关柜或配电箱至消防专业的控制屏及就地控制屏的电源电缆及配件														◆			●		○							
(13) 供应及安装智能化系统、火灾自动报警系统及火灾紧急广播系统的一次结构暗敷管及明丝管下工作														◆			●	○								
(14) 动力电源电缆接至热力站内电源箱														◆			●					○				
(15) FCU电源线的配管、配线及接驳														◆			●									
(16) VAV电源线的配管、配线及接驳														◆	○		●									
(17) 分体空调、VRV空调及精密空调机组电源的提供及接驳														◆	○		●									
4.3.2　变配电																										
(1) 柴油发电机柜的高压出线电缆电线、桥架的供货、安装、调试														◆			○						●			
(2) 高压冷冻机组电源柜高压进线电缆、桥架供货、安装、调试														◆			○						●			○
(3) 变配电室内外10kV电缆及槽盒供货、安装、调试														◆			○						●			
(4) 电力监控系统实施，提供数据接口														◆			○	○								
4.4　消防工程																										
4.4.1　消防水子项工程																										
(1) 室内装修II-B、III电梯厅及大堂的消火栓门供应及安装	■					●	●							◆					○							
(2) 大空间智能灭火系统、气体灭火系统的设备、电动阀、系统自带控制柜、扫描设备及相关配件的供应及安装，以及本系统的二次深化设计及送审盖章														◆					●							
(3) 气体灭火系统至控制柜的信号线的供应及接驳、调试														◆					●							
(4) 除室内装修II-B、III电梯厅及大堂之外的消火栓门供应及安装	○				○	○	○							◆					●							
(5) 地下一层消防转输水箱及其内部管道的供应、安装，并与冷却排水预留接口														◆		○			●							
(6) 所有消防水箱（池）的液位传感及溢流报警等设备供应、安装及提供信号接口														◆					●	○						
(7) 厨房内喷淋管道的供应、安装及调试	■													◆					●							
(8) 配合精装修单位完成精装修区域内消防水末端等设施的追位安装	■				○	○	○							◆					●	○						
(9) 预留屋顶停机坪消防管接口	■													◆					●							
4.4.2　电气防火子项工程																										
(1) 消火栓箱上报警按钮的开孔及消防报警按钮的供货、安装、调试														◆					○	●						
(2) 所有消防水箱（池）的液位传感器及溢流报警等设备各信号传感器的配管、配线、接驳及调试	○													◆					○	●						
(3) 消防系统电动阀、电磁阀信号及电源的配管、配线及接驳														◆		○	○			●						
(4) 负责协调消防强切、防火卷帘、电梯迫降等消防联动控制	○								○					◆						●						
(5) 负责与厨房设备自带灭火系统的消防信号接驳	○						○							◆						●						
(6) 按建筑设备管理监控点一览表及图纸要求提供监视接口														◆				○		●						
(7) 负责接驳消防减压阀超压报警信号														◆					○	●						
(8) 负责接驳气体灭火系统控制柜至消防中控室的信号线及控制线并调试														◆					○	●						
(9) 消防设备的手动直接启动控制线路由从各消防设备接至明配箱接至消防中控室														◆					○	●						
(10) 电梯机房电梯控制柜内消防联动接驳点（应提供满足电梯所需信号要求）至消防控制室的线管、线缆压接及明配保护管供应及安装	■													◆						●						
4.5　智能化系统工程																										
(1) 智能化系统所需的线槽、明敷线管、二次结构配管、穿线的供应、安装等工作														◆				●								
(2) 火灾自动报警系统通讯接口的信号管线接驳由智能化专业实施														◆				●		○						
(3) 停车场管理系统、速通门系统、门禁系统接受消防联动要求，用于紧急疏散														◆				●								
(4) 给水、中水各水池（箱）的液位传感器、溢流报警设备信号线、控制线的接驳及调试														◆		○		●								
(5) 风机盘管的控制器和温控器的供应、穿接线、安装及调试														◆	○			●								
(6) VAV BOX箱的控制器和温控器的供应、穿接线、安装及调试														◆	○			●								
(7) VAV BOX箱的供应														◆	○			●								
(8) 暖通系统的电动风阀、电动水阀的供应、指导安装及调试（不含冰蓄冷机房、热力站、防排烟的所有风阀、空调通风系统所有风阀及消防/通风空调双重控制的电动风阀）														◆	○			●								
(9) 暖通电动风阀的执行机构、电动水阀的执行机构的供应、安装及调试														◆	○			●								
(10) 传感器（温湿度传感器、水流开关、低温断路控制器、压差开关、二氧化碳传感器、一氧化碳传感器、电磁流量计、压力传感器等）的供应、安装及调试（热力站、蓄冰机房除外）														◆	○			●								
(11) 传感器（温湿度传感器、水流开关、低温断路控制器、压差开关、二氧化碳传感器、一氧化碳传感器、电磁流量计、压力传感器等）配套配件的供应（热力站、蓄冰机房除外）														◆	○			●								
(12) 暖通水泵运行状态、故障、启停、手/自动转换开关等监视报警配管、配线及接驳（热力站、蓄冰机房除外）														◆	○	○		●								
(13) 给排水系统所有水泵运行状态、故障、启停、手/自动转换开关等监视报警配管、配线及接驳至控制柜														◆		○		●								
(14) 提供暖通及给排水系统的电动阀、电磁阀信号线和220V以下电源线并接驳														◆	○	○		●								

4.4 施工双总包管理体系概况
Overview Of Construction Double General Contractor Management System

中信大厦的建造采用施工总承包和机电总承包的双总包管理模式。施工总承包承担土建、钢结构、幕墙、装饰、景观工程等内容，同时统一配置施工共享资源；机电总承包负责暖通、电气、智能化、消防和给排水五大类机电工程内外的协调和实施，发挥双总包各方优势，是实现总工期目标的主要方法。

施工总承包地下阶段采取直线型组织结构，便于问题的快速解决；进入地上阶段施工涉及专业多、专业单位多，采取以建造归口部门管理为主的矩阵式管理架构，成立各类型施工管理小组，虽然一定程度上增加了总包内部协调与沟通难度，但提高了与分包间的沟通效率，提高项目管理效率。

同时，施工总承包成立专门的 BIM 管理部、计划部、设计协调部。其中 BIM 管理部统筹管理项目施工中 BIM 的实施及应用；计划部负责组织编制并发布多层次计划，包括总体控制计划、年度计划、月度计划、周计划、日计划、专项计划以及交安交装计划等；设计协调部负责统筹管理各专业深化设计的事宜。

机电总承包实行企业保障层、机电总承包管理层、专业分包管理层和施工作业管理层四级管理，北京市首次为机电总承包单独颁发施工许可证。

CITIC Tower carries out the double-general contractor management mode of general construction Contractor and electromechanical general Contractor. The general construction contractor is responsible for civil work, steel structure, curtain wall, decoration, landscape engineering, and uniformly allocates construction and shared resources; The electromechanical general contractor is responsible for the coordination and implementation of HVAC, electrical, intelligent firefighting and water supply and drainage, five types of mechanical and electrical projects, and giving full play to the advantages of all parties in double general contractor is the main method to achieve the target of the total construction period.

Linear organizational structure is adopted in the underground stage of general construction contractor, which is convenient to solve the problem quickly. There are many specialties and units involved in the construction on the ground. The matrix management structure, which mainly focuses on the management of the construction centralized department, is adopted and various types of construction management groups are established. Although the difficulty of coordination and communication within the general contractor is increased to a certain extent, the communication efficiency between the general contractor and the subcontractor is improved and the project management efficiency is enhanced.

At the same time, special BIM management department, planning department and design coordination department are established in the construction general contractor. The BIM management department manages the implementation and application of BIM in the project construction as a whole. The planning department is responsible for organizing the preparation and issuance of multi-level plans, including general control plans, annual plans, monthly plans, weekly plans, daily plans, special plans and delivery and loading plans; The design coordination department is responsible for coordinating and managing the deepening of design in various professions.

Electromechanical general contractor is managed at four levels: the guarantee level of the enterprise, the management level of the general mechanical and electrical contractor, the professional subcontractor management level and the construction management level. Beijing municipality has for the first time issued a separate construction permit for the general mechanical and electrical contractor.

机电总承包直线型组织架构
Line organization structure of electromechanical general contractor

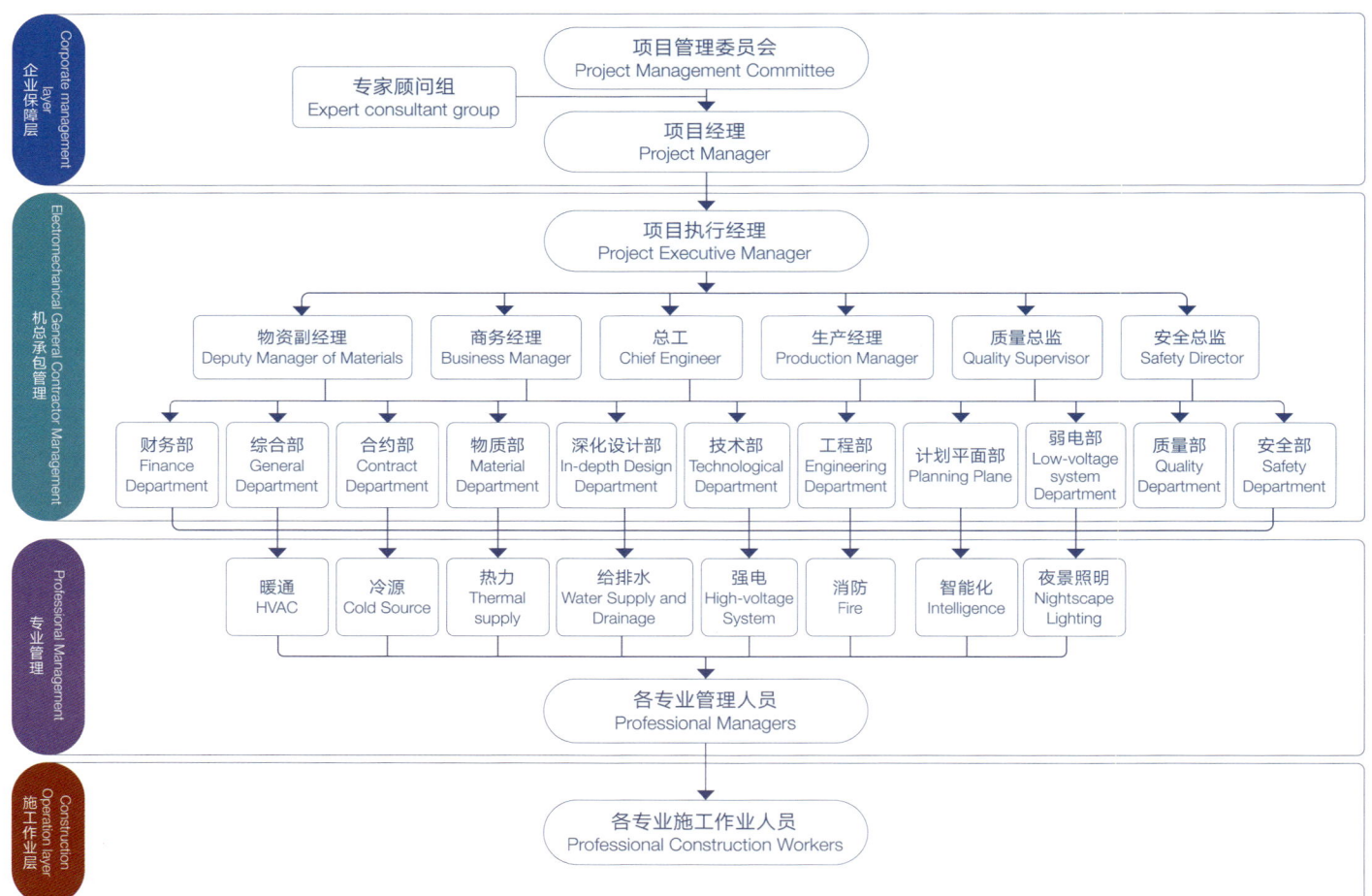

施工总承包矩阵式组织架构
Matrix organization structure of construction general contractor

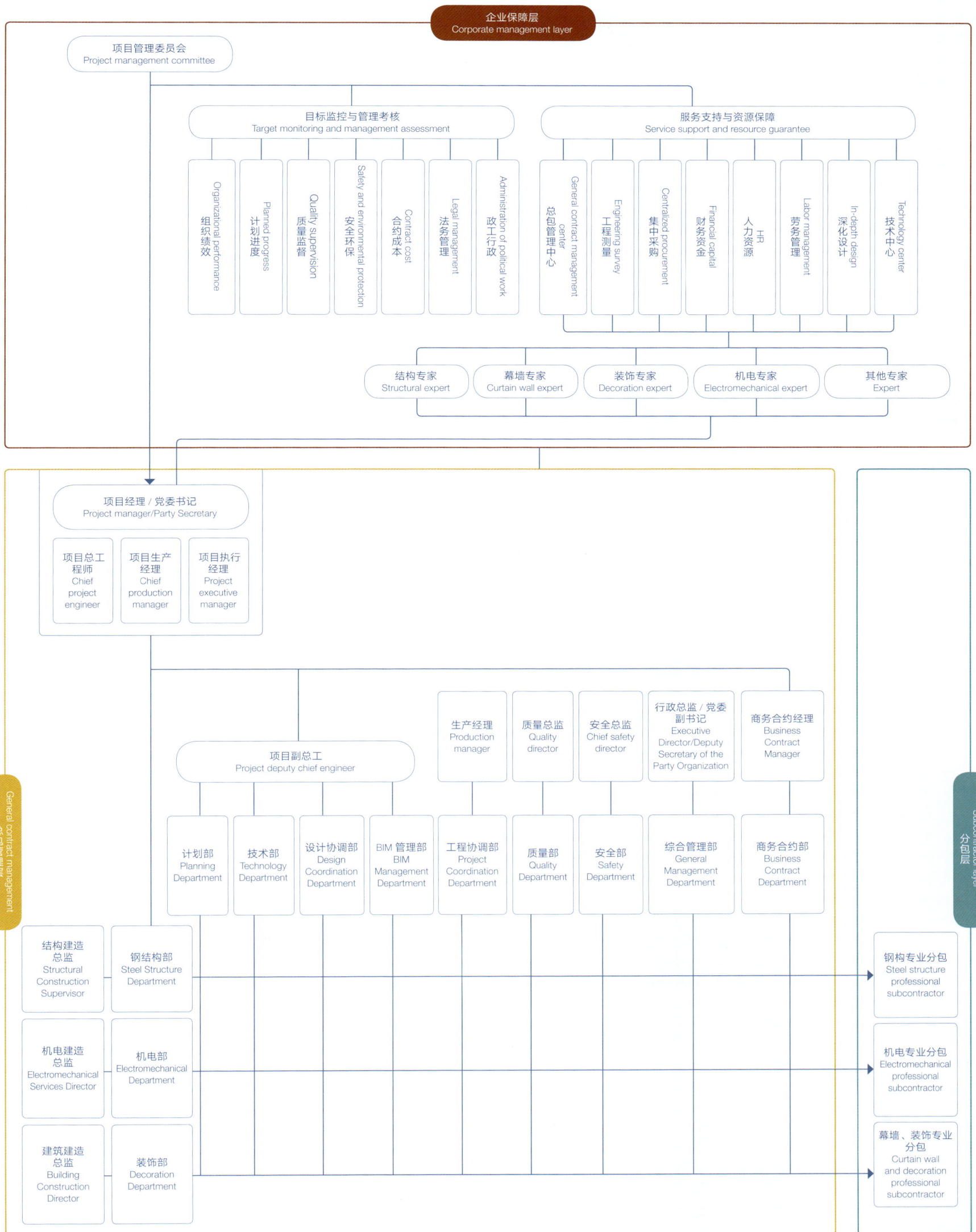

企业保障层
Corporate management layer

项目管理委员会
Project management committee

目标监控与管理考核
Target monitoring and management assessment

| 组织绩效 Organizational performance | 计划进度 Planned progress | 质量监督 Quality supervision | 安全环保 Safety and environmental protection | 合约成本 Contract cost | 法务管理 Legal management | 政工行政 Administration of political work |

服务支持与资源保障
Service support and resource guarantee

| 总包管理中心 General contract management center | 工程测量 Engineering survey | 集中采购 Centralized procurement | 财务资金 Financial capital | 人力资源 HR | 劳务管理 Labor management | 深化设计 In-depth design | 技术中心 Technology center |

| 结构专家 Structural expert | 幕墙专家 Curtain wall expert | 装饰专家 Decoration expert | 机电专家 Electromechanical expert | 其他专家 Expert |

总包管理层
General contract management

项目经理 / 党委书记
Project manager/Party Secretary

| 项目总工程师 Chief project engineer | 项目生产经理 Chief production manager | 项目执行经理 Project executive manager |

| 生产经理 Production manager | 质量总监 Quality director | 安全总监 Chief safety director | 行政总监 / 党委副书记 Executive Director/Deputy Secretary of the Party Organization | 商务合约经理 Business Contract Manager |

项目副总工
Project deputy chief engineer

| 计划部 Planning Department | 技术部 Technology Department | 设计协调部 Design Coordination Department | BIM 管理部 BIM Management Department | 工程协调部 Project Coordination Department | 质量部 Quality Department | 安全部 Safety Department | 综合管理部 General Management Department | 商务合约部 Business Contract Department |

结构建造总监 Structural Construction Supervisor	钢结构部 Steel Structure Department
机电建造总监 Electromechanical Services Director	机电部 Electromechanical Department
建筑建造总监 Building Construction Director	装饰部 Decoration Department

分包层
Subcontractor layer

钢构专业分包
Steel structure professional subcontractor

机电专业分包
Electromechanical professional subcontractor

幕墙、装饰专业分包
Curtain wall and decoration professional subcontractor

4.5

进度管理
Progress Management

大数据 *Data*

2013.7.29
正式开工日期
Formal commencement date

2018.12.28
实际竣工日期
Actual completion date

1.4 倍
开发速度超同类
Development speed is 1.4 times
faster than that of the similar type

1,978 天
实现交钥匙工期目标
Achievement of turnkey
schedule target 1,978 days

延伸阅读 *Links*

P314. P316

在项目建设过程中，影响超高层建筑成本最敏感的因素是工期，严控工期成为中信大厦开发建设成败的重要因素。中信大厦于2013年7月29日正式开工，2018年12月27日单位工程竣工验收，并于2018年12月28日完成初步移交，2019年2季度投入运营。开发工期93个月，开发速度（㎡/月）是中国已建成同类超高层建筑平均开发速度的约1.4倍。

按照正常的报建报批流程，中信大厦项目要到2014年6月才能拿到施工许可证。在北京市政府与中信集团的支持下，我们对施工许可证的办理流程进行"再造"，将一个施工许可证申请分解成了4个，大大缩短了办证周期，为施工方提前施工创造了条件。

施工中采用区段、楼层、工作面、工时、交接日期、关键节点6个参数为依据编制总进度计划，建立三级进度管理体系，有效组织土建、机电、装饰等30多个承包商的工作面协调和穿插配合，科学合理配置各项施工资源，执行严格工期进度考核制度，实现了本工程1,978天"交钥匙"的工期目标。

In the process of project construction, the most sensitive factor affecting the cost of super high-rise building is the construction period, and strict control of the construction period has become an important factor in the development and construction of CITIC Tower. With commenced operation on July 29, 2013, completed acceptance of the unit works on December 27, 2018, completed preliminary handover on December 28, 2018, CITIC Tower was put into operation in the second quarter of 2019. The development period is 93 months, and the speed of development (m²/month) is about 1.4 times of the average development speed of similar super high-rise buildings in China.

According to the normal construction approval process, CITIC Tower would not get a construction permit until June 2014. With the support of the Beijing Municipal Government and the Group, we "re-made" the process of construction permit processing, splitting one construction permit application into four parts to shorten the period of permit processing and create conditions for the construction party to advance construction.

During the construction, six parameters including section, floor, working face and man-hour, handover date and key node are used to compile the overall progress plan, establish three-level progress management system, effectively make the coordination and interpolation of working face of more than 30 contractors, such as civil work, electromechanical, decoration, scientifically and rationally allocate various construction resources, and strictly implement them. The system of checking the schedule of the project has achieved the target of turning key in 1,978 days.

工期管理控制工作流程
Duration management control workflow

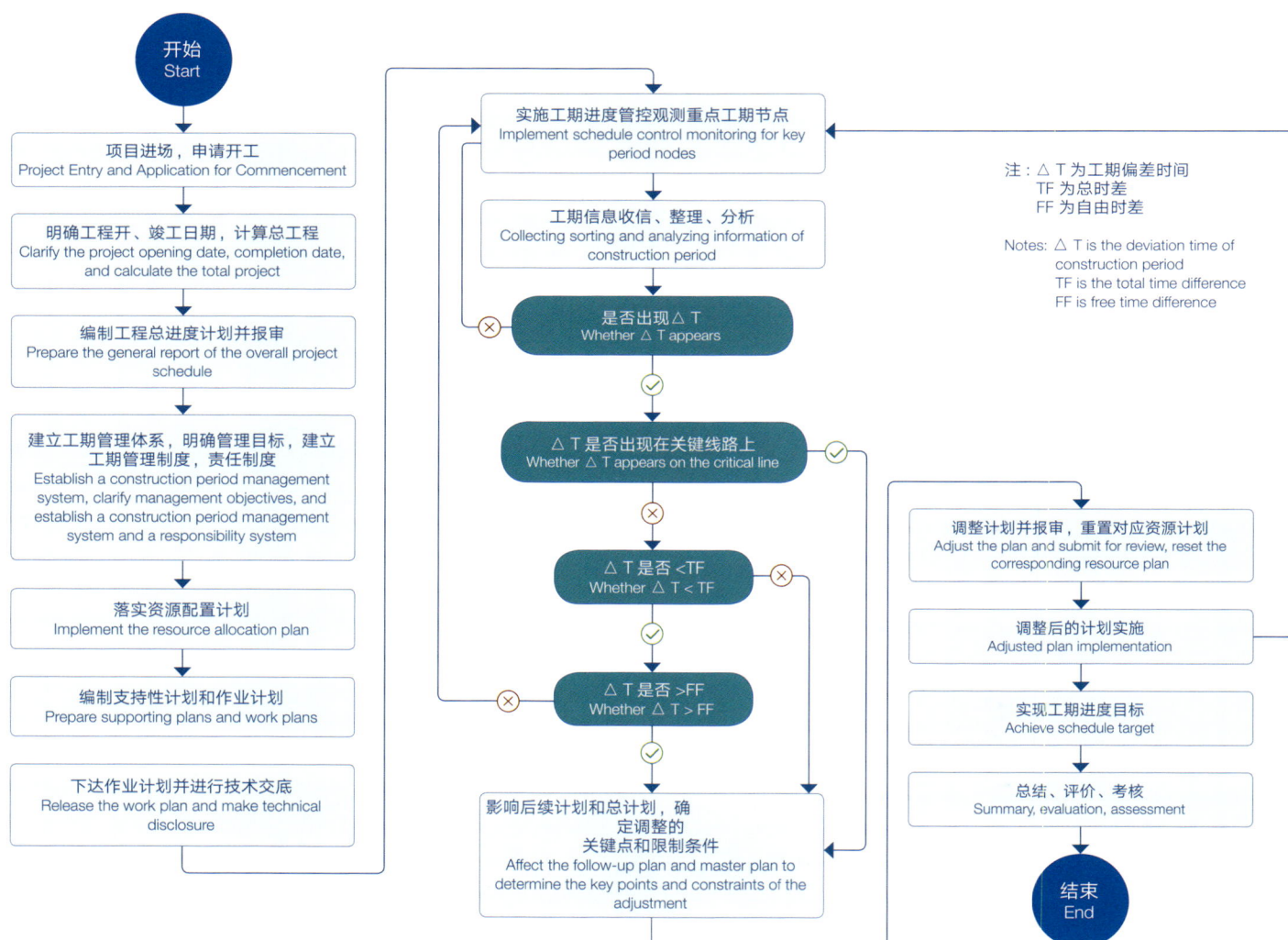

开始
Start

项目进场，申请开工
Project Entry and Application for Commencement

明确工程开、竣工日期，计算总工程
Clarify the project opening date, completion date, and calculate the total project

编制工程总进度计划并报审
Prepare the general report of the overall project schedule

建立工期管理体系，明确管理目标，建立工期管理制度，责任制度
Establish a construction period management system, clarify management objectives, and establish a construction period management system and a responsibility system

落实资源配置计划
Implement the resource allocation plan

编制支持性计划和作业计划
Prepare supporting plans and work plans

下达作业计划并进行技术交底
Release the work plan and make technical disclosure

实施工期进度管控观测重点工期节点
Implement schedule control monitoring for key period nodes

工期信息收信、整理、分析
Collecting sorting and analyzing information of construction period

是否出现△T
Whether △T appears

△T是否出现在关键线路上
Whether △T appears on the critical line

△T是否 <TF
Whether △T < TF

△T是否 >FF
Whether △T > FF

影响后续计划和总计划，确定调整的关键点和限制条件
Affect the follow-up plan and master plan to determine the key points and constraints of the adjustment

注：△T为工期偏差时间
TF为总时差
FF为自由时差

Notes: △T is the deviation time of construction period
TF is the total time difference
FF is free time difference

调整计划并报审，重置对应资源计划
Adjust the plan and submit for review, reset the corresponding resource plan

调整后的计划实施
Adjusted plan implementation

实现工期进度目标
Achieve schedule target

总结、评价、考核
Summary, evaluation, assessment

结束
End

常规超高层建筑施工许可办理流程（约 24 个月）
Processing Process of Conventional Construction Permit of High-rise Building (approx. 24 months)

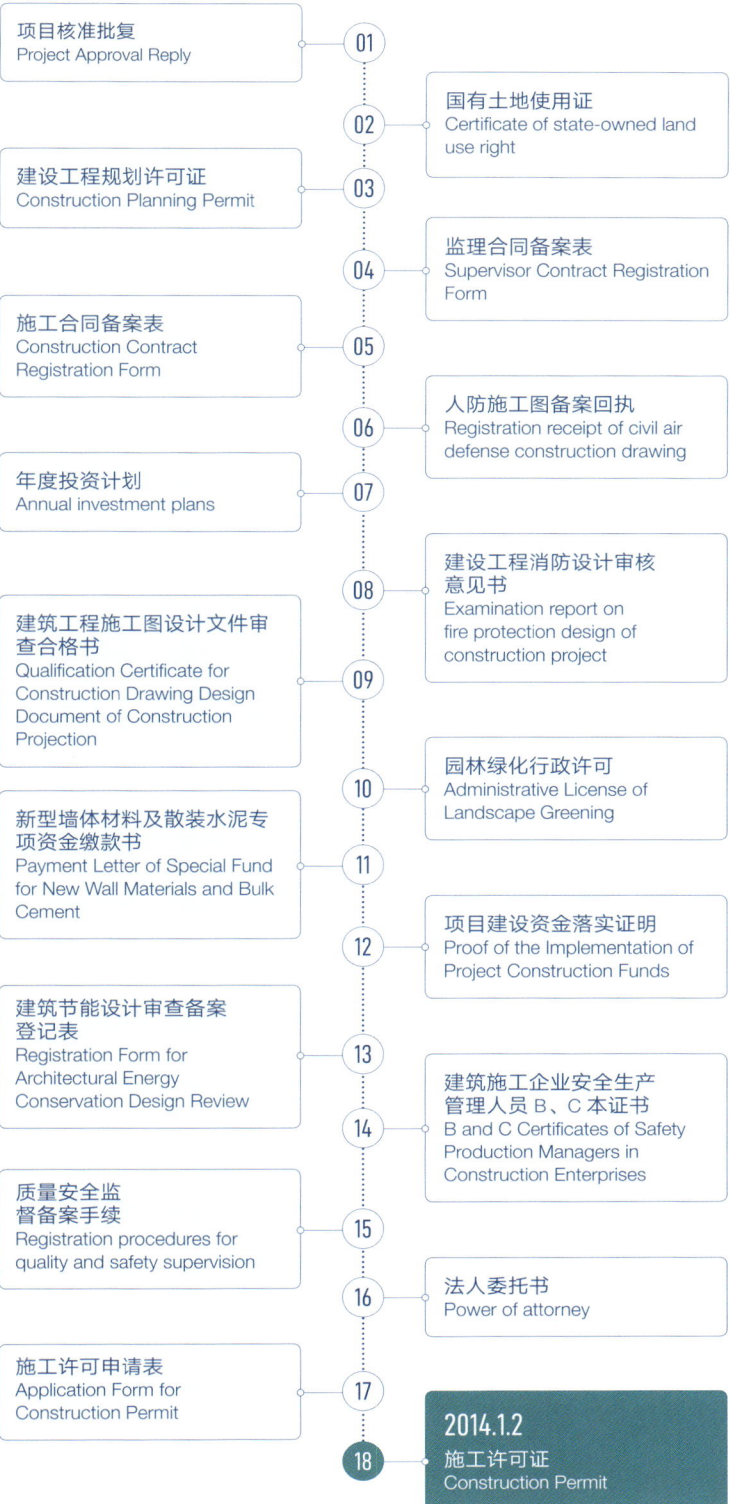

01 项目核准批复
Project Approval Reply

02 国有土地使用证
Certificate of state-owned land use right

03 建设工程规划许可证
Construction Planning Permit

04 监理合同备案表
Supervisor Contract Registration Form

05 施工合同备案表
Construction Contract Registration Form

06 人防施工图备案回执
Registration receipt of civil air defense construction drawing

07 年度投资计划
Annual investment plans

08 建设工程消防设计审核意见书
Examination report on fire protection design of construction project

09 建筑工程施工图设计文件审查合格书
Qualification Certificate for Construction Drawing Design Document of Construction Projection

10 园林绿化行政许可
Administrative License of Landscape Greening

11 新型墙体材料及散装水泥专项资金缴款书
Payment Letter of Special Fund for New Wall Materials and Bulk Cement

12 项目建设资金落实证明
Proof of the Implementation of Project Construction Funds

13 建筑节能设计审查备案登记表
Registration Form for Architectural Energy Conservation Design Review

14 建筑施工企业安全生产管理人员 B、C 本证书
B and C Certificates of Safety Production Managers in Construction Enterprises

15 质量安全监督备案手续
Registration procedures for quality and safety supervision

16 法人委托书
Power of attorney

17 施工许可申请表
Application Form for Construction Permit

18 2014.1.2
施工许可证
Construction Permit

中信大厦开创性分段申办施工许可证（第 1 个施工许可证申办耗时较常规缩短了 17 个月）
CITIC Tower applies for construction permit in phases (Time consumed in the first construction permit application is 17 months shorter than usual.)

01 2013.1.23
项目核准批复
Project approval reply

02 2013.7.10
施工合同备案表
Construction Contract Registration Form

03 2013.7.11
监理合同备案表
Registration form for supervising contract

04 2013.7.12
项目建设资金落实证明
Proof of project construction fund implementation

05 2013.7.17
建设工程规划许可证（地下部分）
Construction Planning Permit (underground part)

06 2013.7.19
建筑工程施工图设计文件审查合格书（底板及以下部分）
Qualification Certificate for Construction Drawing Design Document of Construction Projection (bottom plate and below part)

07 2013.7.19
施工企业安全管理人员 B、C 本证书
B and C Certificates of Safety Production Managers in Construction Enterprises

08 2013.7.19
施工许可申请表（底板及以下部分）
Application Form for Construction Permit (Bottom Plate and Below)

09 2013.7.19
法人委托书
Power of attorney

10 2013.7.24
施工许可证 1（底板及以下部分）
Construction Permit 1 (Bottom Plate and Below)

11 2013.8.27
新型墙体材料及散装水泥专项资金缴款书
Payment Letter of Special Fund for New Wall Materials and Bulk Cement

12 2013.12.13
建设工程消防设计审核意见书（±0.00 以下部分）
Examination report on fire protection design of construction project (below ±0.00)

13 2013.12.20
建筑工程施工图设计文件审查合格书（±0.00 以下部分）
Qualification Certificate for Construction Drawing Design Document of Construction Projection (below ±0.00)

14 2014.1.2
施工许可证 2（±0.00 以下部分）
Construction Permit 2 (below ±0.00)

15 2014.6.4
人防工程异地建设证明书
Certificate of civil air defense construction in different places

16 2014.11.2
年度投资计划
Annual investment plans

17 2014.11.14
建设工程消防设计审核意见书（±0.00 以上部分）
Examination report on fire protection design of construction project (above ±0.00)

18 2014.11.20
建筑工程施工图设计文件审查合格书（±0.00 以上部分）
Qualification Certificate for Construction Drawing Design Document of Construction Projection (above ±0.00)

19 2014.11.28
施工总承包合同备案
Registration of Construction General Contract

20 2014.12.8
施工许可证 3（±0.00 以上部分）
Construction Permit 3 (above ±0.00)

21 2015.4.23
年度投资计划
Annual investment plans

22 2015.5.19
机电总承包合同备案
Registration of Electromechanical General Contractor

23 2015.5.26
施工企业安全管理人员 B、C 本证书
B and C Certificates of Safety Production Managers in Construction Enterprises

24 2015.7.2
施工许可申请表（机电总承包）
Application Form for Construction Permit (electromechanical general contractor)

25 2015.7.7
施工许可证 4（机电总承包）
Construction Permit 4 (electromechanical general contractor)

中信大厦与同类超高层建筑开发建设全周期对比图
Comparison Chart of Development and Construction Period between CITIC Tower and Similar Super High-rise Projects

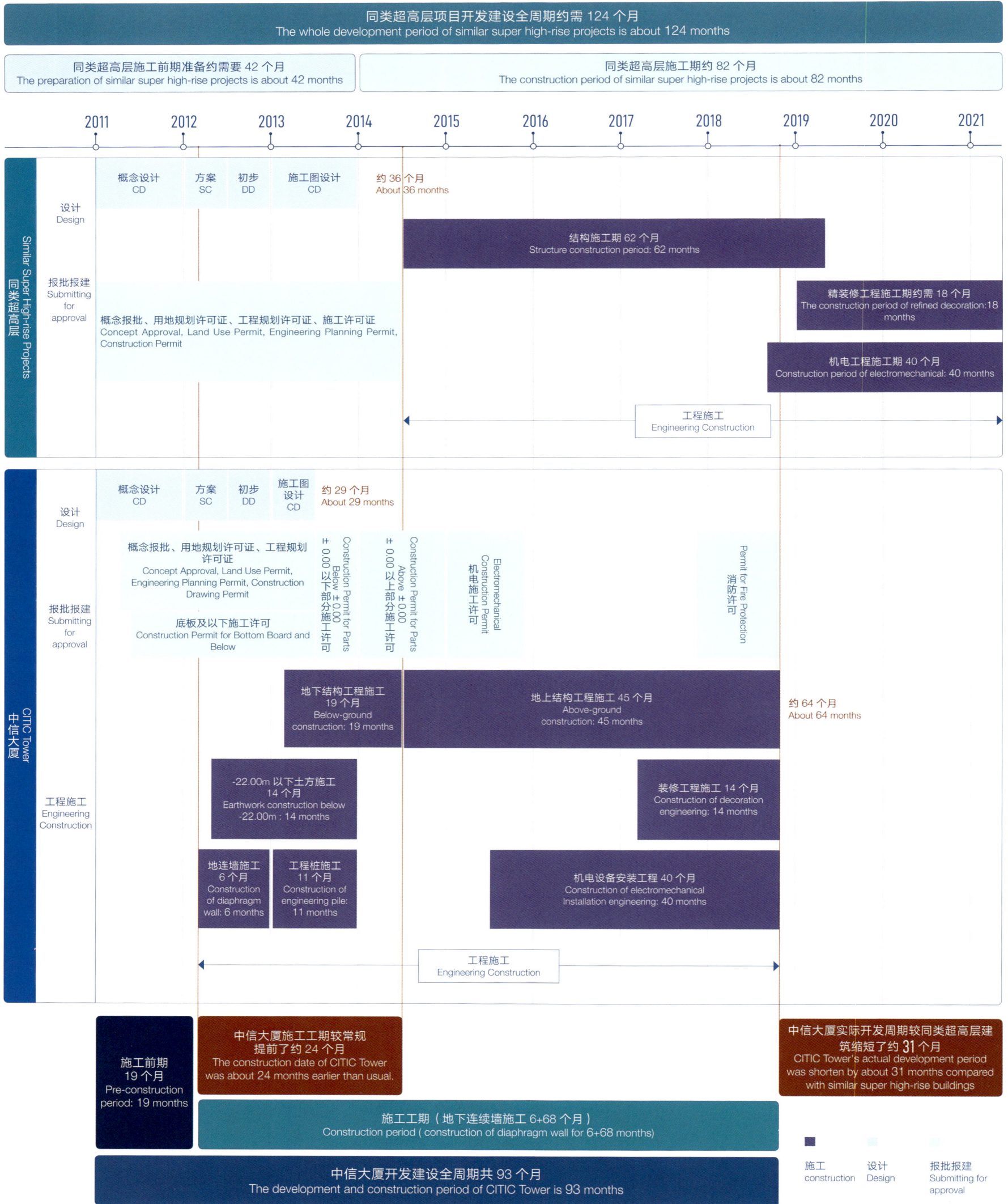

同类超高层项目开发建设全周期约需 124 个月
The whole development period of similar super high-rise projects is about 124 months

同类超高层施工前期准备约需要 42 个月
The preparation of similar super high-rise projects is about 42 months

同类超高层施工期约 82 个月
The construction period of similar super high-rise projects is about 82 months

2011　2012　2013　2014　2015　2016　2017　2018　2019　2020　2021

Similar Super High-rise Projects 同类超高层

设计 Design

概念设计 CD　方案 SC　初步 DD　施工图设计 CD

约 36 个月 About 36 months

报批报建 Submitting for approval

概念报批、用地规划许可证、工程规划许可证、施工许可证
Concept Approval, Land Use Permit, Engineering Planning Permit, Construction Permit

结构施工期 62 个月
Structure construction period: 62 months

精装修工程施工期约需 18 个月
The construction period of refined decoration: 18 months

机电工程施工期 40 个月
Construction period of electromechanical: 40 months

工程施工
Engineering Construction

CITIC Tower 中信大厦

设计 Design

概念设计 CD　方案 SC　初步 DD　施工图设计 CD

约 29 个月 About 29 months

报批报建 Submitting for approval

概念报批、用地规划许可证、工程规划许可证
Concept Approval, Land Use Permit, Engineering Planning Permit, Construction Drawing Permit

底板及以下施工许可
Construction Permit for Bottom Board and Below

Below ±0.00 ±0.00 以下部分施工许可 Construction Permit for Parts

±0.00 以上部分施工许可 Above ±0.00 Construction Permit for Parts

机电施工许可 Electromechanical Construction Permit

消防许可 Permit for Fire Protection

工程施工 Engineering Construction

地下结构工程施工 19 个月
Below-ground construction: 19 months

地上结构工程施工 45 个月
Above-ground construction: 45 months

约 64 个月 About 64 months

-22.00m 以下土方施工 14 个月
Earthwork construction below -22.00m: 14 months

装修工程施工 14 个月
Construction of decoration engineering: 14 months

地连墙施工 6 个月
Construction of diaphragm wall: 6 months

工程桩施工 11 个月
Construction of engineering pile: 11 months

机电设备安装工程 40 个月
Construction of electromechanical Installation engineering: 40 months

工程施工
Engineering Construction

施工前期 19 个月
Pre-construction period: 19 months

中信大厦施工工期较常规提前了约 24 个月
The construction date of CITIC Tower was about 24 months earlier than usual.

中信大厦实际开发周期较同类超高层建筑缩短了约 31 个月
CITIC Tower's actual development period was shorten by about 31 months compared with similar super high-rise buildings

施工工期（地下连续墙施工 6+68 个月）
Construction period (construction of diaphragm wall for 6+68 months)

中信大厦开发建设全周期共 93 个月
The development and construction period of CITIC Tower is 93 months

施工 construction　设计 Design　报批报建 Submitting for approval

中信大厦工期形象进度折线图
CITIC Tower's progress image line chart

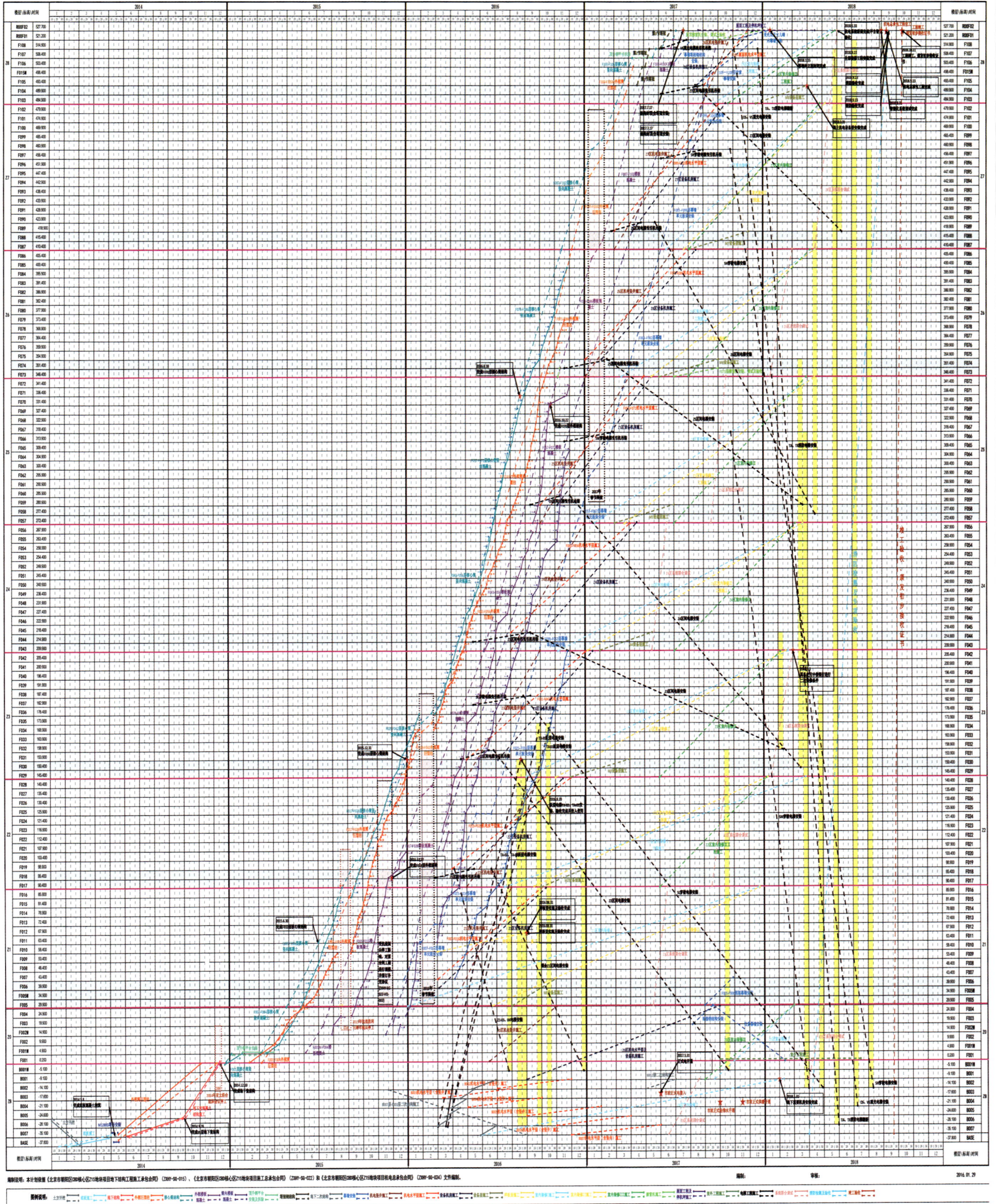

4.6　工程质量管理
Engineering Quality Management

大数据 Data

449 项
建筑类及机电类材料
检测、复试
Items of inspection, re-
examination of building and
electromechanical materials

50 项
品质检测项目
Items of quality inspection

119 份
第三方检测报告总计
Total third-party testing reports

10 个
涉及分部工程
Branch projects involved

75 个
涉及子分部工程
Sub-branch projects involved

117 个
涉及子分项工程
Sub-item projects involved

16,765 次
检验批次
Inspection batches

>50,000 次
工序过程验收次数
Process acceptance times

95%
一次验收合格率
The one-time acceptance rate

在中信大厦建设全过程中，中信和业投资有限公司秉承"全生命周期"的建造理念，坚持对工程使用的设备、材料，进行独立的第三方检验、测试，以确保工程品质符合设计要求；中信大厦按规完成建筑类及机电工程设备材料类检测复试 449 项：其中建筑类材料检测复试 330 项，机电类设备材料检测复试 119 项；共计获得检测报告 119 份，工程类及机电设备材料类品质检测 50 项。

中信大厦施工期间涉及地基与基础、主体结构、建筑装饰装修、屋面、给排水及采暖、通风与空调、建筑电气、智能建筑、建筑节能、电梯等分部工程 10 个，子分部工程 75 个，分项工程 117 个，检验批 16,765 次，全部一次验收合格，开展工序过程验收 50,000 余次，一次验收合格率 95% 以上。

推行样板验收、首件验收、举牌验收等制度。成立以项目经理为第一责任人，囊括各专业分包商项目经理在内的项目质量管理领导小组，并建立以项目质量管理小组为核心的质量管理组织架构，明确各组成机构管理职责；通过合同、总分包质量管理协议落实质量管理责任分解，保障质量目标的实现。项目建立"源头控制"和"过程控制"相结合的施工质量管理体系，实现质量的全过程管控。根据项目实施进展、资源配备的动态变化，在不同阶段识别质量控制重点，动态调整质量管理计划，对不同阶段的关键工序及特殊部位实施重点监控，保证整体工程质量。工程项目已获北京市结构工程长城杯金质奖、亚太地区地产领袖峰会 2018 重大项目金奖、中国钢结构金奖杰出工程大奖等奖项。

中信和业

中国尊大厦
第三方试验检验报告

中信和业投资有限公司
2018 年 12 月 17 日

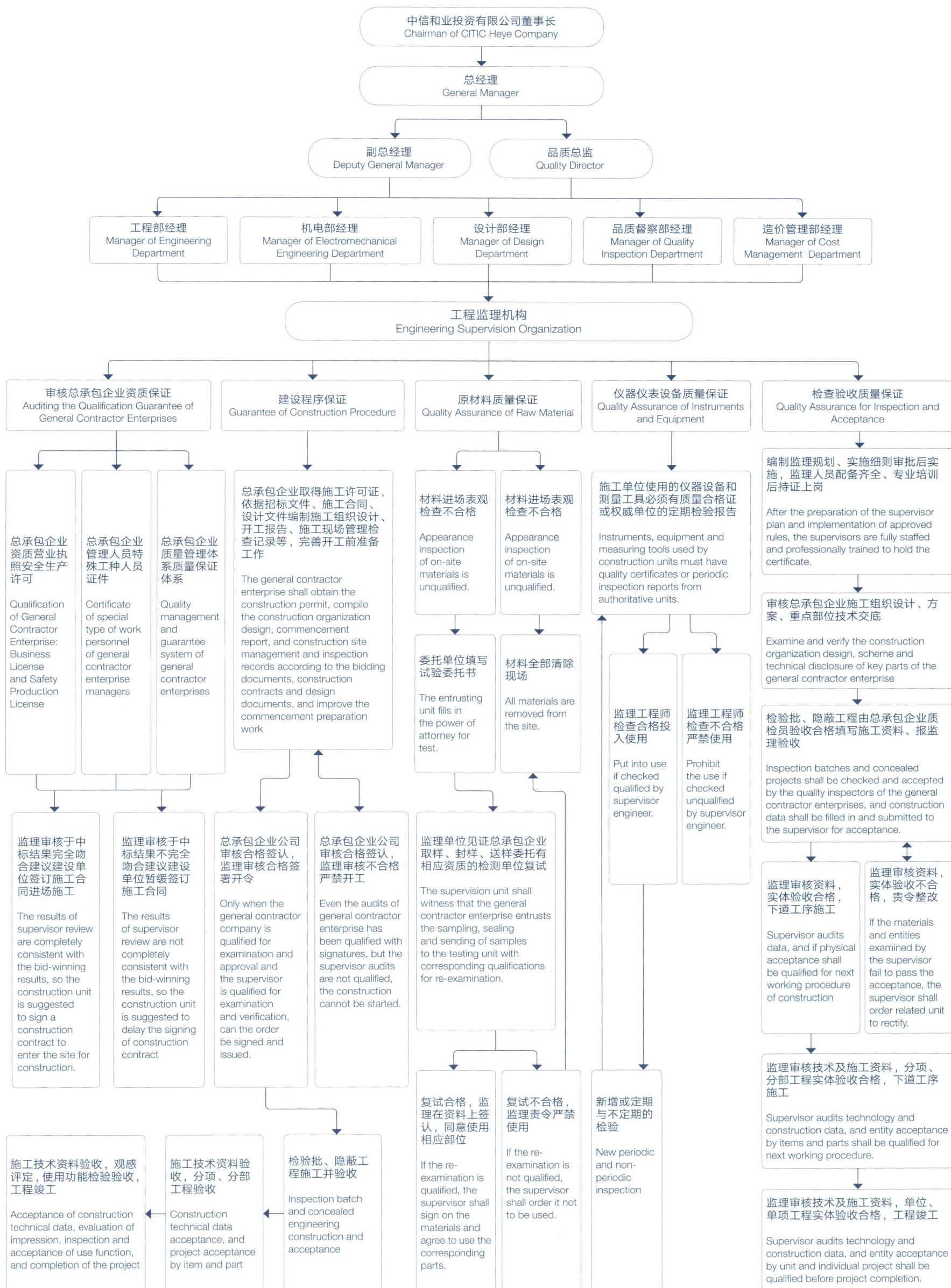

中信和业投资有限公司董事长
Chairman of CITIC Heye Company

总经理
General Manager

副总经理
Deputy General Manager

品质总监
Quality Director

工程部经理
Manager of Engineering Department

机电部经理
Manager of Electromechanical Engineering Department

设计部经理
Manager of Design Department

品质督察部经理
Manager of Quality Inspection Department

造价管理部经理
Manager of Cost Management Department

工程监理机构
Engineering Supervision Organization

审核总承包企业资质保证
Auditing the Qualification Guarantee of General Contractor Enterprises

建设程序保证
Guarantee of Construction Procedure

原材料质量保证
Quality Assurance of Raw Material

仪器仪表设备质量保证
Quality Assurance of Instruments and Equipment

检查验收质量保证
Quality Assurance for Inspection and Acceptance

总承包企业资质营业执照安全生产许可
Qualification of General Contractor Enterprise: Business License and Safety Production License

总承包企业管理人员特殊工种证件
Certificate of special type of work personnel of general contractor enterprise managers

总承包企业质量管理体系质量保证体系
Quality management and guarantee system of general contractor enterprises

总承包企业取得施工许可证，依据招标文件、施工合同、设计文件编制施工组织设计、开工报告、施工现场管理检查记录等，完善开工前准备工作
The general contractor enterprise shall obtain the construction permit, compile the construction organization design, commencement report, and construction site management and inspection records according to the bidding documents, construction contracts and design documents, and improve the commencement preparation work

材料进场表观检查不合格
Appearance inspection of on-site materials is unqualified.

材料进场表观检查不合格
Appearance inspection of on-site materials is unqualified.

施工单位使用的仪器设备和测量工具必须有质量合格证或权威单位的定期检验报告
Instruments, equipment and measuring tools used by construction units must have quality certificates or periodic inspection reports from authoritative units.

编制监理规划、实施细则审批后实施，监理人员配备齐全、专业培训后持证上岗
After the preparation of the supervisor plan and implementation of approved rules, the supervisors are fully staffed and professionally trained to hold the certificate.

审核总承包企业施工组织设计、方案、重点部位技术交底
Examine and verify the construction organization design, scheme and technical disclosure of key parts of the general contractor enterprise

委托单位填写试验委托书
The entrusting unit fills in the power of attorney for test.

材料全部清除现场
All materials are removed from the site.

检验批、隐蔽工程由总承包企业质检员验收合格填写施工资料、报监理验收
Inspection batches and concealed projects shall be checked and accepted by the quality inspectors of the general contractor enterprises, and construction data shall be filled in and submitted to the supervisor for acceptance.

监理工程师检查合格投入使用
Put into use if checked qualified by supervisor engineer.

监理工程师检查不合格严禁使用
Prohibit the use if checked unqualified by supervisor engineer.

监理审核于中标结果完全吻合建议建设单位签订施工合同进场施工
The results of supervisor review are completely consistent with the bid-winning results, so the construction unit is suggested to sign a construction contract to enter the site for construction.

监理审核于中标结果不完全吻合建议建设单位暂缓签订施工合同
The results of supervisor review are not completely consistent with the bid-winning results, so the construction unit is suggested to delay the signing of construction contract

总承包企业公司审核合格签认，监理审核合格签署开令
Only when the general contractor company is qualified for examination and approval and the supervisor is qualified for examination and verification, can the order be signed and issued.

总承包企业公司审核合格签认，监理审核不合格严禁开工
Even the audits of general contractor enterprise has been qualified with signatures, but the supervisor audits are not qualified, the construction cannot be started.

监理单位见证总承包企业取样、封样、送样委托有相应资质的检测单位复试
The supervision unit shall witness that the general contractor enterprise entrusts the sampling, sealing and sending of samples to the testing unit with corresponding qualifications for re-examination.

监理审核资料，实体验收合格，下道工序施工
Supervisor audits data, and if physical acceptance shall be qualified for next working procedure of construction

监理审核资料，实体验收不合格，责令整改
If the materials and entities examined by the supervisor fail to pass the acceptance, the supervisor shall order related unit to rectify.

监理审核技术及施工资料，分项、分部工程实体验收合格，下道工序施工
Supervisor audits technology and construction data, and entity acceptance by items and parts shall be qualified for next working procedure.

监理审核技术及施工资料，单位、单项工程实体验收合格，工程竣工
Supervisor audits technology and construction data, and entity acceptance by unit and individual project shall be qualified before project completion.

施工技术资料验收，观感评定，使用功能检验验收，工程竣工
Acceptance of construction technical data, evaluation of impression, inspection and acceptance of use function, and completion of the project

施工技术资料验收，分项、分部工程验收
Construction technical data acceptance, and project acceptance by item and part

检验批、隐蔽工程施工并验收
Inspection batch and concealed engineering construction and acceptance

复试合格，监理在资料上签认，同意使用相应部位
If the re-examination is qualified, the supervisor shall sign on the materials and agree to use the corresponding parts.

复试不合格，监理责令严禁使用
If the re-examination is not qualified, the supervisor shall order it not to be used.

新增或定期与不定期的检验
New periodic and non-periodic inspection

中信大厦质量管理组织结构
Project quality management organizational structure of CITIC Tower

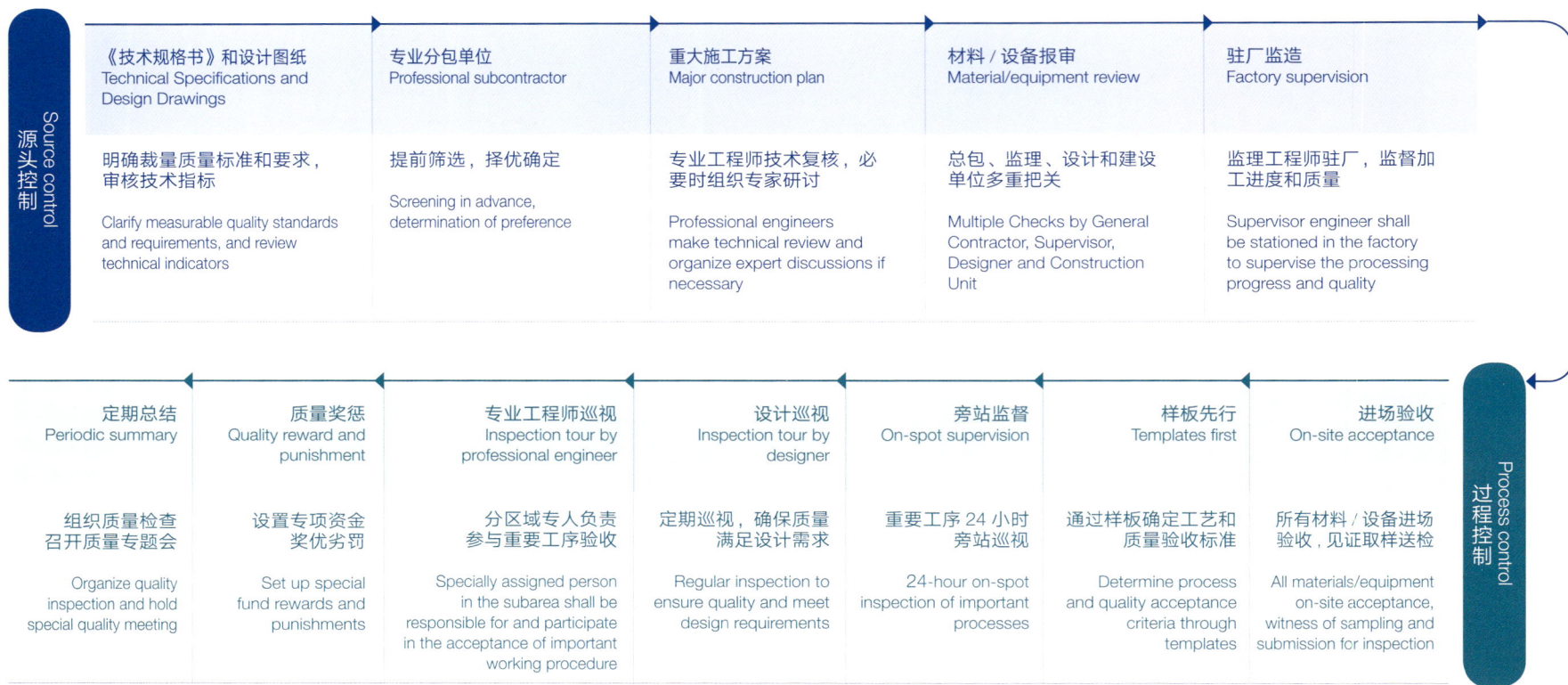

源头控制 / Source control

《技术规格书》和设计图纸 Technical Specifications and Design Drawings	专业分包单位 Professional subcontractor	重大施工方案 Major construction plan	材料 / 设备报审 Material/equipment review	驻厂监造 Factory supervision
明确裁量质量标准和要求，审核技术指标 Clarify measurable quality standards and requirements, and review technical indicators	提前筛选，择优确定 Screening in advance, determination of preference	专业工程师技术复核，必要时组织专家研讨 Professional engineers make technical review and organize expert discussions if necessary	总包、监理、设计和建设单位多重把关 Multiple Checks by General Contractor, Supervisor, Designer and Construction Unit	监理工程师驻厂，监督加工进度和质量 Supervisor engineer shall be stationed in the factory to supervise the processing progress and quality

过程控制 / Process control

定期总结 Periodic summary	质量奖惩 Quality reward and punishment	专业工程师巡视 Inspection tour by professional engineer	设计巡视 Inspection tour by designer	旁站监督 On-spot supervision	样板先行 Templates first	进场验收 On-site acceptance
组织质量检查召开质量专题会 Organize quality inspection and hold special quality meeting	设置专项资金奖优劣罚 Set up special fund rewards and punishments	分区域专人负责参与重要工序验收 Specially assigned person in the subarea shall be responsible for and participate in the acceptance of important working procedure	定期巡视，确保质量满足设计需求 Regular inspection to ensure quality and meet design requirements	重要工序 24 小时旁站巡视 24-hour on-spot inspection of important processes	通过样板确定工艺和质量验收标准 Determine process and quality acceptance criteria through templates	所有材料 / 设备进场验收，见证取样送检 All materials/equipment on-site acceptance, witness of sampling and submission for inspection

中信大厦施工总承包质量管理组织结构
CITIC Tower management organizational structure of construction general contractor quality

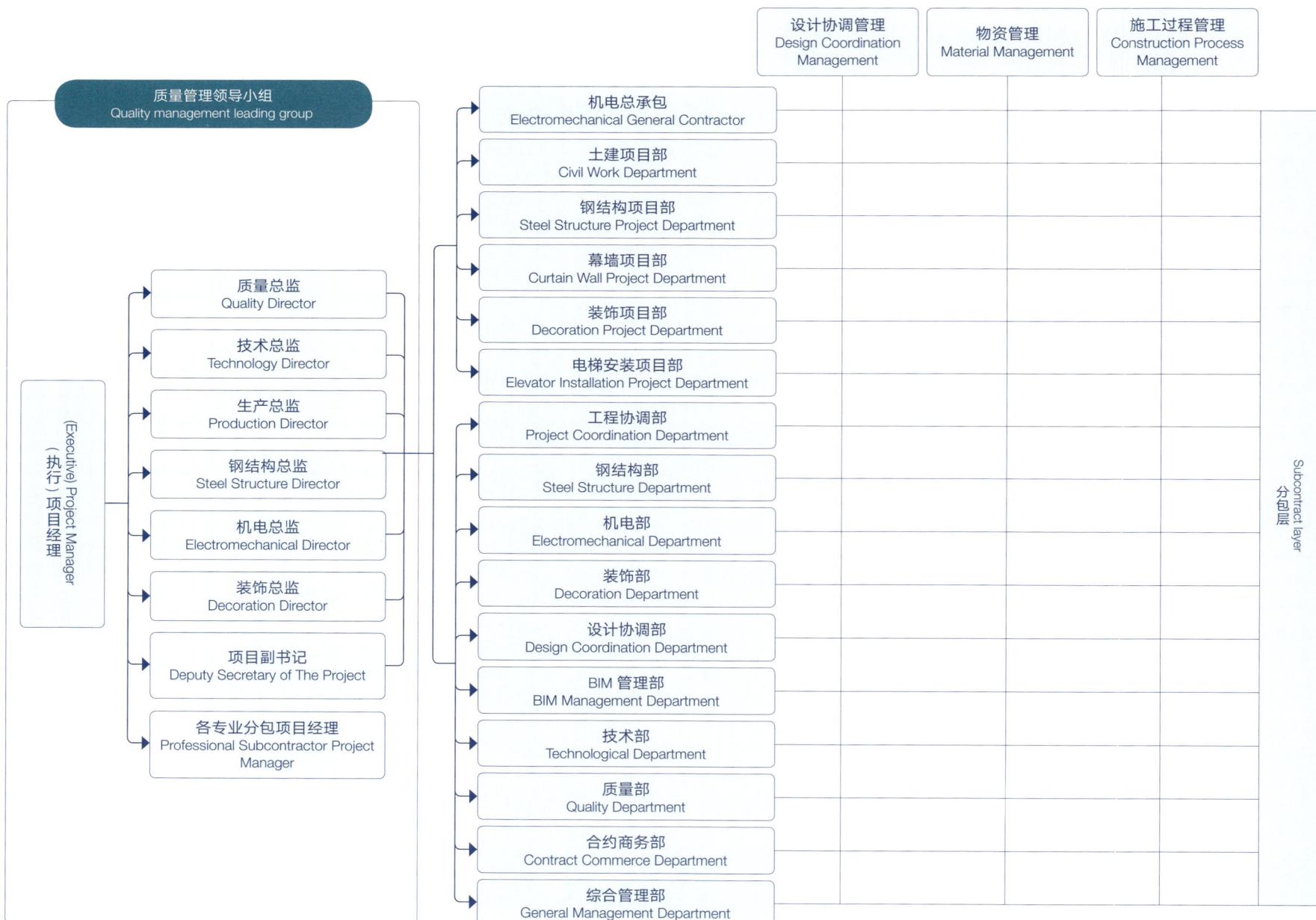

(Executive) Project Manager（执行）项目经理		设计协调管理 Design Coordination Management	物资管理 Material Management	施工过程管理 Construction Process Management

质量管理领导小组
Quality management leading group

质量总监 Quality Director
技术总监 Technology Director
生产总监 Production Director
钢结构总监 Steel Structure Director
机电总监 Electromechanical Director
装饰总监 Decoration Director
项目副书记 Deputy Secretary of The Project
各专业分包项目经理 Professional Subcontractor Project Manager

机电总承包 Electromechanical General Contractor
土建项目部 Civil Work Department
钢结构项目部 Steel Structure Project Department
幕墙项目部 Curtain Wall Project Department
装饰项目部 Decoration Project Department
电梯安装项目部 Elevator Installation Project Department
工程协调部 Project Coordination Department
钢结构部 Steel Structure Department
机电部 Electromechanical Department
装饰部 Decoration Department
设计协调部 Design Coordination Department
BIM 管理部 BIM Management Department
技术部 Technological Department
质量部 Quality Department
合约商务部 Contract Commerce Department
综合管理部 General Management Department

分包层 / Subcontract layer

During the whole process of construction of the CITIC Tower, CITIC Heye Investment Co., Ltd. adheres to the construction concept of "Full Life Cycle" and carries out independent third-party inspection and test on the equipment and materials used in the project to ensure that the quality of the project meets the design requirements; Up to now, CITIC Tower has completed 449 items of re-examination for construction and electromechanical engineering equipment and materials according to regulations: among them, there are 330 items of re-examination for building materials and 119 items of re-examination for electromechanical equipment materials. A total of 119 inspection reports have been made, and 50 items of quality inspection of engineering and electromechanical equipment materials have been made.

During the construction period of the CITIC Tower, 10 branch projects including foundation and foundation, main structure, building decoration and decoration, roof, water supply and drainage, heating, ventilation and air conditioning, building electricity, intelligent building, building energy conservation, elevator, etc. are involved. There are 75 sub-branch projects and 117 sub-item projects, with 16,765 inspection batches, all of which have passed the acceptance at one time, and more than 50,000 inspections have been carried out in the process, with the acceptance rate of more than 95% at one time. Carry out the system of template acceptance, first article acceptance, brand acceptance and so on. They establish a project quality management leading group with the project manager as the first responsible person, including the project managers of various professional subcontractors, and establish a quality management organizational structure with the project quality management group as the core, and clarify the management responsibilities of each organization; implement the decomposition of quality management responsibilities through contractors and general subcontractor quality management agreements to ensure the realization of quality objectives. The project is ensured by a construction quality management system combining "source control" and "process control" to realize the whole process control of quality. According to the progress of project implementation and the dynamic change of resource allocation, the key points of quality control can be identified in different stages, the quality management plan can be adjusted dynamically, and the key procedures and special parts in different stages can be monitored to ensure the overall engineering quality. The project has won the Gold Prize in Beijing Structure Great Wall Cup, Major Projects Gold Award in 2018 Asia-Pacific Real Estate Leaders' Summit and Outstanding Engineering Award of China Steel Structure Gold Award.

施工质量管理流程
Construction quality management process

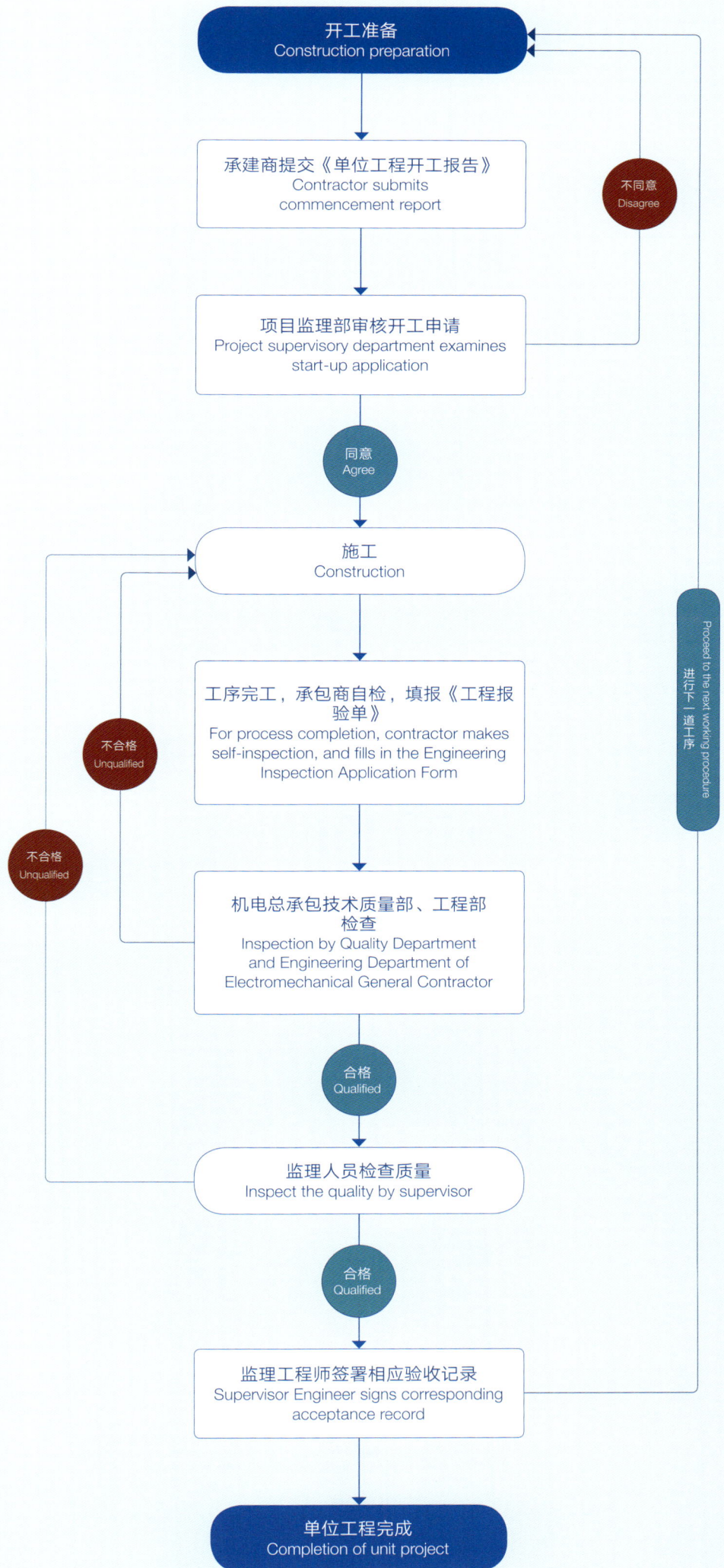

开工准备
Construction preparation

承建商提交《单位工程开工报告》
Contractor submits commencement report

不同意
Disagree

项目监理部审核开工申请
Project supervisory department examines start-up application

同意
Agree

施工
Construction

不合格
Unqualified

工序完工，承包商自检，填报《工程报验单》
For process completion, contractor makes self-inspection, and fills in the Engineering Inspection Application Form

不合格
Unqualified

机电总承包技术质量部、工程部检查
Inspection by Quality Department and Engineering Department of Electromechanical General Contractor

合格
Qualified

监理人员检查质量
Inspect the quality by supervisor

合格
Qualified

监理工程师签署相应验收记录
Supervisor Engineer signs corresponding acceptance record

单位工程完成
Completion of unit project

Proceed to the next working procedure
进行下一道工序

工程项目管理
Project Management

大数据 Data

>28 家
施工单位项下分包商
Subcontractors under
construction unit

>4,000 人
高峰期作业人员
Construction operation personnel

延伸阅读 Links

P311. P330

4.7 专业工程分包管理
Professional Engineering Subcontractor Management

为规范建设项目相关方的管理行为，给专业分包商提供优质管理与服务，提升工程整体履约水平，施工总承包特编制《中信大厦项目专业分包商管理行为准则》。从分包商进场管理、文档管理、会议管理、综合事务管理、施工现场协调、计划管理、技术与深化设计管理、施工质量管理、安全文明施工管理、合约计量管理、工程交付管理、竣工结算管理等方面对总分包管理行为进行约束。

施工总承包项下有 28 余家专业分包商和劳务分包，通过市场招标和企业集团内部招标两种方式，选择行业优秀专业分包商和劳务分包商，建立以施工单位为管理主体、分包合同履约为管理目标的新型承包模式，充分发挥总承包单位的统筹、协调、监督、服务等管理职能，确保工程顺利实施。高峰期双总包管理人员逾 480 人，施工作业人员逾 4,000 人。

In order to standardize the management behavior of relevant parties of construction project, provide high-quality management and service to professional subcontractors, and enhance the overall project performance level, the construction unit specially compiles the Code of Management Conduct for Professional Subcontractors of CITIC Tower. Constraints are imposed on subcontractor management from the following aspects: subcontractor entry management, document management, meeting management, comprehensive affairs management, construction site coordination, planning management, technical and in-depth design management, construction quality management, safe and civilized construction management, contractor measurement management, project delivery management and completion settlement management.

There are more than 28 professional subcontractors and labor subcontractors under the construction unit. By means of market bidding and internal bidding of enterprise groups, excellent professional subcontractors and labor subcontractors in the industry are selected, and a new contract mode is established with the construction unit as the main body of management and the subcontractor performance as the management objective, giving full play to the management functions of overall planning, coordination, supervision and service of electromechanical general contractor, so as to ensure the smooth implementation of the project. During the peak period, there were more than 480 double-general contractor managers and over 4,000 construction operators.

机电工程各分包管理模式
Electromechanics engineering subcontract management mode

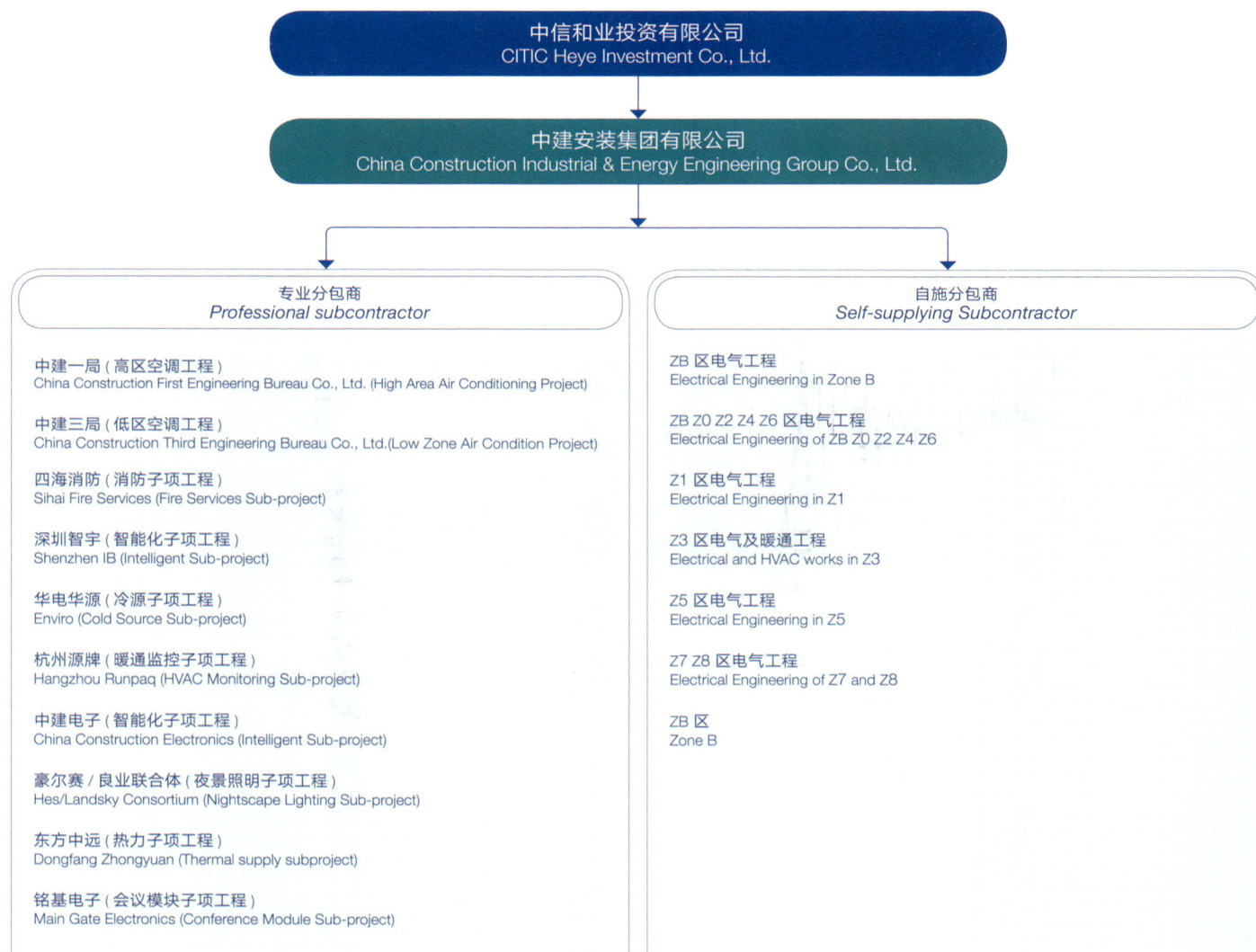

中信和业投资有限公司
CITIC Heye Investment Co., Ltd.

中建安装集团有限公司
China Construction Industrial & Energy Engineering Group Co., Ltd.

专业分包商 Professional subcontractor	自施分包商 Self-supplying Subcontractor
中建一局（高区空调工程）China Construction First Engineering Bureau Co., Ltd. (High Area Air Conditioning Project)	ZB 区电气工程 Electrical Engineering in Zone B
中建三局（低区空调工程）China Construction Third Engineering Bureau Co., Ltd.(Low Zone Air Condition Project)	ZB Z0 Z2 Z4 Z6 区电气工程 Electrical Engineering of ZB Z0 Z2 Z4 Z6
四海消防（消防子项工程）Sihai Fire Services (Fire Services Sub-project)	Z1 区电气工程 Electrical Engineering in Z1
深圳智宇（智能化子项工程）Shenzhen IB (Intelligent Sub-project)	Z3 区电气及暖通工程 Electrical and HVAC works in Z3
华电华源（冷源子项工程）Enviro (Cold Source Sub-project)	Z5 区电气工程 Electrical Engineering in Z5
杭州源牌（暖通监控子项工程）Hangzhou Runpaq (HVAC Monitoring Sub-project)	Z7 Z8 区电气工程 Electrical Engineering of Z7 and Z8
中建电子（智能化子项工程）China Construction Electronics (Intelligent Sub-project)	ZB 区 Zone B
豪尔赛／良业联合体（夜景照明子项工程）Hes/Landsky Consortium (Nightscape Lighting Sub-project)	
东方中远（热力子项工程）Dongfang Zhongyuan (Thermal supply subproject)	
铭基电子（会议模块子项工程）Main Gate Electronics (Conference Module Sub-project)	

施工总承包分包商管理流程
Subcontractor management process of construction general contractor

開始
Start
↓
分包商招标、合同签订
Subcontractor bidding, contract signing
↓
分包商进场审批
Approval for subcontractor entering the field
↓
月度考核
Monthly assessment
↓
分包商使用调查、评价
Survey and evaluation for subcontractor
↓
分包商履约调查、评价
Survey and evaluation for subcontractor performance
↓
完工考核
Completion Assessment
↓
考核结果应用
Application of assessment results
↓
分包商退场审批
Approval for subcontractor exiting
↓
结束
End

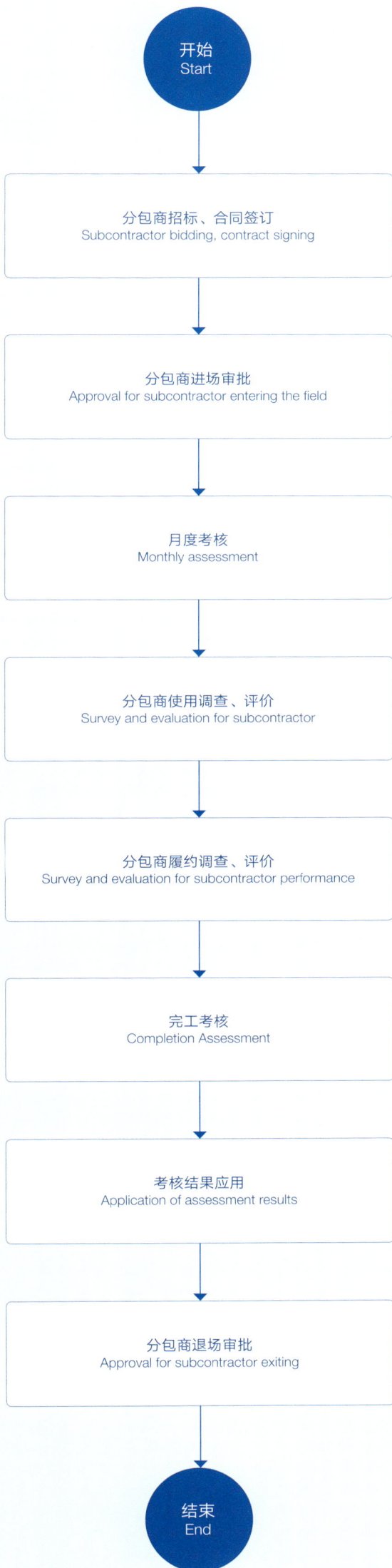

施工总承包专业分包名录		
序号	单位	施工范围
1	中建钢构有限公司	钢结构制作、安装
2	江苏沪宁钢机股份有限公司	钢结构制作
3	宝钢钢构有限公司	钢结构制作
4	中建东方装饰有限公司	室内装饰工程 I
		F004 精装修
5	中国建筑装饰集团有限公司	室内装饰工程 II（Z1、Z4 区）
6	深圳市建筑装饰（集团）有限公司	室内装饰工程 II（Z2、Z5 区）
7	中建深圳装饰有限公司	室内装饰工程 II（Z3、Z6 区）、Z3 区标准层精装修
8	北京江河幕墙系统工程有限公司	幕墙工程
9	北京华美装饰工程有限责任公司	室内装饰工程 II（Z7B 区）
10	浙江亚厦装饰股份有限公司	Z3、Z5 空中大堂精装修工程
		地下大堂精装修工程
		Z3 区（F32-F33）精装修工程
11	苏州金螳螂建筑装饰股份有限公司	F003 会议中心精装修工程
		首层大堂精装修工程
		Z7 区精装修工程
		Z2 区精装修工程
		首层大堂样板（A/C 区）精装修工程
12	北京弘高建筑装饰设计工程有限公司	Z1 区精装修工程
		首层大堂样板（B 区）精装修工程
		F010 样板装修工程
13	通力电梯有限公司	跃层电梯工程

CONSTRUCTION GENERAL CONTRACT PROFESSIONAL SUBCONTRACTOR LIST		
NO.	UNIT	CONSTRUCTION SCOPE
1	China Construction Steel Co., Ltd.	Steel structure fabrication and installation
2	Jiangsu Huning Steel Machinery Co., Ltd.	Steel structure
3	Baosteel Steel Structure Co., Ltd.	Steel structure
4	Zhongjian Oriental Decoration Co., Ltd.	Interior Decoration Engineering I
		F004 layer decoration
5	China Building Decoration Group Co., Ltd.	Interior Decoration Engineering II (Z1, Z4 District)
6	Shenzhen Building Decoration (Group) Co., Ltd.	Interior Decoration Engineering II (Z2, Z5 District)
7	China Construction Shenzhen Decoration Co., Ltd.	Interior decoration project II (Z3, Z6 area), Z3 area standard layer decoration
8	Beijing Jianghe Curtain Wall System Engineering Co., Ltd.	Curtain wall engineering
9	Beijing Huamei Decoration Engineering Co., Ltd.	Interior Decoration II (Z7B District)
10	Zhejiang Yaxia Decoration Co., Ltd.	Z3, Z5 sky lobby decoration project
		Underground lobby decoration project
		Z3 District (F32-F33) fine decoration project
11	Suzhou Jinhao Building Decoration Co., Ltd.	F003 floor conference center decoration project
		First floor lobby decoration project
		Z7 District Fine Decoration Project
		Z2 District Fine Decoration Project
		First floor lobby model (A/C area) fine decoration project
12	Beijing Honggao Architectural Decoration Design Engineering Co., Ltd.	Z1 District Fine Decoration Project
		First floor lobby model (B area) fine decoration project
		F010 layer model decoration project
13	KONE Elevator Co.	Elevator project

大数据 *Data*

751 项
双总包共同组织编制施工方案
Total construction schemes have been organized and compiled by the double general contractor

23 项
双总包共重点创新技术方案
Key innovative technical schemes of double general contractor

33 项次
双总包共重要专家论证、设备选型
Important expert demonstrations and equipment selection of double general contractor

<5 mm
同心（轴）度精度
Concentricity (axis) accuracy

273 种
材料考察、实验、检测和审核
Investigation, testing, detecting and review of 273 materials

延伸阅读 *Links*

P298. P300

4.8	施工技术管理
	Technical Work In Construction

中信大厦施工方案由施工总承包单位组织编制，由监理单位审核、建设单位复核后，监理单位审批执行。施工、机电双总包共同组织编制施工方案 751 项，重点创新技术方案 23 项，重要专家论证、设备选型 33 项次。

期间施工总承包组织编制了 623 项施工方案，其中对 15 项重点方案进行专家论证。针对超高层建筑多因素下易导致施工中主体结构倾斜的问题，施工总承包以折点坐标算法、等高比修正补偿技术，通过建立三维测量控制网，采用激光投影贯通测量技术，经复核，结构封顶后，同心（轴）度精度控制在 5mm 以内，满足规范及设计要求。

机电总承包组织编制了 128 项施工方案，对 273 种主要材料报审组织召开了专家技术考察、试验、检测和审核，对 3 个重点施工方案和 15 种重要材料进行了专家论证会。

针对中信大厦新、特、高、难、美的特点，项目实施初期对项目的施工重点、难点进行分析，中信和业投资有限公司、施工总承包和机电总承包在施工中积极研发并应用创新技术，旨在减少施工污染，缩短建造工期，提升工程质量。期间联合清华大学等高校，进行相应技术的研究，形成产学研一体的研究模式，并将相应成果形成专利、工法、论文、标准等，以期为其他超高层建筑的施工提供可借鉴的技术经验，引领行业发展。

中信大厦施工方案报审流程（不涉及造价变更）
Application procedure for approval of construction scheme of CITIC Tower (not involved in cost change)

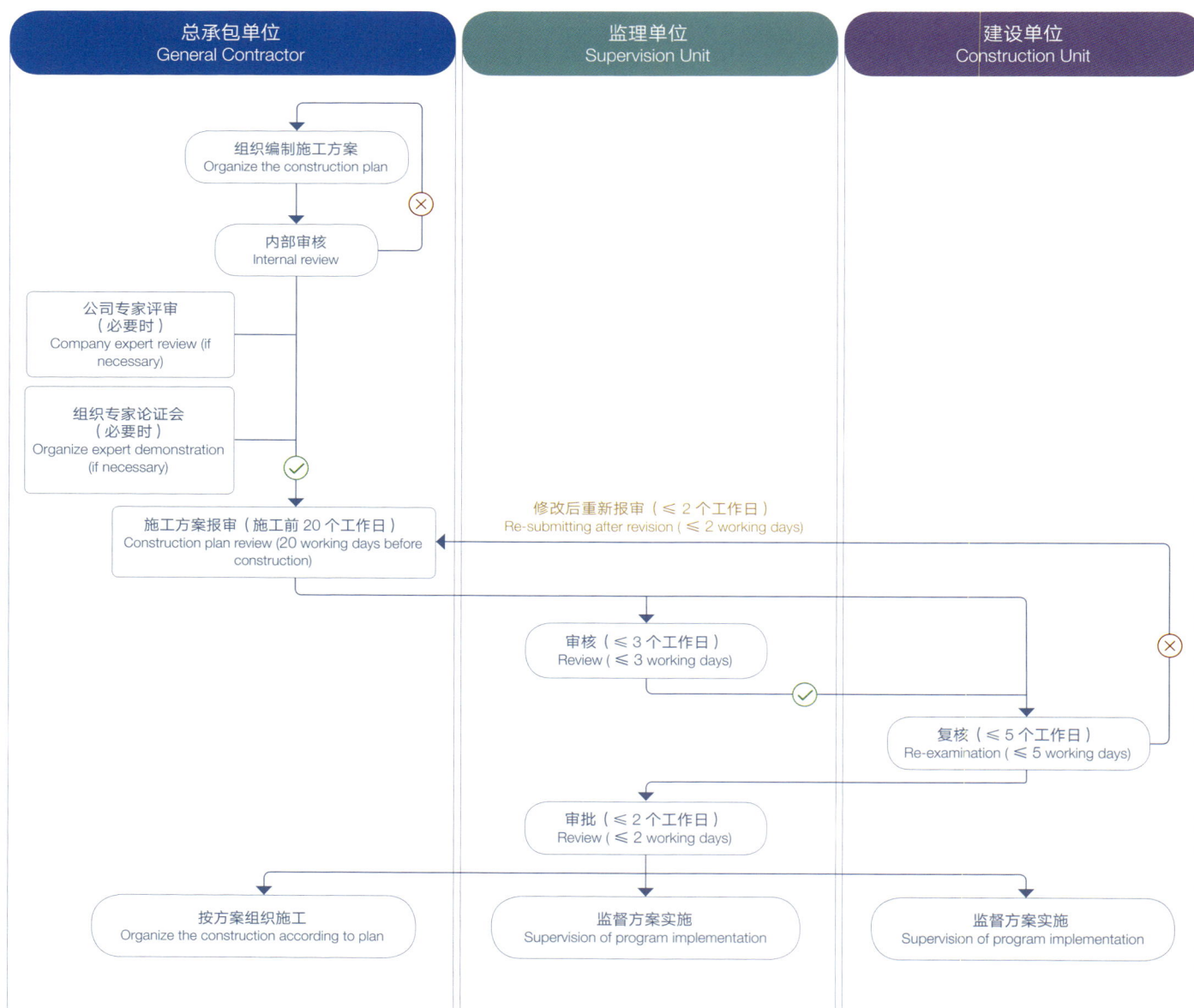

总承包单位 General Contractor	监理单位 Supervision Unit	建设单位 Construction Unit
组织编制施工方案 Organize the construction plan		
内部审核 Internal review		
公司专家评审（必要时）Company expert review (if necessary)		
组织专家论证会（必要时）Organize expert demonstration (if necessary)		
施工方案报审（施工前 20 个工作日）Construction plan review (20 working days before construction)	修改后重新报审（≤ 2 个工作日）Re-submitting after revision (≤ 2 working days)	
	审核（≤ 3 个工作日）Review (≤ 3 working days)	复核（≤ 5 个工作日）Re-examination (≤ 5 working days)
	审批（≤ 2 个工作日）Review (≤ 2 working days)	
按方案组织施工 Organize the construction according to plan	监督方案实施 Supervision of program implementation	监督方案实施 Supervision of program implementation

The construction plan of CITIC Tower is prepared by the general contractor and reviewed by the supervision unit. After the construction unit reexamines, the supervision unit examines and approves the implementation. 761 construction schemes, 23 key innovative technical schemes, 33 important expert demonstrations and equipment selection have been jointly compiled by the construction and electromechanical double general contractor unit.

During the construction period, 623 construction schemes have been compiled by the construction general contractor organization, of which 15 key schemes have been verified by experts. Aiming at the inclination of the main structure in the construction of super high-rise building under multi-factors, the general construction contractor adopts the algorithm of fold point coordinate and the correction and compensation technology of equal height ratio, establishes the 3D measurement control network, adopts the laser projection-through measurement technology, and controls the concentricity (axis) accuracy within 5mm after the structure is checked and the roof is sealed, which meets the requirements of the code and design.

Electromechanical general contractor organization has prepared 128 construction schemes, conducted technical investigation, testing, testing and review of 273 main materials for examination and approval, and conducted expert demonstration on 3 key construction schemes and 15 important materials.

In view of the new, special, high, difficult and beautiful characteristics of CITIC Tower, the key and difficult points of the project construction were analyzed in the early stage of the project implementation. CITIC Heye Investment Co., Ltd., General construction contractor and Electromechanical general contractor actively developed and applied innovative technologies in the construction, aiming at reducing construction pollution, shortening construction period and improving project quality. During this period, we joined Tsinghua University and other universities to study the corresponding technology and form a research model of integration of production, teaching and research. And the corresponding results have been formed into patents, construction methods, papers, standards, in order to provide technology experience reference for the construction of other super high-rise building, leading the development of the industry.

施工总承包	
序号	重点创新技术
1	超深超厚大体积基础底板施工技术
2	超大面积地脚锚栓群施工关键技术及应用
3	超大型结构底板灌浆施工关键技术及应用
4	智能化超高层建筑施工集成平台关键技术及应用
5	混凝土内灌外包多腔体巨型柱施工技术
6	超高层建筑跃层电梯的应用技术
7	巨柱集成式操作平台的应用技术
8	智能行车吊系统关键技术及应用
9	超高层建筑施工期间临时永久结合系统
10	BIM 技术全生命周期应用技术
11	超高层建筑巨型钢结构现场焊接机器人研发与应用
序号	重要专家论证方案列表
1	《土方开挖及内支撑结构施工方案》
2	《大体积混凝土底板施工方案》
3	《高大模板支撑架安全专项施工方案》
4	《地上结构测量方案》
5	《智能顶升钢平台专项施工方案》
6	《M900D 塔吊安装及 M1280D、M900D 塔吊爬升方案》
7	《钢结构安装方案》
8	《核心筒硬质防护与行车吊系统设计方案》
9	《巨柱集成式操作平台方案》
10	《建筑幕墙安装工程安全专项施工方案》
11	《电梯井脚手架安全专项施工方案》
12	《建筑幕墙安装工程悬臂吊施工安全专项施工方案》
13	《大型动臂塔机 M1280D 和 M900D 高空移位及拆卸施工方案》
14	《塔冠幕墙工程安全专项施工方案》
15	《塔冠区核心筒临时结构拆除工程施工方案》

机电总承包	
序号	重点创新技术
1	中信大厦项目核心内筒预制立管成套施工技术
2	施工现场地下室"永临结合"排水及排污技术
3	中信大厦大厦窗台系统优化及模块化
4	三维扫描点云技术与三维模型放样技术应用研究
5	超高层建筑"临永结合"消防水系统施工技术
6	BIM 在超高层项目机电工程中的实施应用
7	中信大厦智慧物联云平台建筑能源管理系统技术研究
8	超高层给排水一体化成套技术应用
9	中信大厦大厦组合式预制立管设计、安装的 BIM 应用
10	超高层管井模块化制作研发
11	机电模块化预制全生命周期管理精品工程
12	中信大厦大厦空调机组消声一体化
13	LED 灯具 DMX512+RDM 智能控制技术
14	主动式有源电力滤波谐波治理
序号	重要设备选型和重要方案专家论证
1	《冷却塔选型》
2	《预制立管施工方案》
3	《制冷机组选型》
4	《蓄冰盘管选型》
5	《内衬风管选型》
6	《自动扫描喷水灭火装置选型》
7	《火灾报警装置选型》
8	《高压配电柜选型》
9	《低压配电柜选型》
10	《矿物绝缘电缆选型》
11	《智能照明控制系统选型》
12	《高压垂吊电缆选型》
13	《柴油发电机组选型》
14	《暖通监控现场控制器选型》
15	《综合能源管理系统选型》
16	《网络交换设备选型》
17	《夜景照明定制吊篮和擦窗机措施方案》
18	《夜景照明专项安全施工方案》

工程项目管理
Project Management

CONSTRUCTION CONTRACT	
NO.	Name Of Key Innovation Technology
1	Construction technology of ultra-deep and super-thick mass foundation bottom plate
2	Key technology and application of super large area anchor bolt group construction
3	Key technology and application of bottom plate grouting for super large structure
4	Key technology and application of intelligent integrated platform for construction of super high-rise building
5	Construction technology of multi-cavity colossal column with inner concrete filling and outer cladding
6	Application technology of super high-rise jump lift
7	Application technology of giant column integrated operating platform
8	Key technology and application of intelligent traveling crane system
9	Temporary permanent bonding system for super high-rise building during construction
10	Full life cycle application technology of BIM technology
11	Development and application of field welding robot for super high-rise giant steel structure
NO.	List Of Important Expert Demonstration Proposals
1	Construction Scheme for Earth Excavation and Internal Support Structure
2	Construction Scheme of Mass Concrete Bottom Plate
3	Special Construction Scheme for Safety of Large Formwork Support Frame
4	Above-ground Structure Measurement Scheme
5	Special Construction Scheme of Intelligent Lifting Steel Platform
6	Installation of M900D Tower Crane and Climbing Scheme of M1280D and M900D Tower Cranes
7	Installation Scheme of Steel Structure
8	Design Scheme of Core Tube Hard Protection and Traveling Crane System
9	Scheme of Giant Column Integrated Operating Platform
10	Special Construction Scheme for Safety of Building Curtain Wall Installation Project
11	Special Construction Scheme for Safety of Elevator Shaft Scaffold
12	Special Construction Scheme for Safety of Cantilever Crane for Building Curtain Wall In-stallation Project
13	Construction Scheme for High Altitude Displacement and Disassembly of Large Boom Tower Cranes M1280D and M900D
14	Special Construction Scheme for Safety of Curtain Wall Project of Tower Crown
15	Construction Scheme for Temporary Structure Demolition of Core Tube in Tower Crown Area

MECHANICAL AND ELECTRICAL GENERAL CONTRACTOR	
NO.	Name Of Key Innovation Technology
1	CITIC Tower core internal cylinder precast complete vertical pipes construction technology
2	Construction site basement " permanent and temporary combination " drainage and sewage technology
3	Optimization and modularization of window system of CITIC Tower
4	Research on application of 3D point cloud technology and 3D pattern layout technology
5	Super high building " permanent and temporary combination " firefighting water system construction technology
6	Application of BIM in mechanical and electrical engineering of super high building project
7	Research on technology of building energy management system in CITIC Tower intelligent internet of things cloud platform
8	Application of complete technology on super high building water supply and drainage integration
9	BIM application on CITIC Tower combined precast vertical pipe design and installation
10	Development and research on super high building pipe shaft
11	Excellent course of mechanical and electrical modularization precast product lifecycle management
12	CITIC Tower air conditioner silencer integration
13	LED DMX512+RDM smart control technology
14	Active power filtering and harmonic treatment
NO.	Model Selection Of Main Equipment And Expert Argumentation On Important Schemes
1	Model selection of cooling tower
2	Construction scheme of precast vertical pipes
3	Model selection of refrigerating unit
4	Model selection of ice storage coil
5	Model selection of lining air duct
6	Model selection of automatic spray water and firefighting device
7	Model selection of fire alarm device
8	Model selection of HV distribution cabinet
9	Model selection of LV distribution cabinet
10	Model selection of mineral isolating cable
11	Model selection of smart lighting control system
12	Model selection of HV suspended cable
13	Model selection of diesel generator
14	Model selection of site monitor of heating and ventilation system
15	Model selection of comprehensive energy management system
16	Model selection of network switch
17	Measurement scheme of nightscape lighting customized basket and window cleaning equipment
18	Construction scheme for nightscape lighting special safety

施工安全管理
Construction safety management

本工程规模超大、施工周期长，现场高峰时段的施工作业人员超过 4,000 人。为进一步提高工人的安全意识，创新性地采用多媒体安全培训工具箱 + 安全体验馆 + 管理平台 + 手机 APP 查询与监管的方式辅助开展项目的安全培训工作，提高工人安全意识，消除事故隐患，保障生命安全，并实现对工人的规范管理。

中信大厦工程以施工总承包项目经理为第一责任人，各关键岗位、职能人员为组员的安全生产领导小组。各施工单位专职安全管理人员共计 54 人，总分包兼职安全员、消防巡查员共计 110 余人，形成了总包领导，分包支撑，专业配合，横向到边、纵向到底的牢固安全监管体系，有效落实项目安监责任。建立各部门、各级管理人员安全生产责任制，分层级签订年度安全生产责任状，每月对安全责任落实情况进行考核。

严格落实新入场工人三级安全教育，保障 50 小时培训时长，工人培训上岗合格率 100%。施工单位开展入场安全教育、周安全教育、季节性安全教育、节后安全教育 400 余次，累计教育施工人员达 2,672,000 人次，安全技术交底 54,000 余次，发现并消除各类安全隐患 15,600 余处。中信大厦项目获得"全国建设工程项目施工安全生产标准化工地"（原全国"AAA级安全文明样板工地"）和 2018 年度北京市安全最高荣誉北京市绿色安全样板工地荣誉。

>4,000 人
高峰施工作业人员
On-site workers during peak construction period

54 人
专职施工安全管理人员
Full-time safety management personnel

>110 人
兼职安全员、消防巡查员
Part-time safety inspector and fire inspector

100%
工人培训上岗合格率
Qualification rate of workers' training on duty

2,672,000 人次
累计教育施工人员
Cumulative training activities for construction personnel of the construction unit

54,000 次
安全技术交底
Safety technical disclosures by the construction unit

>15,600 处
发现并消除各类安全隐患
Discover and eliminate hidden dangers

与时俱进的施工安全管理模式
Construction safety management mode that keeps pace with the times

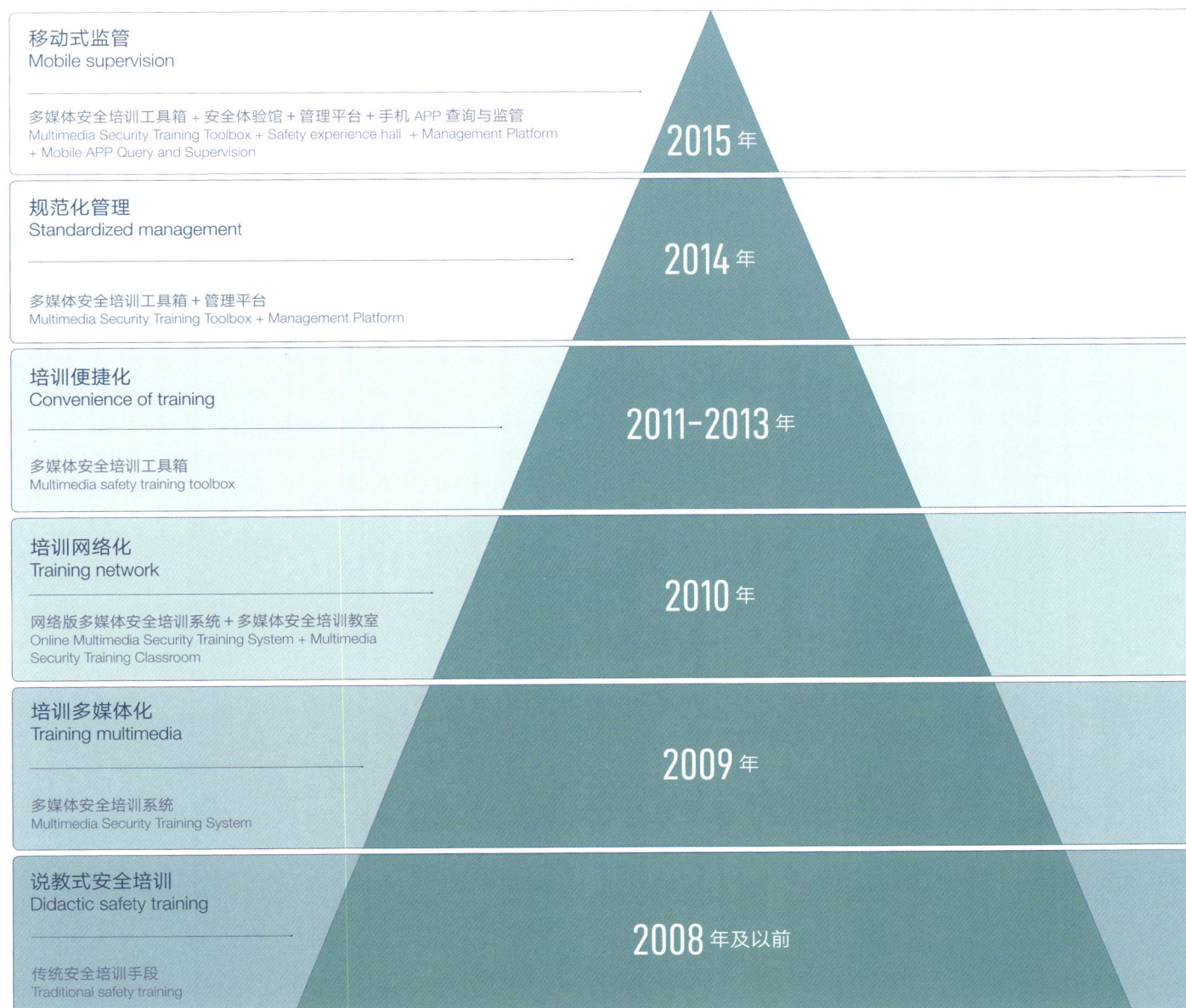

移动式监管
Mobile supervision

多媒体安全培训工具箱 + 安全体验馆 + 管理平台 + 手机 APP 查询与监管
Multimedia Security Training Toolbox + Safety experience hall + Management Platform + Mobile APP Query and Supervision

2015 年

规范化管理
Standardized management

多媒体安全培训工具箱 + 管理平台
Multimedia Security Training Toolbox + Management Platform

2014 年

培训便捷化
Convenience of training

多媒体安全培训工具箱
Multimedia safety training toolbox

2011-2013 年

培训网络化
Training network

网络版多媒体安全培训系统 + 多媒体安全培训教室
Online Multimedia Security Training System + Multimedia Security Training Classroom

2010 年

培训多媒体化
Training multimedia

多媒体安全培训系统
Multimedia Security Training System

2009 年

说教式安全培训
Didactic safety training

传统安全培训手段
Traditional safety training

2008 年及以前

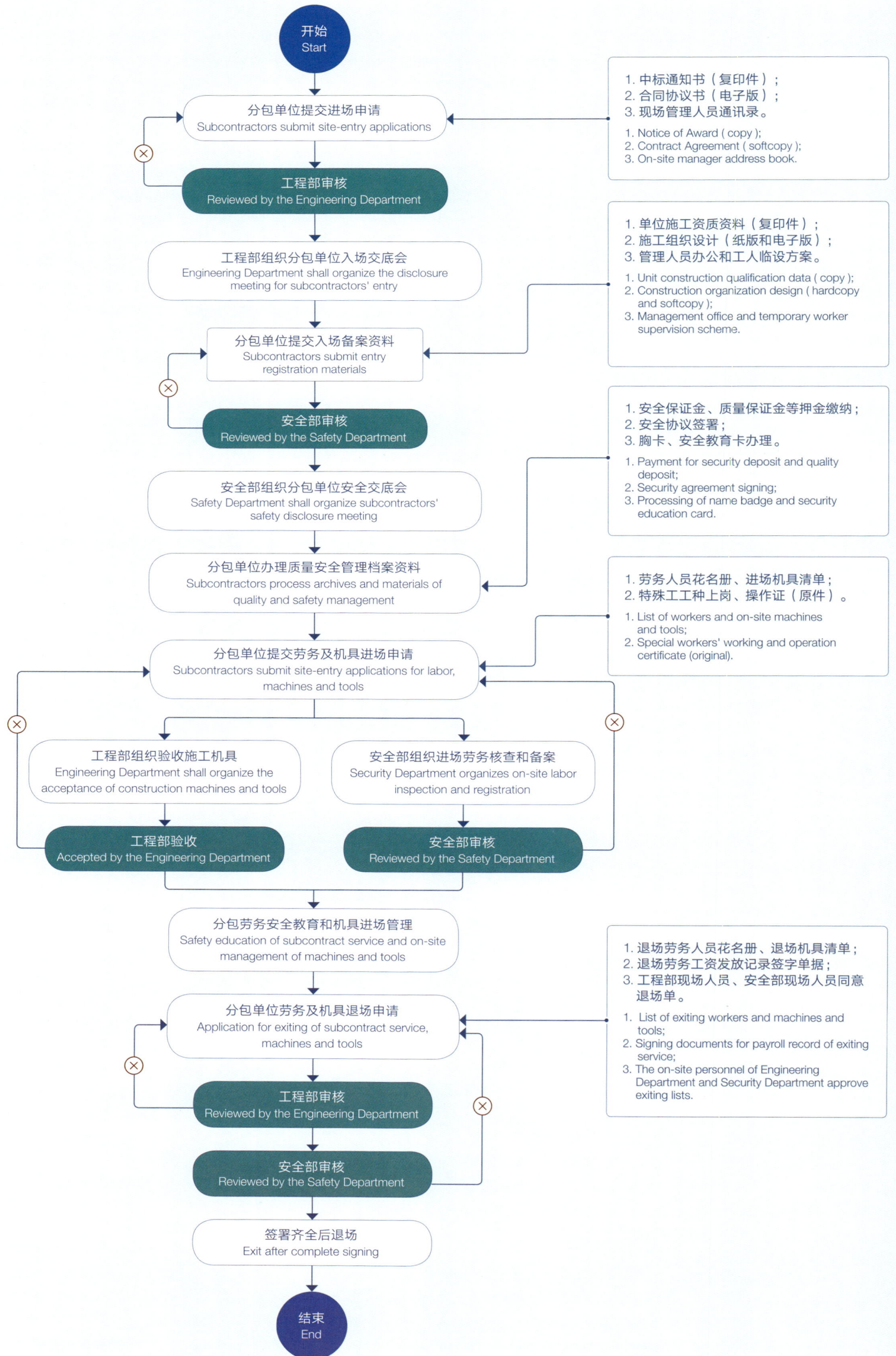

开始 / Start

分包单位提交进场申请
Subcontractors submit site-entry applications

1. 中标通知书（复印件）；
2. 合同协议书（电子版）；
3. 现场管理人员通讯录。

1. Notice of Award (copy);
2. Contract Agreement (softcopy);
3. On-site manager address book.

工程部审核
Reviewed by the Engineering Department

工程部组织分包单位入场交底会
Engineering Department shall organize the disclosure meeting for subcontractors' entry

1. 单位施工资质资料（复印件）；
2. 施工组织设计（纸版和电子版）；
3. 管理人员办公和工人临设方案。

1. Unit construction qualification data (copy);
2. Construction organization design (hardcopy and softcopy);
3. Management office and temporary worker supervision scheme.

分包单位提交入场备案资料
Subcontractors submit entry registration materials

安全部审核
Reviewed by the Safety Department

1. 安全保证金、质量保证金等押金缴纳；
2. 安全协议签署；
3. 胸卡、安全教育卡办理。

1. Payment for security deposit and quality deposit;
2. Security agreement signing;
3. Processing of name badge and security education card.

安全部组织分包单位安全交底会
Safety Department shall organize subcontractors' safety disclosure meeting

分包单位办理质量安全管理档案资料
Subcontractors process archives and materials of quality and safety management

1. 劳务人员花名册、进场机具清单；
2. 特殊工工种上岗、操作证（原件）。

1. List of workers and on-site machines and tools;
2. Special workers' working and operation certificate (original).

分包单位提交劳务及机具进场申请
Subcontractors submit site-entry applications for labor, machines and tools

工程部组织验收施工机具
Engineering Department shall organize the acceptance of construction machines and tools

安全部组织进场劳务核查和备案
Security Department organizes on-site labor inspection and registration

工程部验收
Accepted by the Engineering Department

安全部审核
Reviewed by the Safety Department

分包劳务安全教育和机具进场管理
Safety education of subcontract service and on-site management of machines and tools

分包单位劳务及机具退场申请
Application for exiting of subcontract service, machines and tools

1. 退场劳务人员花名册、退场机具清单；
2. 退场劳务工资发放记录签字单据；
3. 工程部现场人员、安全部现场人员同意退场单。

1. List of exiting workers and machines and tools;
2. Signing documents for payroll record of exiting service;
3. The on-site personnel of Engineering Department and Security Department approve exiting lists.

工程部审核
Reviewed by the Engineering Department

安全部审核
Reviewed by the Safety Department

签署齐全后退场
Exit after complete signing

结束 / End

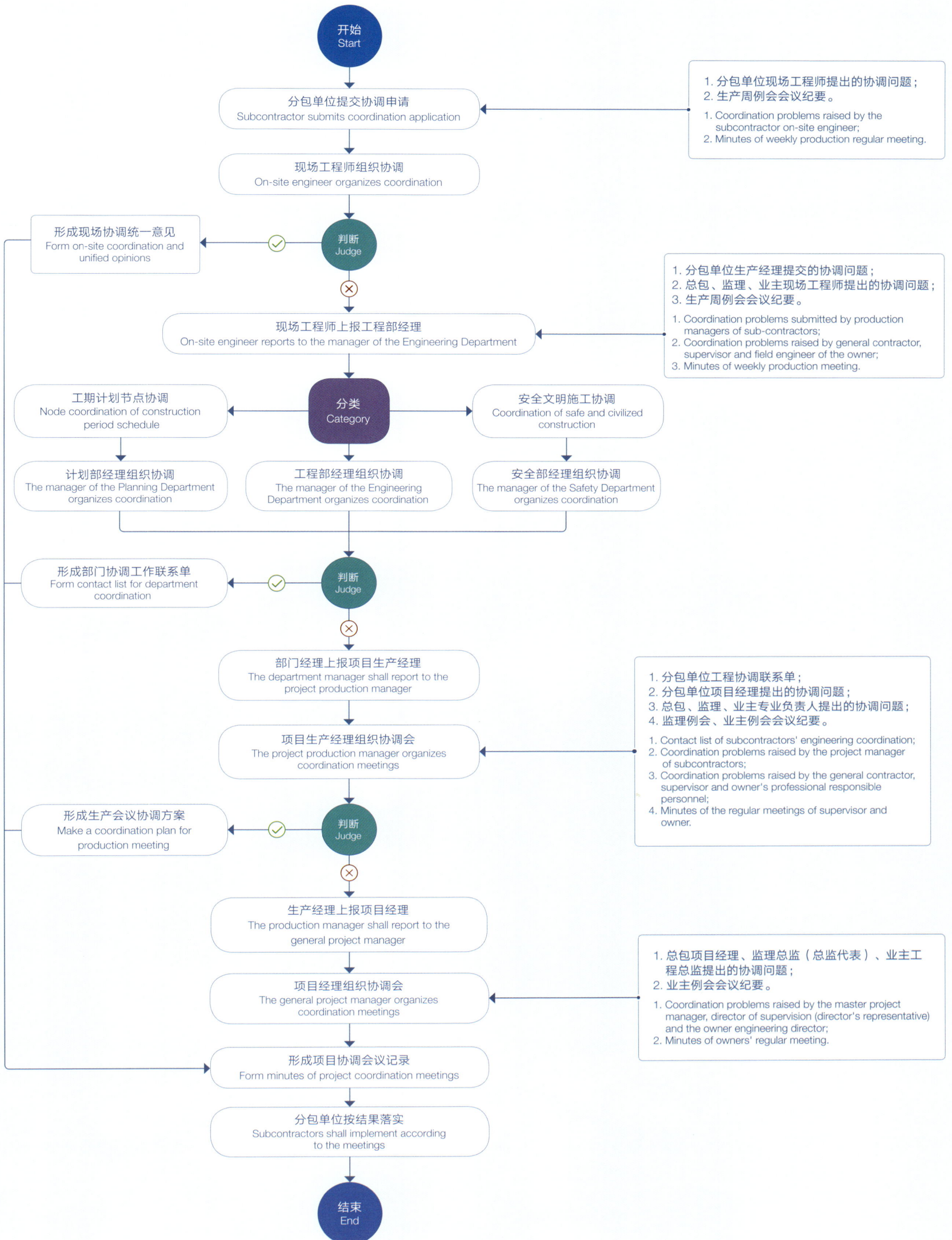

开始
Start

分包单位提交协调申请
Subcontractor submits coordination application

1. 分包单位现场工程师提出的协调问题；
2. 生产周例会会议纪要。

1. Coordination problems raised by the subcontractor on-site engineer;
2. Minutes of weekly production regular meeting.

现场工程师组织协调
On-site engineer organizes coordination

判断
Judge

形成现场协调统一意见
Form on-site coordination and unified opinions

现场工程师上报工程部经理
On-site engineer reports to the manager of the Engineering Department

1. 分包单位生产经理提交的协调问题；
2. 总包、监理、业主现场工程师提出的协调问题；
3. 生产周例会会议纪要。

1. Coordination problems submitted by production managers of sub-contractors;
2. Coordination problems raised by general contractor, supervisor and field engineer of the owner;
3. Minutes of weekly production meeting.

工期计划节点协调
Node coordination of construction period schedule

分类
Category

安全文明施工协调
Coordination of safe and civilized construction

计划部经理组织协调
The manager of the Planning Department organizes coordination

工程部经理组织协调
The manager of the Engineering Department organizes coordination

安全部经理组织协调
The manager of the Safety Department organizes coordination

形成部门协调工作联系单
Form contact list for department coordination

判断
Judge

部门经理上报项目生产经理
The department manager shall report to the project production manager

1. 分包单位工程协调联系单；
2. 分包单位项目经理提出的协调问题；
3. 总包、监理、业主专业负责人提出的协调问题；
4. 监理例会、业主例会会议纪要。

1. Contact list of subcontractors' engineering coordination;
2. Coordination problems raised by the project manager of subcontractors;
3. Coordination problems raised by the general contractor, supervisor and owner's professional responsible personnel;
4. Minutes of the regular meetings of supervisor and owner.

项目生产经理组织协调会
The project production manager organizes coordination meetings

形成生产会议协调方案
Make a coordination plan for production meeting

判断
Judge

生产经理上报项目经理
The production manager shall report to the general project manager

1. 总包项目经理、监理总监（总监代表）、业主工程总监提出的协调问题；
2. 业主例会会议纪要。

1. Coordination problems raised by the master project manager, director of supervision (director's representative) and the owner engineering director;
2. Minutes of owners' regular meeting.

项目经理组织协调会
The general project manager organizes coordination meetings

形成项目协调会议记录
Form minutes of project coordination meetings

分包单位按结果落实
Subcontractors shall implement according to the meetings

结束
End

The project is oversized and the construction period is long, with more than 4,000 construction personnel on site during peak hours. In order to further improve the safety awareness of workers, the multimedia safety training toolbox + Safety experience hall + management platform + mobile APP query and supervision are creatively used to assist in the safety training of the project, improve the safety awareness of workers, eliminate hidden risks of accidents, ensure the safety of life, and realize the standardized management of workers.

The master project manager of the general construction contractor acts as the first person responsible for the construction of CITIC Tower and the key posts and functional personnel act as the team members. The total number of full-time safety production managers of each construction unit is 54, and more than 110 part-time safety inspectors and fire inspectors of sub-contractors in total, which forms a solid safety supervision system of general contractor leadership, subcontracts support, professional cooperation, ensuring safety horizontally and vertically, and effectively implementing the

project safety supervision responsibilities. The project establishes the responsibility system for safety production for managers of all Departments and levels, and annual safety production responsibility commitments are signed at different levels, and the implementation of safety responsibility is appraised monthly.

The project team strictly implement the three-level safety education for newly-admitted workers, guarantee 50 hours of training, and the qualification rate of workers' training is 100%. The construction unit carry out more than 400 times of entrance safety education, weekly, seasonal and post-festival safety education, cumulatively educating the construction personnel 2,672,000 times, disclosing the safety technology 54,000 times and discovering and eliminating more than 15,600 hidden risks. The project has won the "National Construction Project Construction Safety Production Standardization Site" (the original national "AAA Class Safety and Civilization Model Site") and the "Beijing Green Safety Model Site" in 2018, the highest honor for safety in Beijing.

化解垂直运输瓶颈难题新途径 —— 跃层电梯技术运用
A New Way To Break The Vertical Transport Bottleneck--Application Of Jump Lift Technology

在国内传统工程施工模式下，超高层建筑基本采用附墙型临时施工电梯作为人员运输的主要工具。临时施工电梯运行速度慢、运输效率低，且现场可安装临时施工电梯空间有限，不能保证满足超高层建筑施工现场的运力需求。因其占用正式电梯的井道时间长，影响后期正式电梯的安装，会对工程的整体工期造成不利的影响。

中信大厦高峰期施工作业人员逾 4,000 人，为缓解现场垂直运输压力，创新应用了国际先进的跃层电梯技术。跃层电梯是在核心筒结构施工处于低楼层阶段时，在井道内安装部分正式电梯部件，作为运输施工人员及货物的施工电梯投入使用。其机房是可自行跃升的临时机房，通过自爬升机构上的可伸缩支撑梁，将机房临时固定于核心筒墙体上。随着上部井道的施工，逐步向上跃升，在结构封顶前将最终电梯使用的曳引机和控制柜运输至正式机房内，更换电气部件和损耗

部件，电梯重新调试验收后即可投入使用，大大缩短了正式电梯的安装时间。

作为施工电梯，跃层电梯的运行速度约是普通临时施工电梯的 4 倍，可达 4m/s，最大行程达到 514m。单台跃层电梯的运力相当于同规格施工电梯的 10 倍左右，最大载重量达 3,600kg。该安装方式可与建筑高度同步递增，有效解决超高层建筑垂直运输瓶颈问题，且运输效率、安全性、可靠性远高于传统建筑施工电梯。

大数据 Data

>4,000 人
高峰期施工作业人员
On-site workers at the peak of construction

4 倍
跃层电梯的速度
约是普通临时施工电梯的
The running speed of the jump lift is about 4 times that of the ordinary temporary construction elevator

4 m/s
跃层电梯施工期间的运行速度
The running speed of jump lift during construction

10 倍
单台跃层电梯的运力
是同规格施工电梯的
The capacity of a single jump lift is 10 times that of a construction elevator of the same specification

3,600 kg
单台跃层电梯的最大运力
The carrying capacity of a single jump lift

图片说明 Comments

1. 跃层电梯示意图
Jump Lift Schematic Diagram

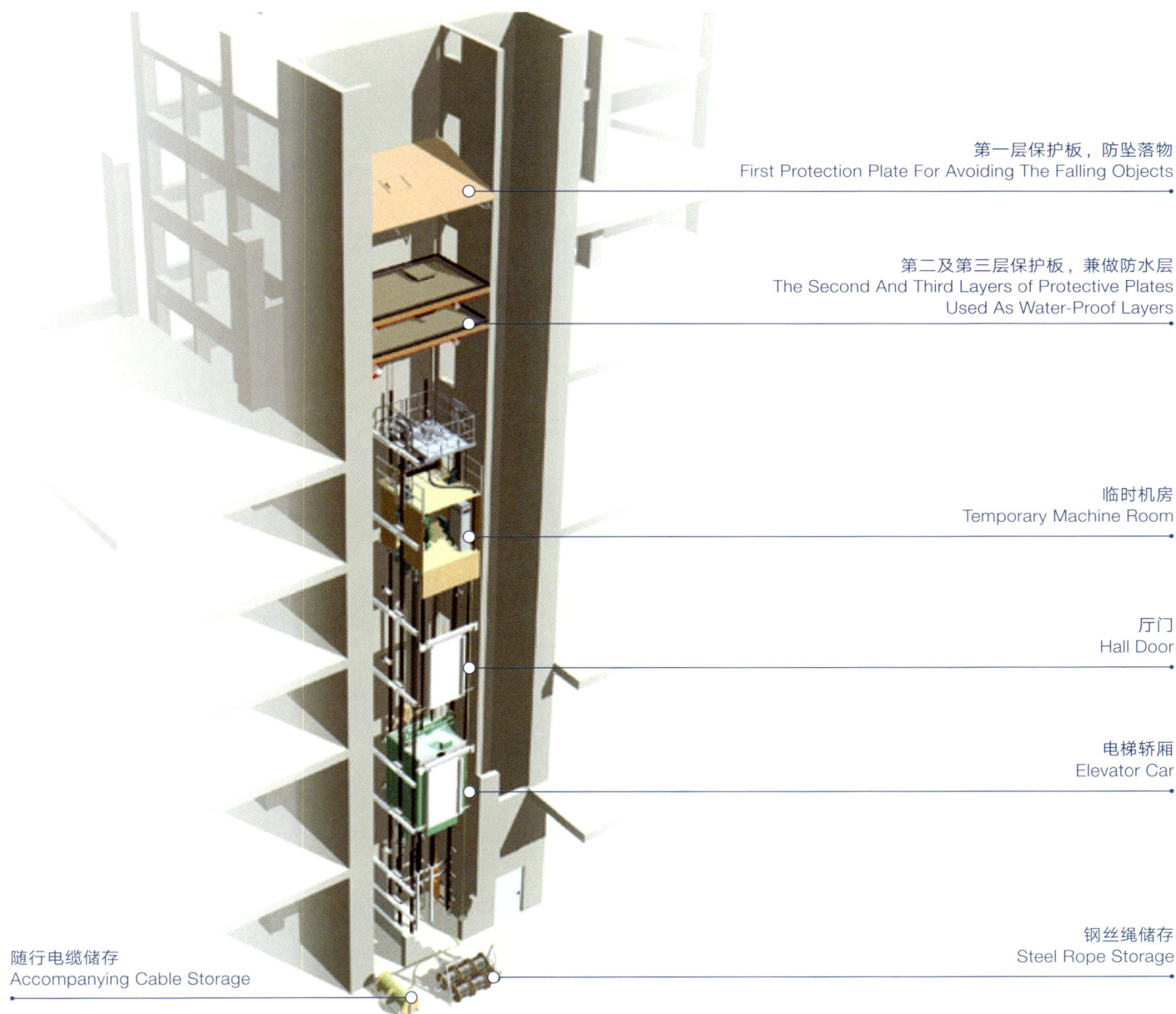

第一层保护板，防坠落物
First Protection Plate For Avoiding The Falling Objects

第二及第三层保护板，兼做防水层
The Second And Third Layers of Protective Plates Used As Water-Proof Layers

临时机房
Temporary Machine Room

厅门
Hall Door

电梯轿厢
Elevator Car

钢丝绳储存
Steel Rope Storage

随行电缆储存
Accompanying Cable Storage

1

1

Under the domestic traditional engineering construction mode, the temporary construction elevator with wall is basically used as the main means of personnel transportation in super high-rise buildings. The temporary construction elevator is slow in operation, low in transportation efficiency, and the space of temporary construction elevator can be installed on site is limited, which cannot meet the transportation demand of super high-rise building construction site. Besides, the temporary construction elevator occupies the shaft of the formal elevator for a long time, so it will affect the installation of the formal elevator in the later period, and have a negative impact on the overall construction period of the project.

During the peak construction period of CITIC Tower, there are more than 4,000 workers on the site. In order to relieve the pressure of vertical transportation on the site, the project creatively applies the jump lift technology. When the low-rise core tube is under construction, the jump lift is adopted to install part of the formal elevator in the shaft and used as the construction elevator for transportation of construction personnel and goods. Its machine room is a temporary machine room which can jump by itself, and the machine room is temporarily fixed on the core tube wall body through the telescopic support beam on the self-climbing mechanism. With the construction of the upper shaft and the jump of temporary machine room, the tractor and control cabinet used by the elevator will be transported to the formal machine room before the structural roof is sealed, and the electrical parts and the worn-out parts will be replaced. After the elevator is readjusted and accepted, the elevator can be put into use, thus greatly shortening the installation time of the formal elevator.

As a construction elevator, the running speed of the jump lift, 4 m/s, is about 4 times of that of the ordinary temporary construction elevator and the maximum stroke can reach 514m. The capacity of a single jump lift is about 10 times of that of the construction elevator with the same specification, which is 3,600 kg. The installation can be added synchronously with the building height, effectively solving the vertical transport bottleneck problem of super high-rise buildings, and the transportation efficiency, safety and reliability are much higher than the traditional building construction elevator.

图片说明 Comments

1. 超高层建筑跃层电梯
Super high-rise jump lift

2. 超高层建筑跃层电梯井
Super high-rise jump lift shaft

3. 超高跃层电梯按钮
Super high-rise jump lift buttons

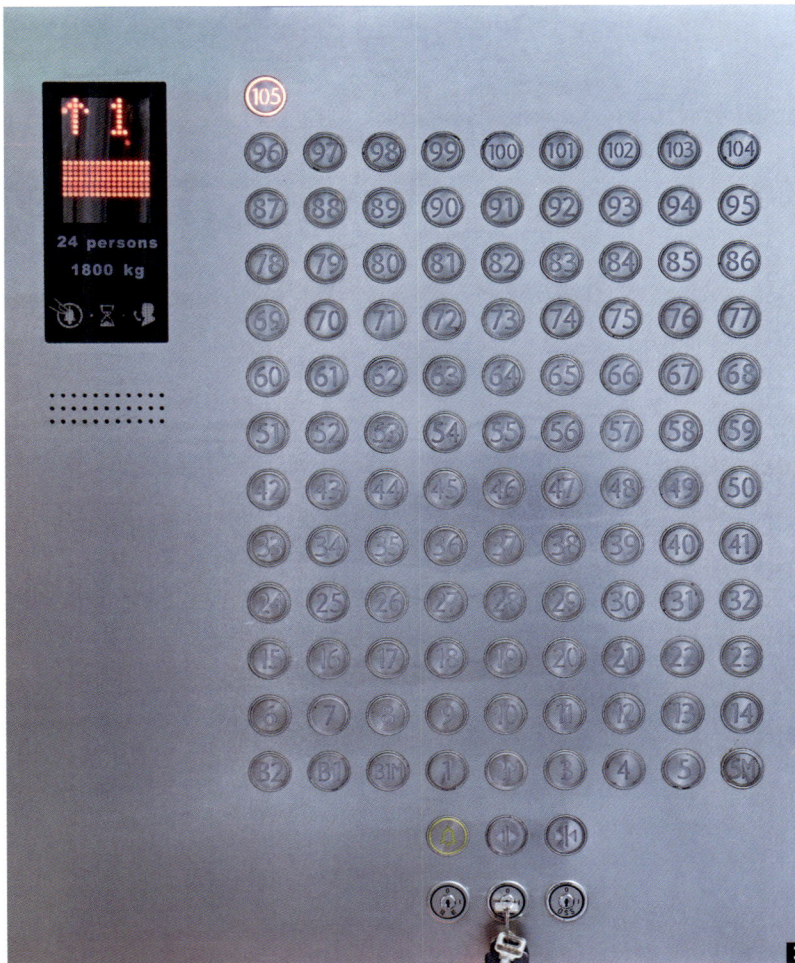

4.11 施工"零场地"条件的应对
Zero Site Construction

大数据 *Data*

-38 m

基坑下搭建钢平台位置
Location of a steel platform
built under foundation pit

130 m × **20** m × **27** m

基坑下搭建钢平台尺寸
Dimension of a steel platform
built under foundation

图片说明 *comments*

1. "生命通道"钢平台
Steel platform of "life channels"

2. 钢平台策略
Strategy for steel platform

中信大厦地处北京市 CBD 核心区，购地红线四周紧邻地下公共空间管廊，现场零施工场地。

土方及地下室施工阶段，在 -38m 的基坑下搭建一座东西长 130m，南北宽 20m，高 27m 的钢平台，为工程提供了物资运输通道、材料堆场和大型机械作业空间。

地下结构施工阶段，在东西两侧纯地下室楼板上设置堆场，供土建及钢结构共同使用。

地上结构施工阶段，将场地模块化分解。策划首层平面，化解零场地困局；充分利用地下室，缓解材料周转压力。同时，合理利用地上楼层作为机电和装饰材料堆场，对楼层平面动态布置和策划，减少材料二次搬运、保障通道通行顺畅，缓解首层和地下室材料周转压力。

施工全过程钢筋、钢结构构件等构件均为场外加工、现场安装，在减小现场用地压力的同时，实现建筑的工业化建造。

CITIC Tower is located in the core area of Beijing CBD core area, close to the underground public space tunnel around the red line of land purchase, and there is no construction site.

In the earthwork and basement construction stage, a steel platform with an east-west length of 130m, south-north width of 20m and height of 27m is built under the negative 38m of foundation pit, which provides the material transportation passage, material yard and large-scale machinery working space for the project.

In the underground structural construction stage, a stacking yard is set up on the basement floors on the east and west sides for both the civil work and the steel structure.

During the aboveground structural construction stage, the site is decomposed into modules. The first floor plane is planned to eliminate the predicament of "No Site"; the basement is fully used to relieve the material turnover pressure. At the same time, the reasonable use of the above-ground floor as an electromechanical and decorative material yard, floor dynamic layout and planning can reduce the secondary material handling, ensure smooth passage, and alleviate the material turnover pressure at first floor and basement.

During the whole construction process, the components such as rebar and steel structure components are processed off-site and installed on-site, which can reduce the pressure on the site while realizing the industrialized construction of the building at the same time.

施工期间立体式各平面作业布置
3D general layout during construction

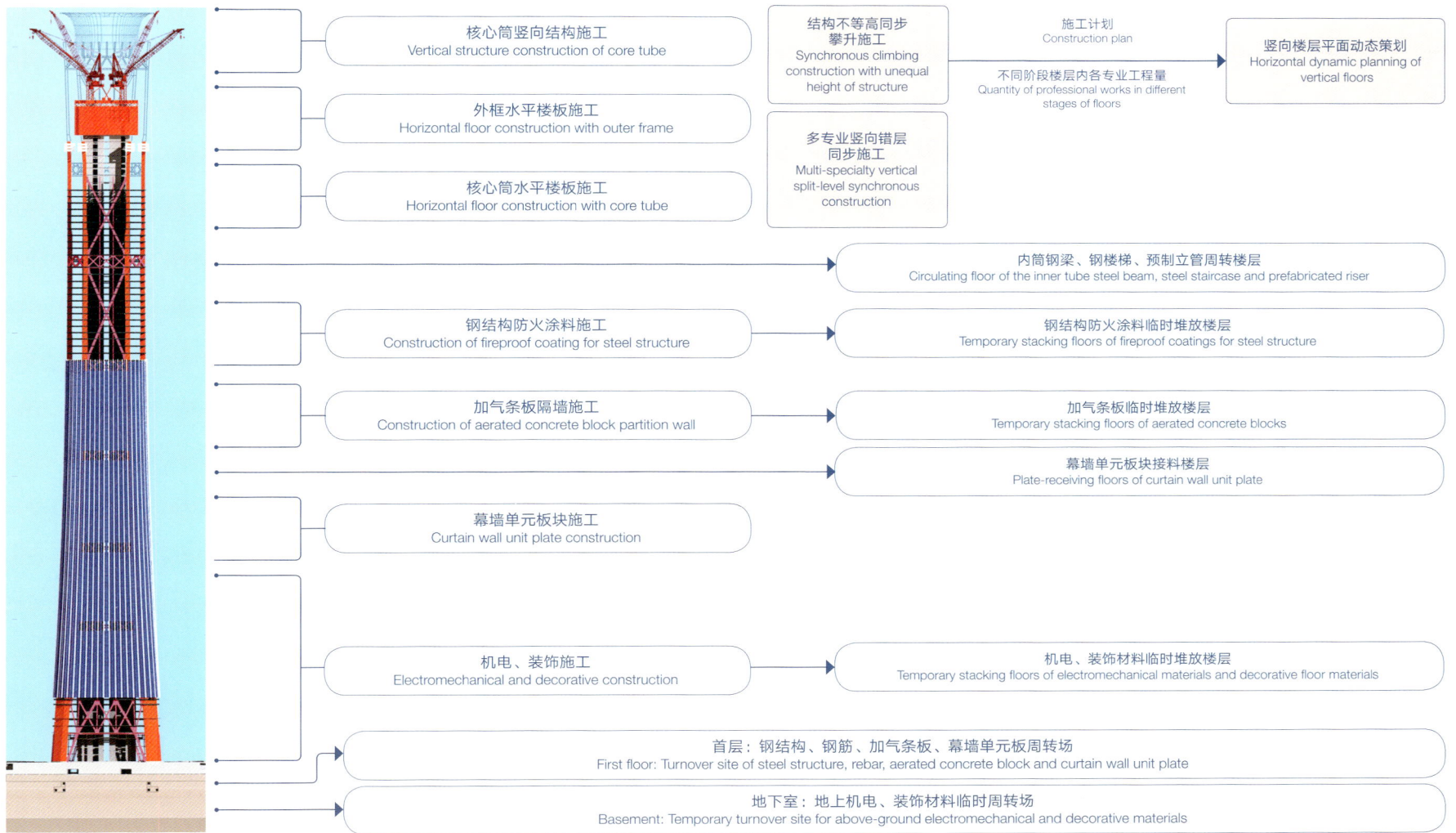

核心筒竖向结构施工
Vertical structure construction of core tube

外框水平楼板施工
Horizontal floor construction with outer frame

核心筒水平楼板施工
Horizontal floor construction with core tube

钢结构防火涂料施工
Construction of fireproof coating for steel structure

加气条板隔墙施工
Construction of aerated concrete block partition wall

幕墙单元板块施工
Curtain wall unit plate construction

机电、装饰施工
Electromechanical and decorative construction

结构不等高同步攀升施工
Synchronous climbing construction with unequal height of structure

多专业竖向错层同步施工
Multi-specialty vertical split-level synchronous construction

施工计划
Construction plan

不同阶段楼层内各专业工程量
Quantity of professional works in different stages of floors

竖向楼层平面动态策划
Horizontal dynamic planning of vertical floors

内筒钢梁、钢楼梯、预制立管周转楼层
Circulating floor of the inner tube steel beam, steel staircase and prefabricated riser

钢结构防火涂料临时堆放楼层
Temporary stacking floors of fireproof coatings for steel structure

加气条板临时堆放楼层
Temporary stacking floors of aerated concrete blocks

幕墙单元板块接料楼层
Plate-receiving floors of curtain wall unit plate

机电、装饰材料临时堆放楼层
Temporary stacking floors of electromechanical materials and decorative floor materials

首层：钢结构、钢筋、加气条板、幕墙单元板周转场
First floor: Turnover site of steel structure, rebar, aerated concrete block and curtain wall unit plate

地下室：地上机电、装饰材料临时周转场
Basement: Temporary turnover site for above-ground electromechanical and decorative materials

1. 中信大厦地块平面图
CITIC Tower plot plan

2. 地上楼层施工平面布置
Construction plane layout
of above-ground floor

3. 中信大厦地块施工平
面布置与周边关系鸟瞰
A bird's-eye view of the relationship
between the construction plane and
the surrounding area of CITIC Tower

4. 通过对公共资源在项目建造
全过程中尤其是复杂条件下的
三维模拟,有利于对现场场地布
置和分配进行直观分析和决策
The 3D simulation of public
resources in the whole construction
process, especially under complex
conditions, is beneficial to the
intuitive analysis and decision-
making of site layout and allocation.

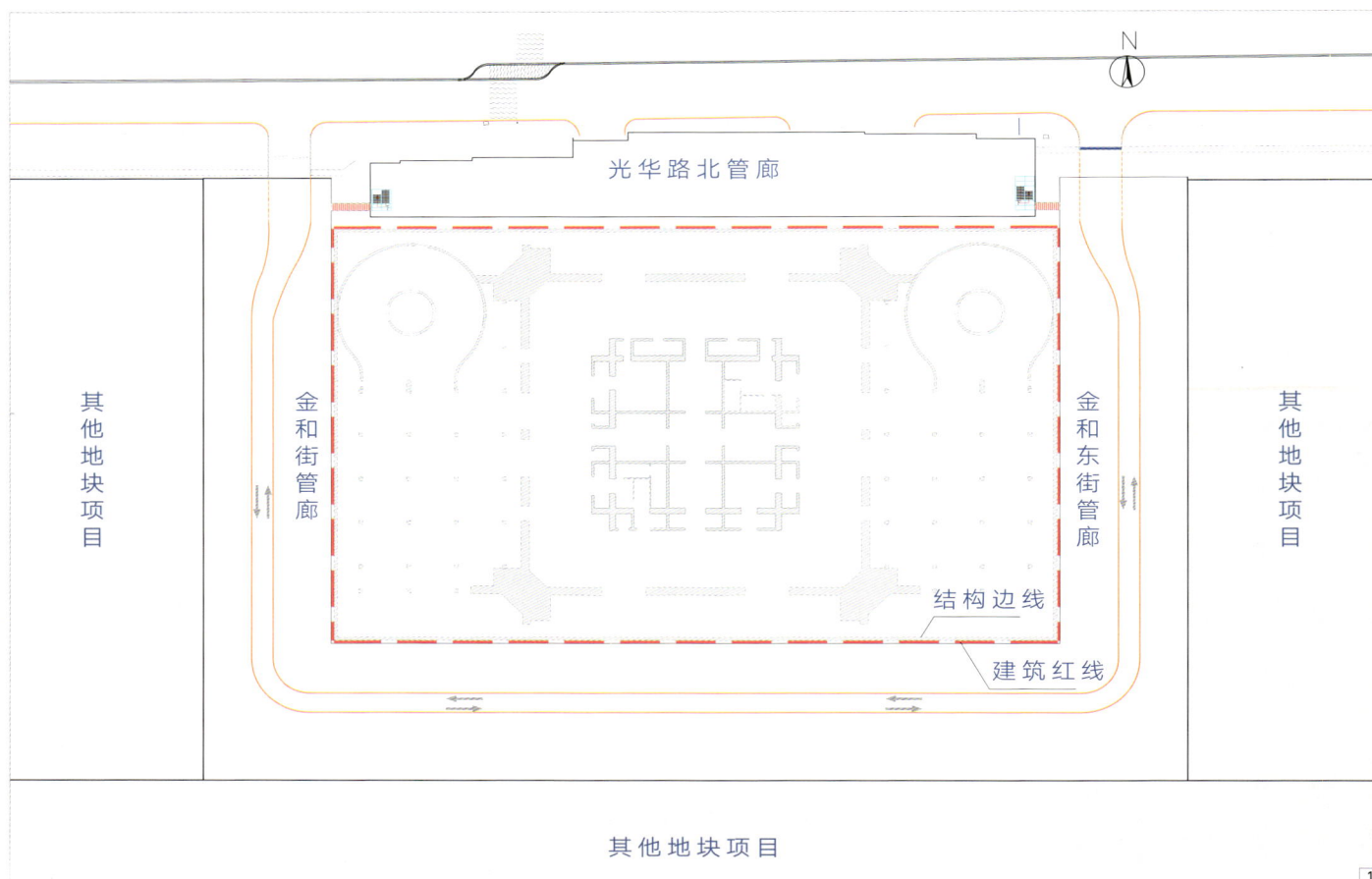

光华路北管廊

其他地块项目

金和街管廊

金和东街管廊

其他地块项目

结构边线

建筑红线

其他地块项目

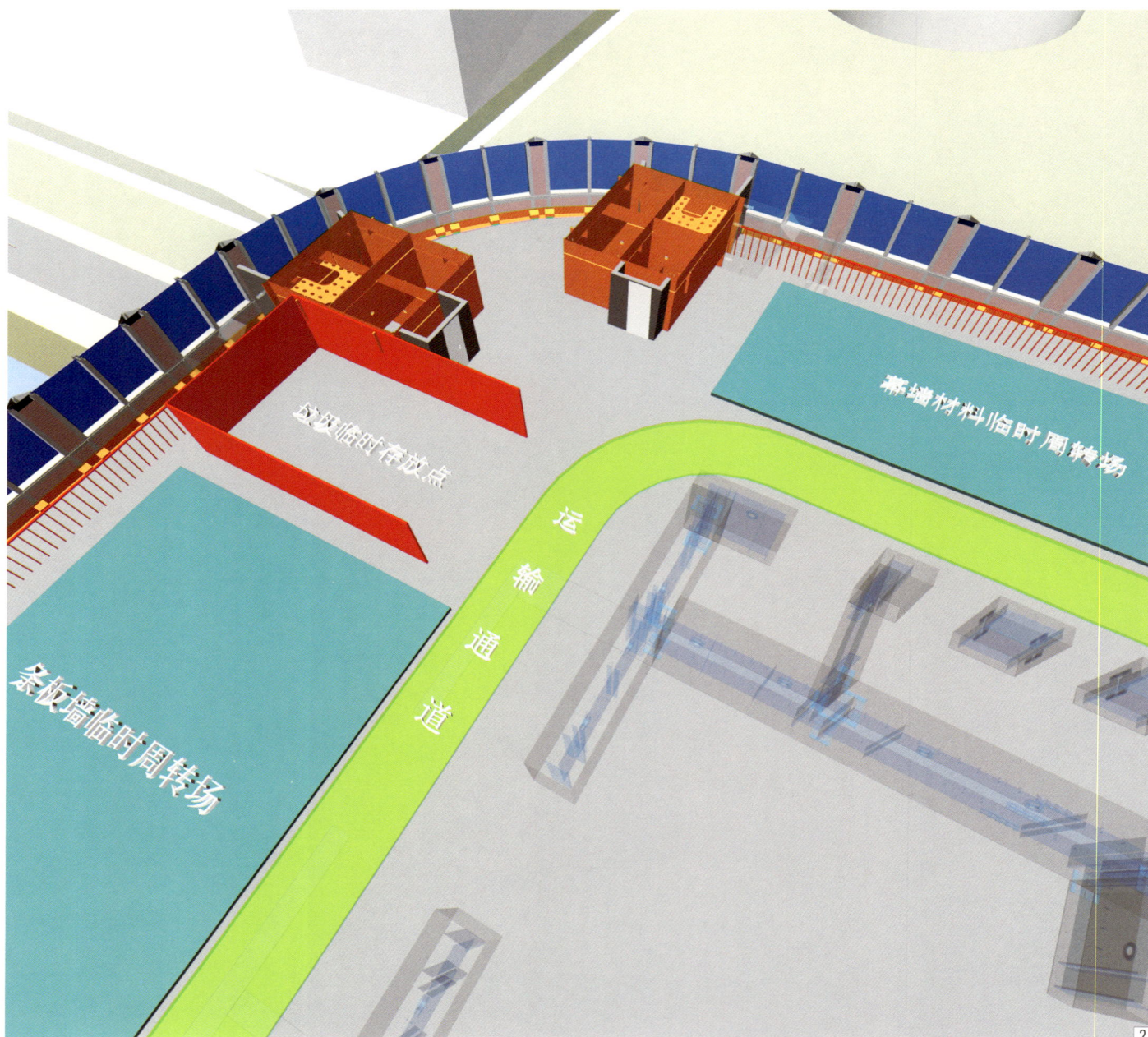

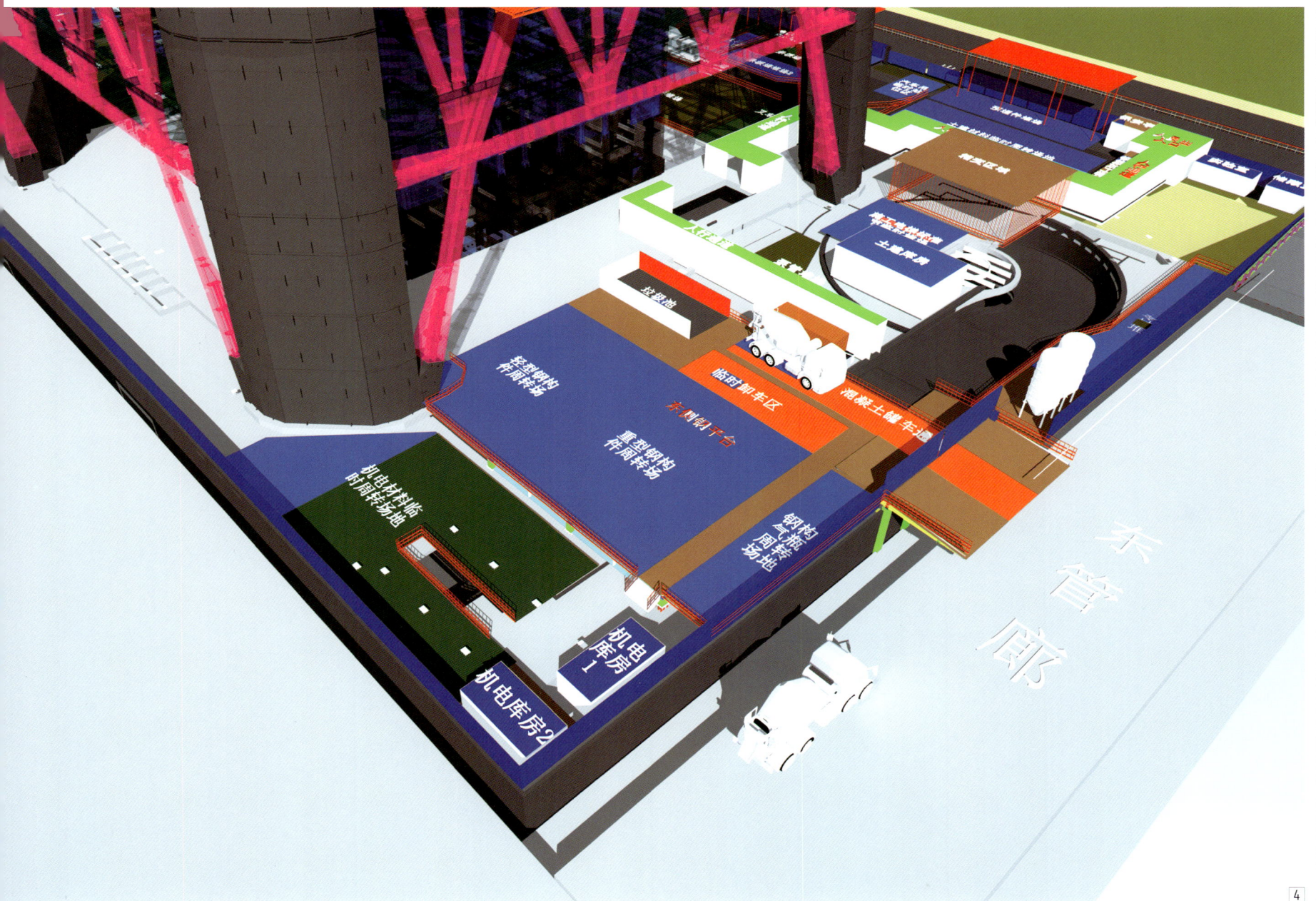

4.12 超高层建筑施工临时供电的特殊性及解决方案
Particularity And Solution Of Temporary Power Supply In Super High-Rise Building Construction

大数据 *Data*

2,890 kVA

大厦临时供电
Temporary power supply
to the tower

图片说明 *Comments*

1. 高低压变配电站
High-voltage and low-voltage
transformer and distribution station

延伸阅读 *Links*

P086. P170

施工总承包负责超高层建筑现场临时供电的实施和管理。随着建筑高度的逐渐增高，临时电缆竖向输送距离越来越长，长期运行中，临时电缆因通电发热极易造成绝缘体与导体脱落，引起电力系统崩溃，特别是垂直电梯和高空吊篮的突发停车，安全风险很大。

工程后期机电总承包正式变配电系统逐步完成，受到临永结合创新的启发，将大厦 2,890kVA 的施工临时供电分步接入正式供电系统，利用正式供电系统的安全可靠性，极大保障了现场施工用电稳定。同时，也可以尽早对正式供电系统进行调试检验，临时供电系统可以尽早拆除，提高工作效率，节约工期。

The general construction contractor is responsible for the implementation and management of the temporary power supply on the super high-rise building site. With the gradual increase of building height, the vertical transmission distance of the temporary cable becomes increasingly longer. During long-term operation, the temporary cable is easy to cause the insulator and conductor to fall off due to the energization and heating, which leads to the collapse of power system, especially the sudden stop of vertical elevator and high-altitude basket, with high security risks.

In the later stage of the project, the formal power transmission and distribution system under the electromechanical general contractor was completed step by step. Inspired by the innovation combining temporary and formal power, the temporary power supply of 2,890 kVA in the Tower was connected to the formal power supply system gradually, and the safety and reliability of the formal power supply system were utilized to ensure the power consumption stability of the construction site. Meanwhile, the formal power supply system can also be debugged and inspected and the temporary power supply system can be dismantled as soon as possible to improve the work efficiency and save time.

中信大厦 2,890kVA 临时永久结合供电方案第一阶段 2018.5.10-2018.8.30
Combined Scheme Phase I of Temporary and Permanent 2890 kVA Power Supply in CITIC Tower from May 10, 2018 to August 30, 2018

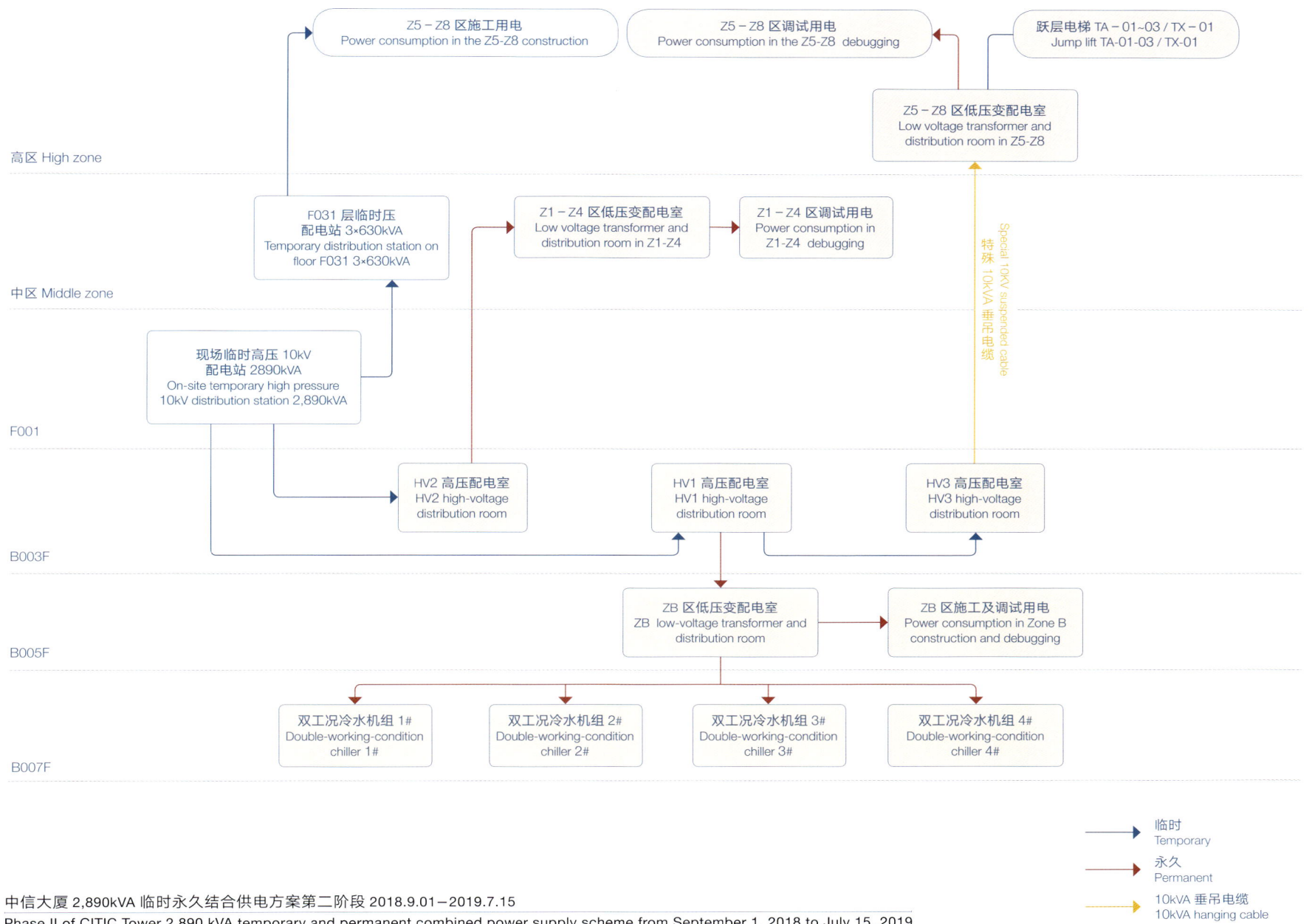

中信大厦 2,890kVA 临时永久结合供电方案第二阶段 2018.9.01-2019.7.15
Phase II of CITIC Tower 2,890 kVA temporary and permanent combined power supply scheme from September 1, 2018 to July 15, 2019

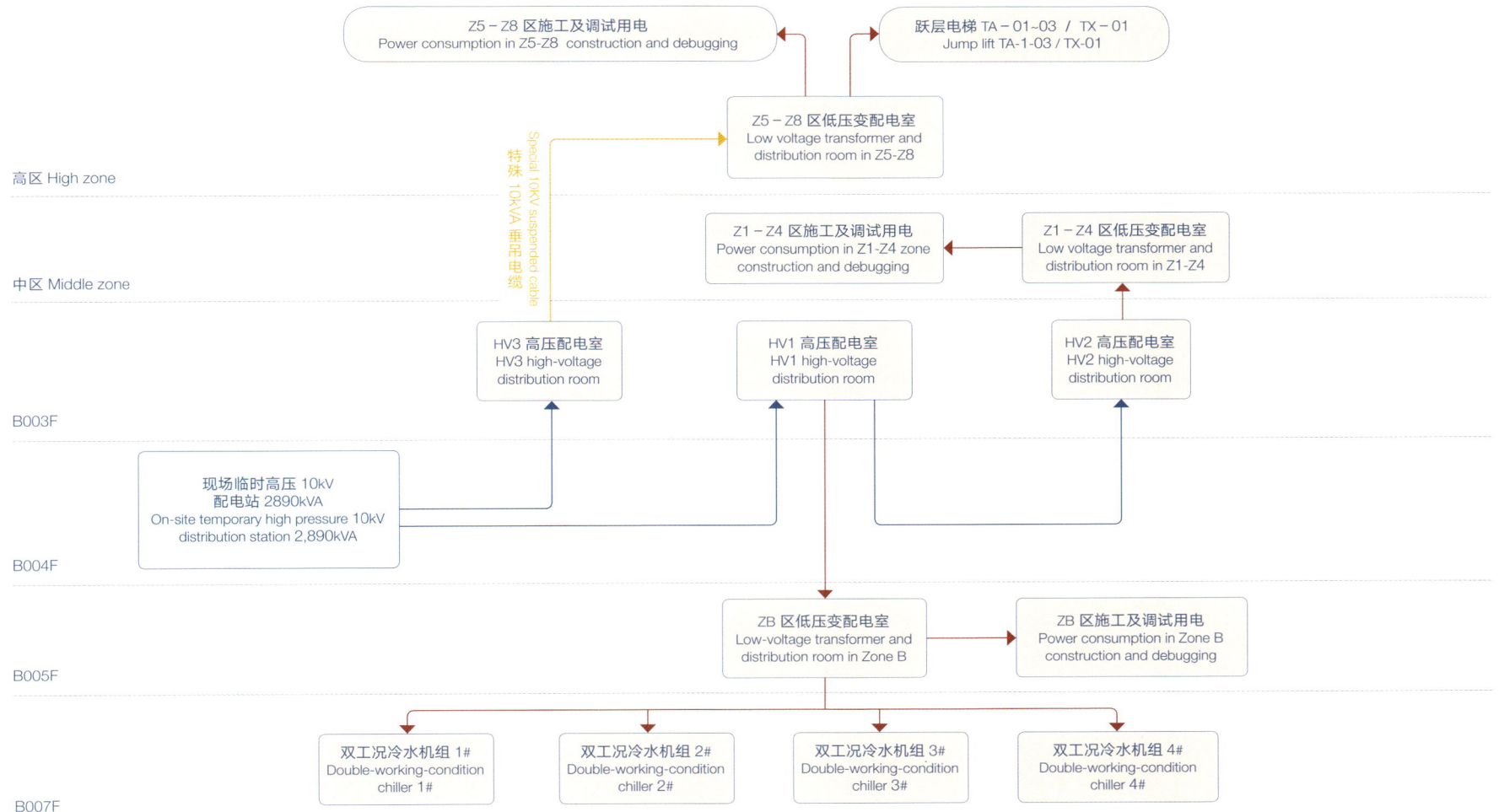

工程项目管理
Project Management

4.13

临永结合的消防自救体系创新
Innovation Of Self-Rescue System Of Fire Protection Based On The Combination Of Temporary And Formal Principle

图片说明 *Comments*

1. 全球首个超 500m 建筑
临永结合消防系统正式启用
The world's first fire protection system with the combination of temporary and formal principle in buildings exceeding 500 m has been put into operation

2. 临永结合消防泵房
The combination of temporary and formal fire-fighting pump houses

延伸阅读 *Links*

P090. P135. P220. P326.

中信大厦"临永结合"消防系统为全球高层建筑首创。实现了"临永"消防系统的无缝对接与科学转换；减少精装修收口工作量；减少临时设施的投入及拆除，节约工期和成本；最重要的是提高了超高层建筑施工现场消防的可靠性。

1. 市政供水至 B1 转输水箱，再由转输水泵加压至 F18 转输水箱，通过连续加压的方式经过 F44、F74，将水最终转输至 F103 的消防水池。

2. 消防水池通过重力方式向 F74、F44、F18 减压水箱供水，减压水箱再向各自分区的消火栓供水，完成 B7 — F96 消火栓的常高压供水系统。

3. F97 至屋顶层为临时高压系统，采用消防贮水池、消防水泵和屋顶消防水箱联合供水形式。

"Combination of Temporary and Formal" fire protection system of CITIC Tower is the first among the country's high-rise buildings. The seamless connection and scientific transformation are realized in the "Temporary and Formal" fire protection system; refined decoration works are simplified; less input and dismantlement of temporary facilities save time and cost; the most important function is to improve the reliability of fire protection in super high-rise building construction site.

1. The municipal water is supplied to the transfer tank on the B1 floor, then pressurized to the tank on the 18th floor by the transfer pump, finally transferred to the fire pool on the F103 through the F44 and F74 by continuous pressurization.

2. The fire pool supplies water to the pressure-reduced tanks on the 74th, 44th and 18th by gravity, and to the hydrants in their respective subareas to complete the normal high-pressure water supply system for the hydrants on the B7 to F96.

3. F97 to the top floor are equipped with temporary high pressure systems, which adopt the form of combined water supply of the fire pool, fire water pump and roof fire water tank.

临永结合消防水系统介绍
Official fire system introduction

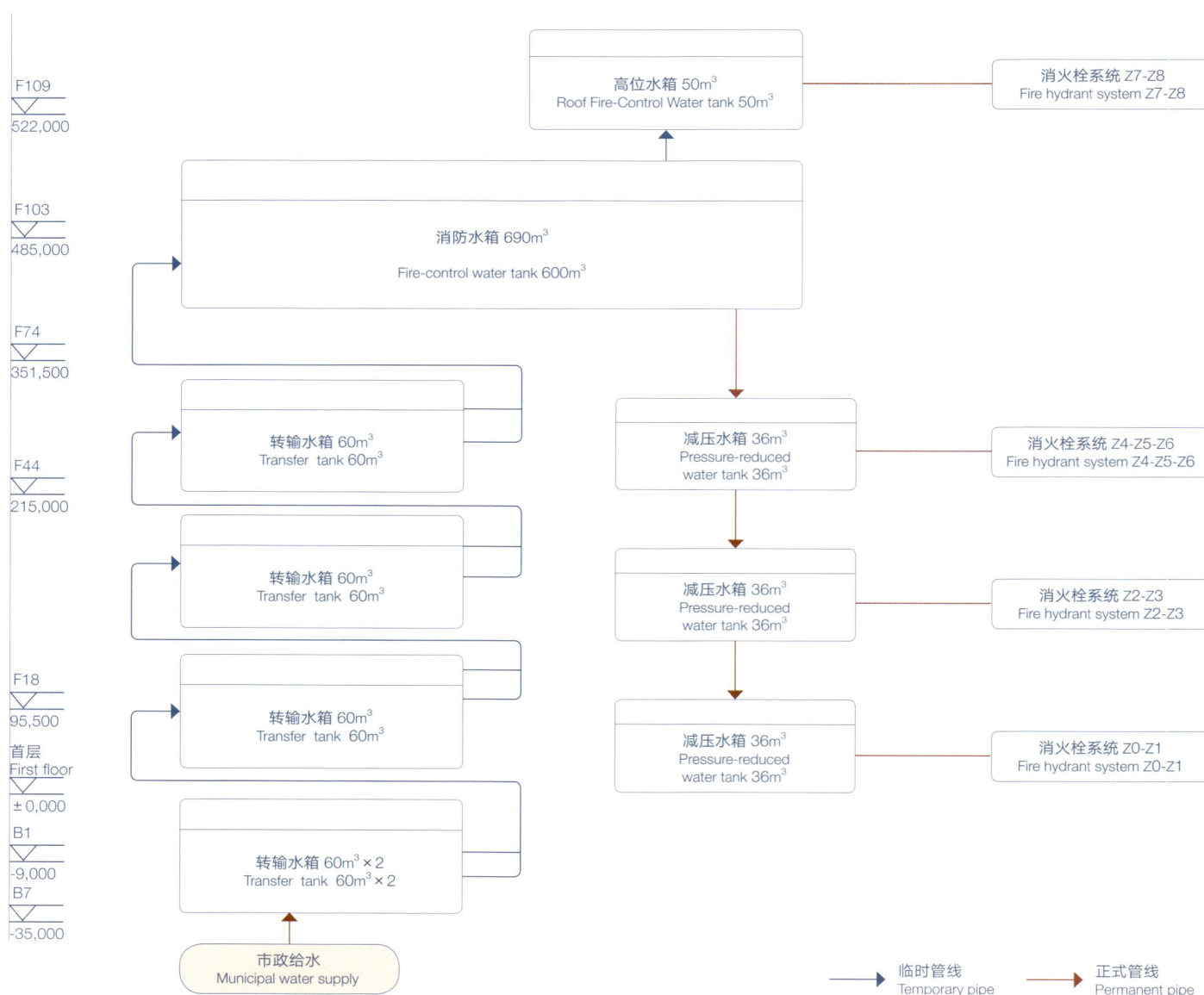

F109 / 522,000	高位水箱 50m³ / Roof Fire-Control Water tank 50m³	消火栓系统 Z7-Z8 / Fire hydrant system Z7-Z8
F103 / 485,000	消防水箱 690m³ / Fire-control water tank 600m³	
F74 / 351,500	转输水箱 60m³ / Transfer tank 60m³	减压水箱 36m³ / Pressure-reduced water tank 36m³ → 消火栓系统 Z4-Z5-Z6 / Fire hydrant system Z4-Z5-Z6
F44 / 215,000	转输水箱 60m³ / Transfer tank 60m³	减压水箱 36m³ / Pressure-reduced water tank 36m³ → 消火栓系统 Z2-Z3 / Fire hydrant system Z2-Z3
F18 / 95,500	转输水箱 60m³ / Transfer tank 60m³	减压水箱 36m³ / Pressure-reduced water tank 36m³ → 消火栓系统 Z0-Z1 / Fire hydrant system Z0-Z1
首层 First floor / ± 0,000		
B1 / -9,000	转输水箱 60m³ × 2 / Transfer tank 60m³ × 2	
B7 / -35,000	市政给水 / Municipal water supply	

临时管线 / Temporary pipe　　正式管线 / Permanent pipe

全球首个超500米建筑临永结合消防系统正式启用

工程项目管理
Project Management

PROJECT WISE 共享数据平台管理
Project Wise Shared Data Platform Management

大数据 Data

797,793 个
累计上传文档
Cumulative upload documents

5.5 TB
总文件大小
Total file size

12 h/次
每 12 小时同步文档
Documents synchronized
every 12 hours

中信大厦作为北京市超大超高的地标性建筑，建设过程中所产生的信息与数据远远超过了一般工程。为承载项目建设过程中的建设信息及超大容量的 BIM 数据，实现参建各方建造信息的高效传递，大厦建造全过程使用 Bentley ProjectWise 平台进行数据的协同管理。平台由业主中信和业、设计院、施工单位各自的 PW 子系统和 PW 云服务系统组成，向参与中信大厦建设的各家单位开放，是整个项目运作的数据库。各单位在一定权限范围内上传或下载数据资料，记录并传递工程建设所需的各类信息。

截止至 2019 年 4 月 23 日，累计上传文档 797,793 个，总文件大小将近 5.5TB，分布式服务，中信大厦 PW 协同平台每 12 小时定时将成果文件同步到业主单位。

As a super-large and super-high landmark building in Beijing, CITIC Tower is built with more information and data than the general project. To carry the construction information and the super-large capacity BIM data in the project construction, and realize the efficient transmission of the construction information of all parties involved in the construction, the Bentley Project Wise platform is used to carry out the collaborative management of the data in the whole construction process. The platform is composed of PW subsystem and PW cloud service system of the owner, design institute and construction unit. It is open to all units involved in the construction of CITIC Tower and is the database of the whole project operation. Each unit shall upload or download data and materials within a certain scope of authority, and record and transmit all kinds of information needed for project construction.

As of April 23, 2019, a total of 797,793 documents had been uploaded, with a total file size of nearly 5.5TB. With distributed services, the PW collaboration platform of CITIC Tower can regularly synchronize the outcome documents to the owner's unit every 12 hours.

文档数统计表
Document count

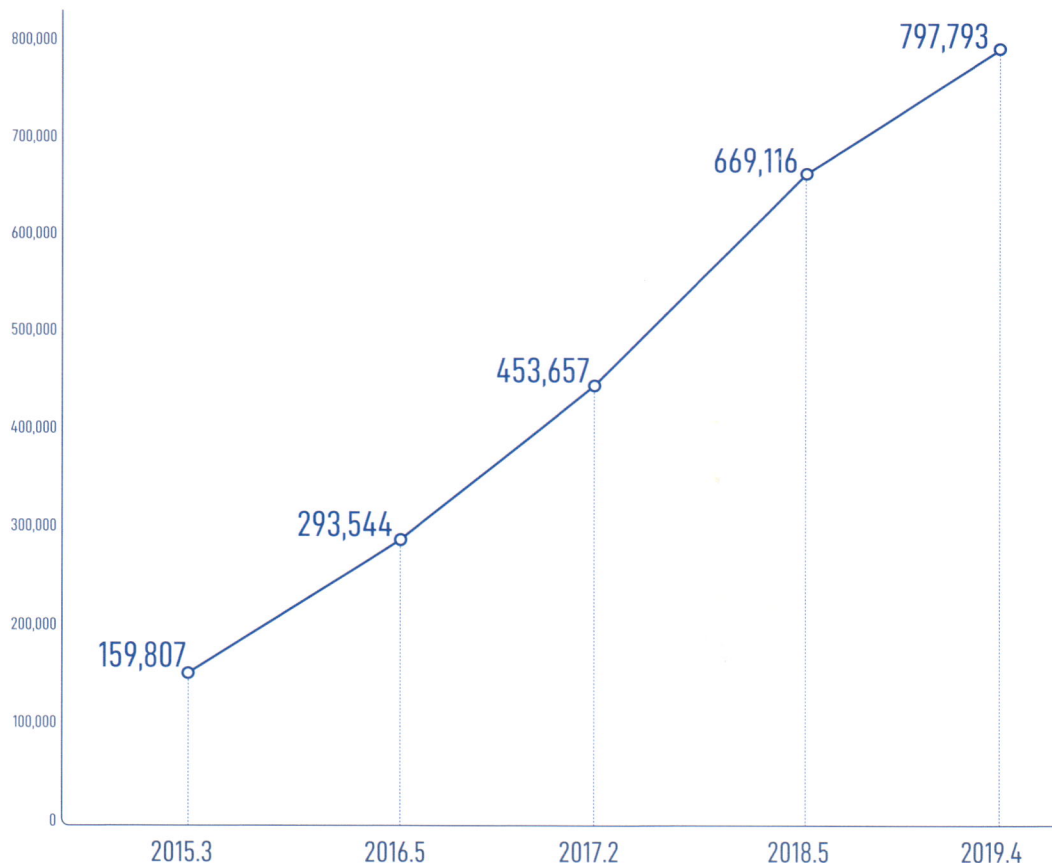

截止至 2019 年 4 月 23 日，累计上传文档 797,793 个，总文件大小约 5.5TB
By April 23, 2019, a total of 797,793 documents had been uploaded, with a total file size of nearly 5.5TB

中信和业 PW 平台架构
PW platform architecture of CITIC Heye

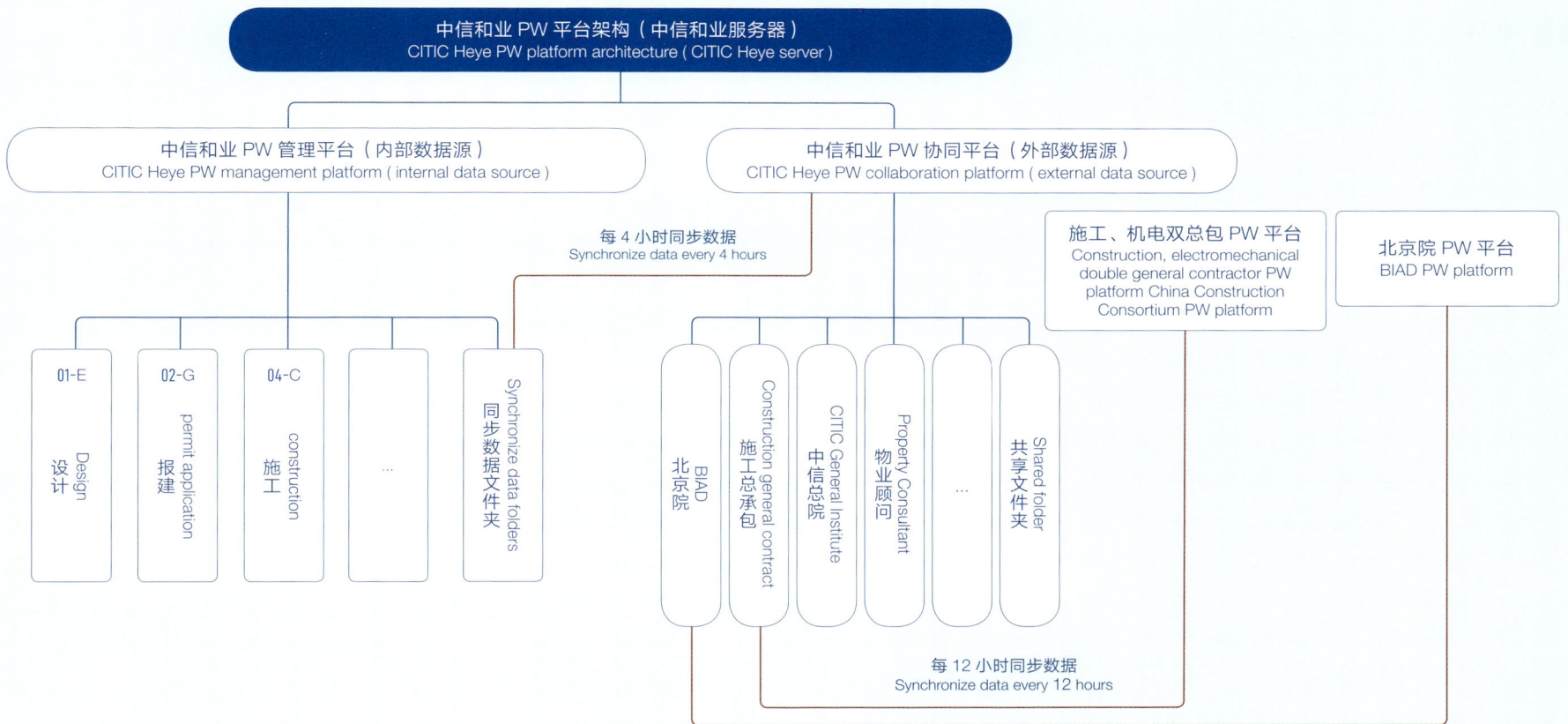

中信大厦 PW 协同平台
CITIC Heye PW collaboration platform

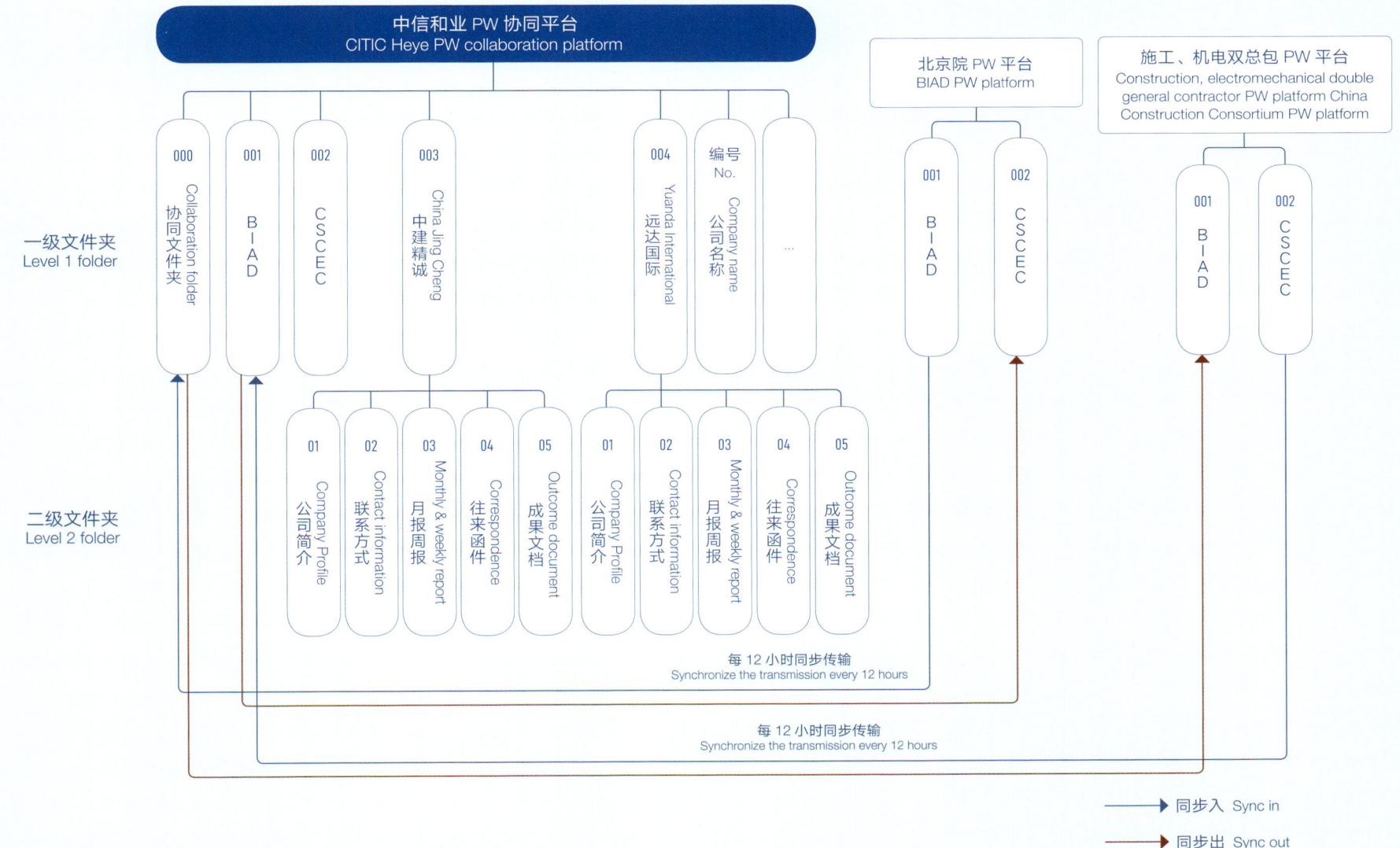

工程项目管理
Project Management

図片说明 *Comments*

1. 中信和业 PW 管理平台
CITIC Heye PW
management platform

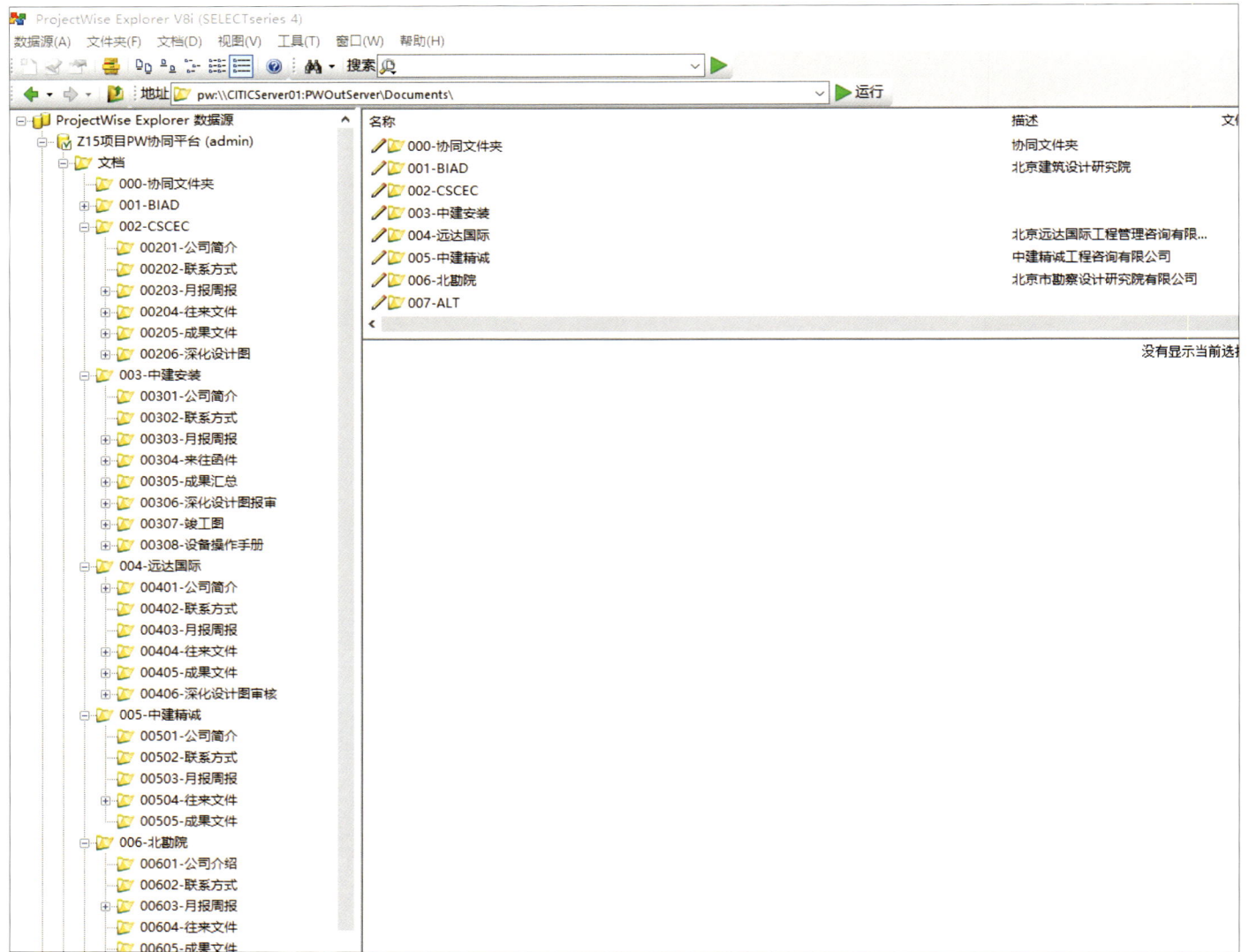

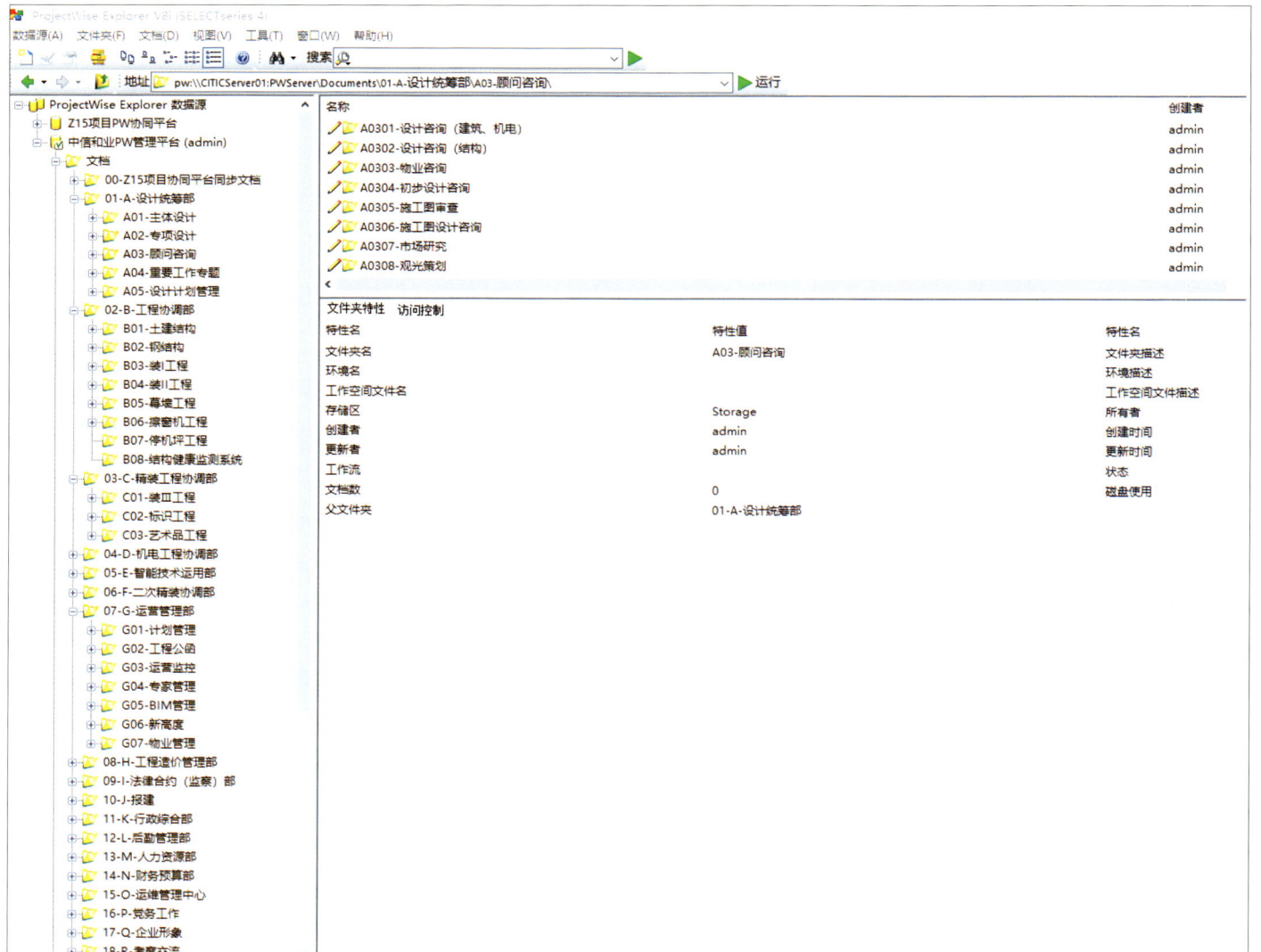

ProjectWise Explorer V8i (SELECTseries 4)

数据源(A) 文件夹(F) 文档(D) 视图(V) 工具(T) 窗口(W) 帮助(H)

搜索

地址 pw:\\CITICServer01:PWOutServer\Documents\ 运行

ProjectWise Explorer 数据源
- Z15项目PW协同平台 (admin)
  - 文档
    - 000-协同文件夹
    - 001-BIAD
    - 002-CSCEC
      - 00201-公司简介
      - 00202-联系方式
      - 00203-月报周报
      - 00204-往来文件
      - 00205-成果文件
      - 00206-深化设计图
    - 003-中建安装
      - 00301-公司简介
      - 00302-联系方式
      - 00303-月报周报
      - 00304-来往函件
      - 00305-成果汇总
      - 00306-深化设计图报审
      - 00307-竣工图
      - 00308-设备操作手册
    - 004-远达国际
      - 00401-公司简介
      - 00402-联系方式
      - 00403-月报周报
      - 00404-往来文件
      - 00405-成果文件
      - 00406-深化设计图审核
    - 005-中建精诚
      - 00501-公司简介
      - 00502-联系方式
      - 00503-月报周报
      - 00504-往来文件
      - 00505-成果文件
    - 006-北勘院
      - 00601-公司介绍
      - 00602-联系方式
      - 00603-月报周报
      - 00604-往来文件
      - 00605-成果文件

名称	描述	文
000-协同文件夹	协同文件夹	
001-BIAD	北京建筑设计研究院	
002-CSCEC		
003-中建安装		
004-远达国际	北京远达国际工程管理咨询有限...	
005-中建精诚	中建精诚工程咨询有限公司	
006-北勘院	北京市勘察设计研究院有限公司	
007-ALT		

没有显示当前选

ProjectWise Explorer V8i (SELECTseries 4)

数据源(A) 文件夹(F) 文档(D) 视图(V) 工具(T) 窗口(W) 帮助(H)

搜索

地址 pw:\\CITICServer01:PWServer\Documents\01-A-设计统筹部\A03-顾问咨询\ 运行

ProjectWise Explorer 数据源
- Z15项目PW协同平台
- 中信和业PW管理平台 (admin)
  - 文档
    - 00-Z15项目协同平台间同步文档
    - 01-A-设计统筹部
      - A01-主体设计
      - A02-专项设计
      - A03-顾问咨询
      - A04-重要工作专题
      - A05-设计计划管理
    - 02-B-工程协调部
      - B01-土建结构
      - B02-钢结构
      - B03-装I工程
      - B04-装II工程
      - B05-幕墙工程
      - B06-擦窗机工程
      - B07-停机坪工程
      - B08-结构健康监测系统
    - 03-C-精装工程协调部
      - C01-装III工程
      - C02-标识工程
      - C03-艺术品工程
    - 04-D-机电工程协调部
    - 05-E-智能技术运用部
    - 06-F-二次精装协调部
    - 07-G-运营管理部
      - G01-计划管理
      - G02-工程公函
      - G03-运营监控
      - G04-专家管理
      - G05-BIM管理
      - G06-新高度
      - G07-物业管理
    - 08-H-工程造价管理部
    - 09-I-法律合约 (监察) 部
    - 10-J-报建
    - 11-K-行政合部
    - 12-L-后勤管理部
    - 13-M-人力资源部
    - 14-N-财务预算部
    - 15-O-运维管理中心
    - 16-P-党务工作
    - 17-Q-企业形象
    - 18-R-考察交流

名称	创建者
A0301-设计咨询 (建筑、机电)	admin
A0302-设计咨询 (结构)	admin
A0303-物业咨询	admin
A0304-初步设计咨询	admin
A0305-施工图审查	admin
A0306-施工图设计咨询	admin
A0307-市场研究	admin
A0308-观光策划	admin

文件夹特性 访问控制

特性名	特性值	特性名	
文件夹名	A03-顾问咨询	文件夹描述	
环境名		环境描述	
工作空间文件名		工作空间文件描述	
存储区	Storage	所有者	
创建者	admin	创建时间	
更新者	admin	更新时间	
工作流		状态	
文档数	0	磁盘使用	
父文件夹	01-A-设计统筹部		

消防工程实施与验收组织
Fire Engineering Implementation And Acceptance Organization

超高层建筑消防性能化设计和消防验收组织是消防工程的关键环节。方案阶段历经 18 轮优化调整，最终获得北京市公安局消防局《建设工程消防设计审核意见书》。临时高压供水优化为重力供水，各项性能化加强措施均提升了大厦消防安全，为消防验收打下良好的基础。

施工阶段正值消防新旧规范更替和消防部门组织管理调整的过渡期，积极主动与消防验收部门做好沟通，邀请其参与过程交流、指导和检查，明确验收依据、要求及流程。

机电总承包牵头组织大厦整体消防调试、消防检测和验收工作。成立消防验收领导小组和各工作小组，由中信和业、机电总承包、施工总承包、设计院和监理单位共同参加，明确各自职责和各小组工作计划。通过大家共同努力，使中信大厦成为北京 CBD 核心区第一个通过消防验收的新建项目。

The design and acceptance of super high-rise fire protection are the key procedure in the fire acceptance. After 18 rounds of optimization tuning and adjustment, the Examination Report of Fire Protection Design for the Construction Project was finally obtained from the Fire Department of the Beijing Municipal Public Security Bureau. The temporary high pressure water supply was optimized for gravity water supply, and all the performance-based strengthening measures enhance the fire safety of the Tower and lay a good foundation for fire acceptance.

The construction stage is in the transition period of replacing the old fire protection standards with the new ones and adjusting the organization and management of the fire control department. It is necessary to actively communicate with the fire acceptance department, invite it to participate in the process of communication, guidance and inspection, and clarify the basis, requirements and procedures of acceptance.

The electromechanical general contractor takes the lead in organizing the overall fire-fighting commissioning, inspection and acceptance of the Tower. A leading group and working groups for fire acceptance are set up, with the participation of CITIC Heye, electromechanical general contractor, general construction contractor, design institute and supervision unit, their respective responsibilities and work plans of each group are clarified. Through joint efforts, CITIC Tower has become the first newly-built project in the CBD core area of Beijing to pass the fire acceptance.

大数据 *Data*

18 轮
消防方案优化调整
Rounds of fire protection scheme optimization tuning and adjustment

延伸阅读 *Links*

P090. P130. P220. P236

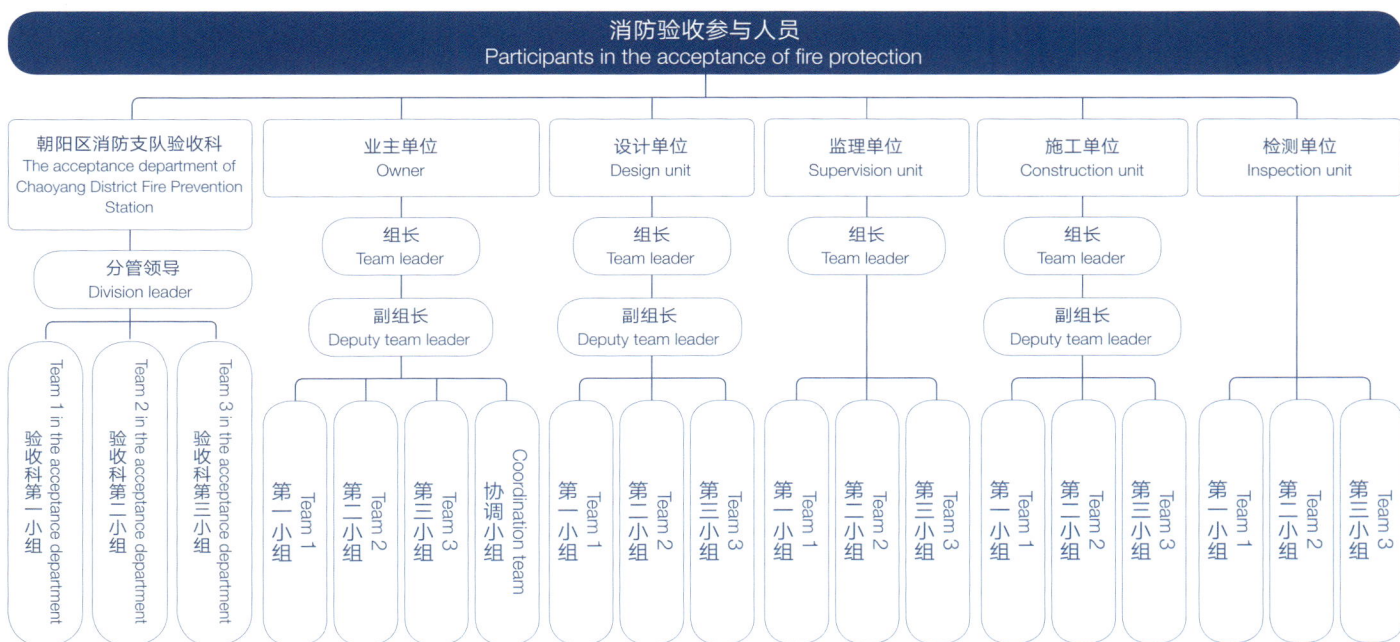

消防验收参与人员
Participants in the acceptance of fire protection

- 朝阳区消防支队验收科 The acceptance department of Chaoyang District Fire Prevention Station
  - 分管领导 Division leader
    - Team 1 in the acceptance department 验收科第一小组
    - Team 2 in the acceptance department 验收科第二小组
    - Team 3 in the acceptance department 验收科第三小组
- 业主单位 Owner
  - 组长 Team leader
    - 副组长 Deputy team leader
      - Team 1 第一小组
      - Team 2 第二小组
      - Team 3 第三小组
      - Coordination team 协调小组
- 设计单位 Design unit
  - 组长 Team leader
    - 副组长 Deputy team leader
      - Team 1 第一小组
      - Team 2 第二小组
      - Team 3 第三小组
- 监理单位 Supervision unit
  - 组长 Team leader
    - Team 1 第一小组
    - Team 2 第二小组
    - Team 3 第三小组
- 施工单位 Construction unit
  - 组长 Team leader
    - 副组长 Deputy team leader
      - Team 1 第一小组
      - Team 2 第二小组
      - Team 3 第三小组
- 检测单位 Inspection unit
  - Team 1 第一小组
  - Team 2 第二小组
  - Team 3 第三小组

消防验收专项工作小组
Special working group on fire acceptance

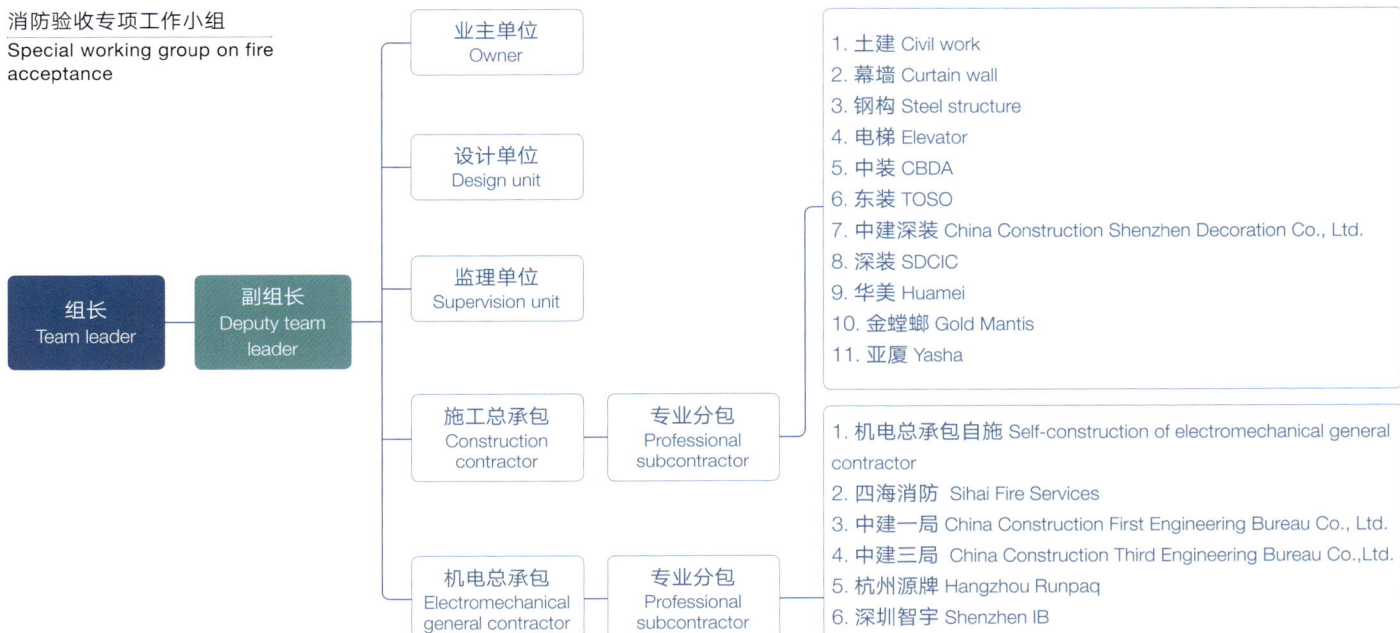

- 组长 Team leader
  - 副组长 Deputy team leader
    - 业主单位 Owner
    - 设计单位 Design unit
    - 监理单位 Supervision unit
    - 施工总承包 Construction contractor
      - 专业分包 Professional subcontractor
    - 机电总承包 Electromechanical general contractor
      - 专业分包 Professional subcontractor

1. 土建 Civil work
2. 幕墙 Curtain wall
3. 钢构 Steel structure
4. 电梯 Elevator
5. 中装 CBDA
6. 东装 TOSO
7. 中建深装 China Construction Shenzhen Decoration Co., Ltd.
8. 深装 SDCIC
9. 华美 Huamei
10. 金螳螂 Gold Mantis
11. 亚厦 Yasha

1. 机电总承包自施 Self-construction of electromechanical general contractor
2. 四海消防 Sihai Fire Services
3. 中建一局 China Construction First Engineering Bureau Co., Ltd.
4. 中建三局 China Construction Third Engineering Bureau Co.,Ltd.
5. 杭州源牌 Hangzhou Runpaq
6. 深圳智宇 Shenzhen IB

5.1 超深基坑后注浆钻孔灌注桩施工技术
Construction Technology Of Post-Grouting Bored Pile In Ultra-Deep Foundation Pit

中信大厦基础形式为桩筏基础，由896根钻孔灌注桩与基础筏板共同作用，撑起整个大楼的重量。有效桩长为54.6m、44.6m、40.1m、26.1m，桩径为1,000mm和1,200mm，桩身为水下混凝土C50、C40。所有工程桩均采用桩侧桩端组合后注浆工艺。

在钻孔灌注桩成桩后，通过埋在桩身的压浆管，以一定压力向桩侧、桩端注入一定量的水泥浆，提高桩侧及桩端的阻力，提高桩基承载性能。

试验桩、锚桩及工程桩均配置3根桩侧压浆管和桩端压浆管。桩侧压浆环管沿整个桩身分别于绝对标高－8.00m、－20.00m、－32.00m左右；桩端压浆导管沿钢筋笼纵向对称设置，底端设三通与环型桩侧压浆阀相连。

注浆流量75L/min，后压浆质量控制采用注浆量和注浆压力双控方法，以水泥注入量控制为主，泵送终止压力控制为辅。

注浆量：
1. 桩端注浆水泥用量：桩径1,000mm桩端为卵石、圆砾⑫层时不小于2.5t/桩；桩径1,200mm桩端为卵石、圆砾⑫层时不小于3.5t/桩；
2. 桩径1,000mm的桩侧每段注浆水泥用量：不小于0.9t/管；
3. 桩径1,200mm的桩侧每段注浆水泥用量：不小于1.1t/管。

注浆压力：
桩端注浆，对于卵石层，注浆压力为5~8MPa；对于细砂中砂层，注浆压力为5~8MPa；

桩侧注浆，终止注浆压力根据土层性质及注浆点深度确定，对于饱和土注浆压力为1.2~4.0MPa，软土取低值，密实性土取高值。

The foundation form of CITIC Tower is piled raft foundation, which consists of 896 bored piles and foundation Bottom Plates to support the weight of the whole building. The effective pile lengths are 54.6m, 44.6m, 40.1m and 26.1m, the pile diameters are 1,000mm and 1,200mm and the piles are made of underwater concrete C50 and C40. All the engineering piles adopt the post grouting technology of pile side and pile end combination.

After the bored pile is completed, a certain amount of cement slurry is injected into the pile side and pile end through the grouting pipe buried in the pile body at a certain pressure, so as to improve the resistance of the pile side and pile end and the bearing capacity of the pile foundation.

The test pile, anchor pile and engineering pile are equipped with 3 pile side and pile end grouting pipes. The absolute elevation of the pile side grouting ring pipe along the pile body is about -8.00m, -20.00m, -32.00m respectively; The pile end grouting catheter is symmetrically arranged along the longitudinal direction of the rebar cage, and the bottom end of the pile end is provided with a tee joint connecting with the ring-shaped pile side grouting valve.

The grouting flow rate is 75L/min, and the post-grouting quality control adopts the double control method of grouting quantity and grouting pressure, mainly controlled by the cement injection quantity and supported by the pumping far-end pressure control.

Grouting volume:
1. Grouting cement amount at the pile end: for a pile diameter of 1,000mm, the pile end made of pebbles and round gravel, not less than 2.5t/pile in 12 layers; for a pile diameter of 1,200mm, the pile end made of pebbles and round gravel, not less than 3.5t/ pile in 12 layers;
2. Grouting cement amount for each section of pile side with pile diameter of 1,000mm: not less than 0.9t/tube;
3. Grouting cement amount for each section of pile side with pile diameter of 1,200mm: not less than 1.1t/tube;

Grouting pressure:
In the pile end, the grouting pressure is 5~8Mpa for the pebble layer; 5~8Mpa for the fine sand and medium sand layer;
In the pile side, the far-end grouting pressure is determined according to the properties of soil layer and the depth of grouting point. For saturated soil, the grouting pressure is 1.2~4.0Mpa, low in soft soil and high in the dense soil.

工程桩汇总表						
桩型	编号	桩径/mm	桩长/m	静载试验最大荷载 P/kN	数量/根	单桩承载力特征值/kN
工程桩 1	P1	1,000	40.1	-	398	14,500（抗压）
工程桩 2	P2	1,200	44.6	-	403	16,000（抗压）
工程桩 3	P3	1,000	26.1	-	40	8,000（抗压）
静载检验桩 1	TP1	1,000	40.1	29,000（抗压）	4	14,500（抗压）
静载检验桩 2	TP2	1,200	44.6	32,000（抗压）	2	16,000（抗压）
静载试验桩 2a	TP2a	1,200	约 54.6	32,000（抗压）	3	16,000（抗压）
试验锚桩 1	AP1	1,000	40.1	-	22	5,800（抗拔）
试验锚桩 2	AP2	1,200	44.6	-	12	6,500（抗拔）
试验锚桩 2a	AP2a	1,200	约 54.6	-	12	6,500（抗拔）
					896（总桩数）	

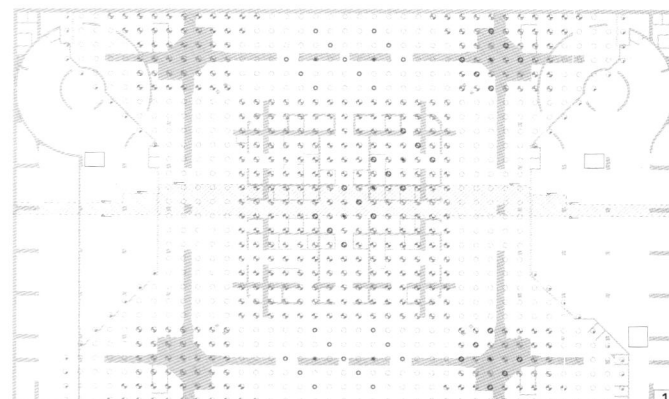

基坑一体化施工技术
Integrated Construction Technology Of Foundation Pit

大数据 data

38 m
最深基坑
Deepest foundation pit

27 m
地下空间基坑深度
Foundation pit depth of
underground space

11 m
后期支护深度
Late support depth

50 mm
结构外皮与红线距离
The distance between the
structural skin and the red line

>7,000 m²
地下室使用面积额外增加
Additional usable area of basement

图片说明 comments

1. 基坑西侧支护详图
Detailed diagram of western
supporting structure
of foundation pit

2. 基坑支护示意图
Schematic diagram of supporting
structure of foundation pit

CBD 核心区内主要道路采用九宫格形式布局，与周边道路衔接，各条道路下为地下管廊。整个地块基坑采用一体化施工方式，即打破各地块间红线的限制，进行整体大开挖、支护、降水，并将深度较大的二级地块基坑简化为"坑中坑"模式，降低支护及降水难度。

中信大厦位于 CBD 核心区，是整个地块中的最高建筑，也是基坑最深项目，达 38m，局部深 40m。地下空间基坑深度约为 27m 左右，采用一体化施工方式，后期支护深度仅约 11m。

大厦 −27.20m 以下部分基坑支护采用"地下连续墙 + 预应力锚杆 + 混凝土内支撑"，标高 −28.60m 以下采用地下连续墙

止水帷幕方案。支护地连墙与止水帷幕布置在公共空间基底以下，可替代部分基础桩，使中信大厦的支护结构能够提前施工，同时其结构外皮距离红线仅 50mm，地下室有效使用面积额外增大超过 7,000m²，提高了土地利用率，多方共赢。

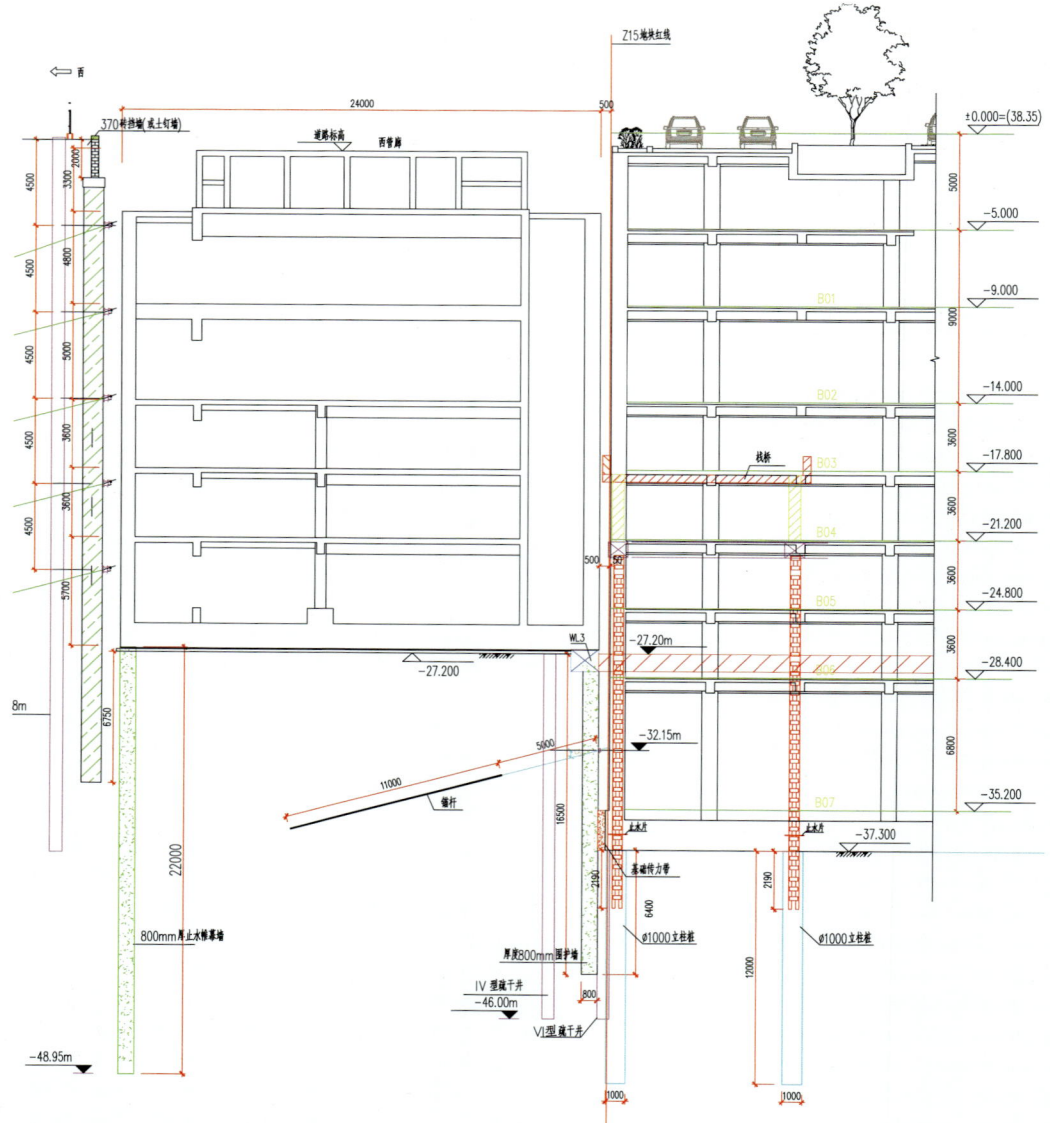

The main roads in the CBD core area are arranged in the form of nine rectangle grid, which is connected with the surrounding roads, and the underground tunnels are laid under each road. The whole block foundation pit is made in the means of integrated construction, namely, breaking the restriction of red lines between blocks, carrying out overall large excavation, support and dewatering, and simplifying the foundation pit of the second-class block with greater depth into a "pit-in-pit" mode to reduce the difficulty of support and dewatering.

Situated in the core area of CBD, CITIC Tower is the tallest building in the whole plot and also the deepest project with 38m of foundation pit,The part of foundation pit 40m. The depth of the foundation pit in underground space is about 27m . The integrated construction mode is adopted, and the late support depth is only about 11m.

"Diaphragm Wall + Pre-stressed Anchor Rod +Concrete Internal Support" shall be adopted for foundation pit support under -27.2m in the building, and underground water below -28.60m in elevation shall use water stopping curtain scheme of the diaphragm wall. The supporting diaphragm wall and water stopping curtain are arranged below the base of the public space, and the supporting structure of CITIC Tower can be constructed ahead of time by replacing some foundation piles. Meanwhile, the structural skin is only 50mm away from the red line, and the effective usable area of the basement is increased by more than 7,000m^2, thus improving the land utilization rate and achieving a win-win situation.

5.3 超深超厚大体积混凝土底板综合施工技术
Integrated Construction Technology Of Ultra-Deep And Ultra-Thick Mass Concrete Bottom Plate

大数据 Data

20,000 t
国内首次采用 HRB500
级钢筋总量
The total amount of HRB500
grade rebar adopted for
the first time in china

40 mm
国内首次采用 HRB500
级钢筋直径
The diameter of HRB500
grade rebar adopted for
the first time in china

0 次
安装地下室柱底板和剪力墙钢
板扩孔
Installing basement sill and
shear wall steel plate reaming

93 h
连续混凝土浇筑时间
Continuous concrete pouring time

56,000 m³
北京一次连续混凝土
浇筑新纪录
A record of one-time continuous
concrete pouring in beijing

602 m³/h
混凝土浇筑速度
Concrete pouring speed

中信大厦基础筏板国内首次采用 HRB500 级直径 40mm 钢筋，总量约 20,000t。钢筋全部场外批量加工，损耗率为 1% 以内。

基础筏板厚度分别为 2,500mm、4,500mm、6,500mm，钢筋支撑架分单元场外加工，现场整吊组拼，减少人工用量，保障了工期。

发明"锚栓套架自适应式设计技术"和"可拆装整体式锚栓套架安装技术"，仅历时 10 天即顺利完成 2,138 根锚栓、165 个锚栓套架的安装，总计起吊次数 253 吊次，仅为逐根安装法起吊次数的 11.8%。混凝土大底板浇筑后，锚栓最大误差为 8mm，远低于预期控制值 13mm，精度提高 1.62 倍，且在安装地下室柱底板和剪力墙钢板的过程中，实现了"零扩孔"的预期目标。

采用大掺量的工业废料粉煤灰配合比水泥，粉煤灰掺量占胶凝材料的 50%，减少了水泥用量（每立方米减少约 70kg），并改善了混凝土拌合料的和易性、密实性，延缓水化速度，减小混凝土因水化热引起的温升，防止混凝土温度裂缝的产生。

发明串管加溜槽组合体系，解决狭小场地深基坑混凝土浇筑问题。利用 12 套泵管体系与 4 套串管加溜槽组合体系共同浇筑，耗时 93h 完成 56,000m³ 混凝土浇筑，平均每小时浇筑混凝土 602m³，创造了北京市一次连续混凝土浇筑的新纪录。

图片说明 comments

1. 串管、溜槽组系统实施现场
Implementation site of string
pipe and chute group system

2. 狭窄高度下混凝土振捣
Concrete vibration at narrow height

3. 底板钢筋绑扎
Bottom plate rebar binding

4. 东北角巨柱锚栓套架
Giant column anchor bolt
sleeve in northeast corner

For the first time in China, HRB500 grade rebar with a diameter of 40mm is adopted for the foundation Bottom Plate in the CITIC Tower, with a total capacity of about 20,000t. All the rebars are processed off-site in batches, and the loss rate is less than 1%.

The thicknesses of the foundation Bottom Plates are 2,500mm, 4,500mm, 6,500mm respectively. The rebar support frame is processed off-site by unit, and is lifted together on site and assembled to reduce the labor consumption and ensure the construction period.

The invention of "Adaptive Design Technology of Anchor Bolt Sleeve" and "Installation Technology of Removable Integral Anchor Bolt Sleeve" successfully make the installation of 2,138 anchor bolts and 165 anchor bolt sleeves completed in 10 days, with a total lifting times of 253, which is only 11.8% of the lifting times of one-by-one installation method. After concrete pouring of mass bottom plate, the maximum deviation of anchor bolt is 8mm, much lower than the expected control value of 13mm, and the precision is increased by 1.62 times. The expected goal of "No Reaming" has been achieved in the process of installing the steel plates of the basement sill and shear wall.

The concrete with a large amount of industrial waste fly ash mixture is adopted, and the amount of fly ash is equal to 5% of cementitious material, which reduces the cement amount (about $70kg/m^3$), improves the workability and compactness of the concrete mixture, retards the hydration speed, decreases the temperature rise of concrete caused by the heat of hydration, and prevents the generation of temperature cracks in concrete.

The combined system of a string pipe and chute group is invented to solve the concrete pouring problem of a deep foundation pit in a narrow and small site. The concreting of $56,000m^3$ was completed in 93 hours, with an average of $602m^3$ per hour, using 12 sets of pump pipe systems and 4 sets of string pipe plus chute combined systems, setting a new record for continuous concrete pouring in Beijing.

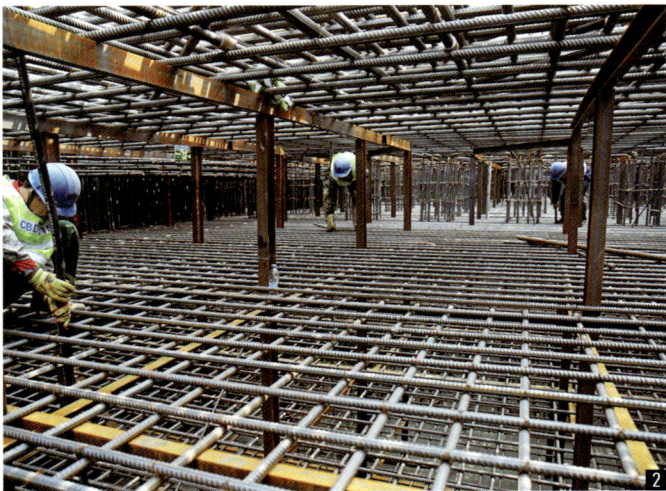

5.6 万 m³ 混凝土基础底板浇筑准备
Prepare to place 56,000m³ of concrete foundation bottom plate

| 5.4 | **超高层建筑智能化施工装备集成平台系统应用技术**
Application Technology Of Intelligent Construction Equipment Integrated Platform System For Super High-Rise Buildings |

大数据 Data

43 m × 43 m × 38 m
智能钢平台尺寸
Intelligent steel platform dimension

1,849 m²
智能钢平台顶部施工面积
Top construction area of intelligent steel platform

9 个
智能钢平台作业层
Operation layer of intelligent steel platform

1,770 m²
外挂架下挂操作层面积
Area of hanging operation layer area under the external rack

5,100 m²
F38 以下核心筒内挂架面积
Rack area in the core tube below 38 floors

4,800 t
智能钢平台顶推力
Intelligent steel platform top thrust

>14 级
智能钢平台可抵御高空强风
The smart steel platform can withstand high-altitude strong wind force

2 台
世界首次将 M900D 塔吊与智能钢平台集成
M900D tower cranes and intelligent steel platforms are integrated for the first time in the world

图片说明 comments

1. 智能化施工装备集成平台顶模剖面图
Sectional drawing of the top mold for intelligent construction equipment integrated platform

2. 高空作业的智能化施工装备集成平台系统
Intelligent steel platform system for intelligent construction equipment integrated platform

项目独创性地采用由中建三局自主研发的第三代集成型自带塔机微凸支点智能化施工装备集成平台体系。

平台长 43m、宽 43m、顶部施工面积约 1,849m²，最大高度 38m。挂架下挂 4.5 个楼层，共 9 个作业层。其中，外挂架下挂操作层面积约 1,770m²；F38 以下核心筒内挂架面积达 5,100m²，施工至 F38 后结构内收拆除了 2,500m²，内挂架下挂操作层面积约 2,600m²。

该平台顶推力达 4,800t，为世界房建施工领域面积最大、承载力最高的施工平台。其强大的空间框架结构体系使其可承受上千吨荷载，可抵御高空 14 级以上的强风。工人们在全封闭平台上作业，如履平地，可同时进行 4 层核心筒立体流水

施工。同时，在世界范围首次创造性地将 2 台 M900D 塔吊（单台自重达 254t）与智能化施工装备平台系统集成，实现塔机平台一体化，采取这项创新技术，节约工期约 56 天。

The project creatively adopts the third generation integrated tower crane micro-convex fulcrum intelligent construction equipment integration platform system independently developed by China Construction Third Engineering Bureau Co., Ltd.

The platform is 43m long and 43m wide. The top construction area is about 1,849m^2 and the maximum height is 38m. Hang 4.5 floors under the rack, with a total of 9 Operation layers. The hanging operation layer area under the external rack is about 1,770m^2; The rack area in the core tube below 38 floors is 5,100m^2, 2,500m^2 have been removed from the structure after the construction to the 38 floors, and the hanging operation layer area under the inner rack is about 2,600m^2.

With a top thrust of 4,800t, the platform has the largest area and highest bearing capacity in the field of building construction in the world. Its strong space frame structure system makes it withstand thousands oft of loads, and high altitude force 14 strong winds. The workers make operations on a fully enclosed platform as easily as walking on firm earth, and can simultaneously carry out the 3D flow repetitive construction operations of four-layer core tubes. Meanwhile, two M900D tower cranes (each weighing 254t) are creatively integrated with the intelligent construction equipment integration platform system to realize the integration of the tower crane platform for the first time in the world.

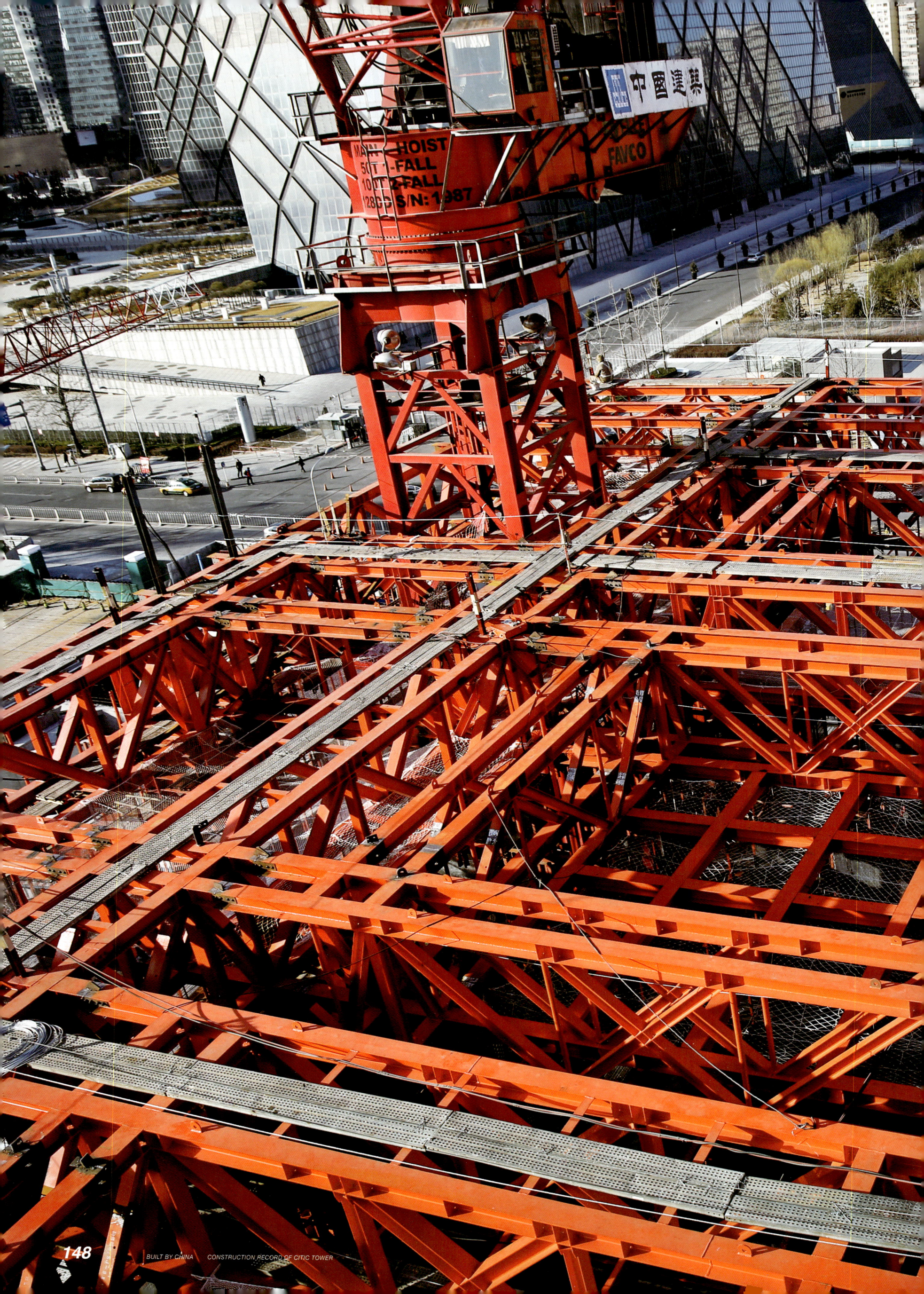

平台长 43m，宽 43m，顶部施工面积约 1,849m²，最大高度 38m。
The platform is 43m long and 43m wide. The top construction area is about 1,849m² and the maximum height is 38m.

5.5 核心筒钢板墙制作及安装技术
Manufacture And Installation Technology Of Core Tube Steel Plate Wall

大数据 *Data*

60 mm
核心筒钢板墙底部区域单钢板厚度
Single steel plate thickness of the bottom area of core tube steel plate wall

8 mm
核心筒钢板墙顶部区域单钢板厚度
Single steel plate thickness of the top area of core tube steel plate wall

30 %
吊装滑梁安全并节省工效
The hoisting sliding beam is safe in operation, and improves the work efficiency

1 d
每节钢板墙节省工期
Savings of 1 day for each steel plate wall

13 m
超厚板超长单条焊缝
Ultra-thick plate and ultra-long single weld seam

图片说明 *Comments*

1. 超大尺度复杂钢板墙体施工阶段
Construction stage of ultra-large scale complex steel plate walls

2. 钢板墙钢筋 BIM 模拟
BIM simulation of steel rebar for steel plate walls

3. 钢板墙吊装
Hoisting of steel plate wall

核心筒钢板墙在底部区域（F046 下）采用的是内置单钢板，厚度最大为 60mm，中间区域（F046－F103）采用的是内置钢暗撑，顶部区域（F103 以上）采用 8mm 厚的内置单钢板。钢结构的材质均为 Q345C。

施工中运用 BIM 技术进行钢板墙施工模拟，深化阶段分段分节避开顶升平台主次桁架影响，构件编号与发运、吊装顺序保持一致，确保安装可行性；安装过程中采用吊装滑梁，规避双机换钩的安全风险，节省工效约 30%，每节钢板墙节省工期 1d。

超厚板超长焊缝，单条焊缝最长 13m，现场采取在钢板墙端部设置约束支撑的措施，制定先立后横、先长后短、先中心后四周的焊接顺序，选用先进的同步对称焊接、分段跳焊焊接工艺，减弱焊接变形。

In the bottom area (under F046), the core tube steel plate wall adopts the built-in single steel plate with the maximum thickness of 60mm; the built-in steel concealed brace is adopted in the middle area (F046-F103), and the built-in single steel plate with the thickness of 8mm is adopted in the top area (above F103). The material of steel structure is Q345C.

During the construction, BIM technology is used to simulate the steel plate wall construction. In the deepening stage, the influence of the primary and secondary truss of the jacking platform is avoided step by step, and the number of the components is consistent with the order of shipment and hoisting to ensure the feasibility of installation; during the installation process, the hoisting sliding beam is adopted to avoid the security risk of double-machine hooks change, which improves the work efficiency by 30% and saves one day for each steel plate wall.

The maximum length of the single super-long weld seam of super-thick plate is 13m. On-site measures are taken to set restraining support at the end of steel plate wall, and the welding sequence is worked out, that is, first standing and then transverse, first long and then short, and first center and then around. Advanced synchronous symmetrical welding and segmented skip welding process are selected to reduce welding deformation.

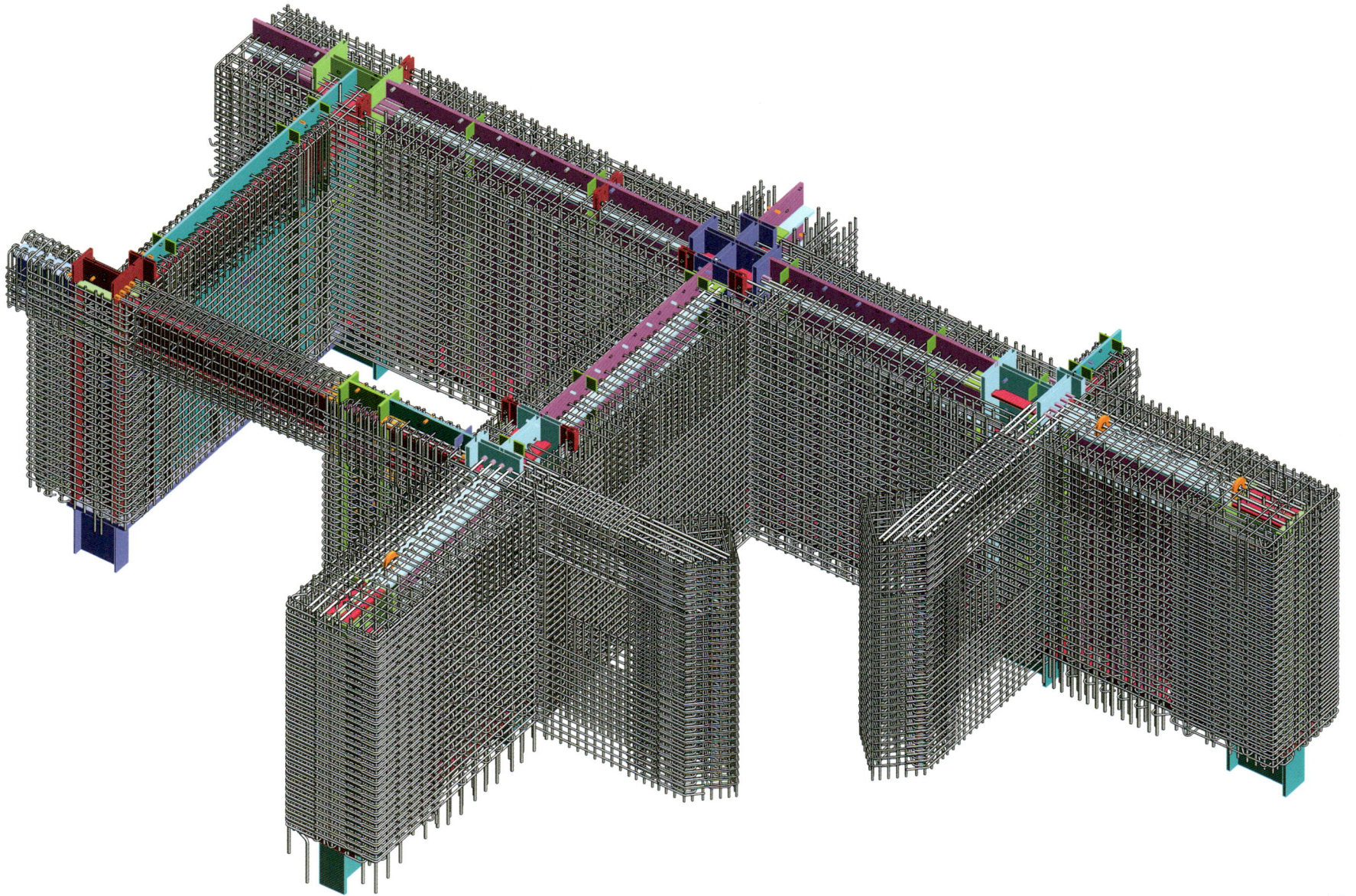

5.6 混凝土内灌外包多腔体巨型柱施工技术
Construction Technology Of Multi-Cavity Giant Column With Inner Concrete Filling And Outer Cladding

大数据 Data

>90m²
世界最大多腔体异形巨柱单层面积
Single-layer area of the largest cavities in the world

13个
最多单层截面腔体数量
Maximum number of single-layer cross-sectional cavities

8.3%
节省巨柱钢筋安装时间
The installation time of giant column rebar saved

图片说明 Comments

1. 巨柱方案模拟
Simulation of the giant column scheme

2. 巨柱浇筑
Giant columns pouring

3. 世界最大 13 腔体巨柱
The world's largest 13-cavity giant column

延伸阅读 Links

P188

中信大厦巨型柱采用了多腔多边形钢管混凝土，尺寸规模和施工难度罕见，腔体内构造复杂，隔板多孔，劲板隔板纵横交错，设有上百根贯穿水平隔板的竖向钢筋构造柱和下人孔处的圆形芯柱。巨柱单层面积最大超过 90m²，是目前世界上最大的多腔体异形巨柱。单层截面最多分为 13 个腔体，内灌 C70/C60/C50 自密实混凝土。地下室部位巨柱外包 C70 高强无收缩自密实混凝土。

施工中科学地对巨柱进行分段分节，通过施工模拟确定"内外组合，横立结合"的焊接顺序，降低焊后矫正时间，节省8.3%安装时间，并显著降低焊接变形。创新采用导管导入法施工技术，确保巨柱内部混凝土的密实度。

The giant columns of CITIC Tower are made of multi-cavity polygonal concrete filled steel tube, which is rare in terms of size and difficulty in construction, complicated in cavity structure, porous in partition plate, crisscrossing in stiffened plate partition plate. The vertical structural rebar and stirrup net for penetrating horizontal partition plate, and reinforced concrete circular core column are arranged. The largest single-layer area of giant columns is over 90m², the largest multi-cavity special-shaped giant columns in the world. The single-layer section can be divided into up to 13 cavities, filled with C70/C60/C50 self-compacting concrete. C70 high-strength shrinkage-free self-compacting concrete is used as cladding around giant columns in the basement.

During the construction, the giant column is segmented scientifically, and the welding sequence of "Inside and Outside, Horizontal and Vertical Combination" is determined through the construction simulation to reduce the post-welding correction time, save installation time by 8.3%, and significantly reduce the welding deformation. The innovative use of conduit leading-in construction technology ensures the density of concrete inside the giant column.

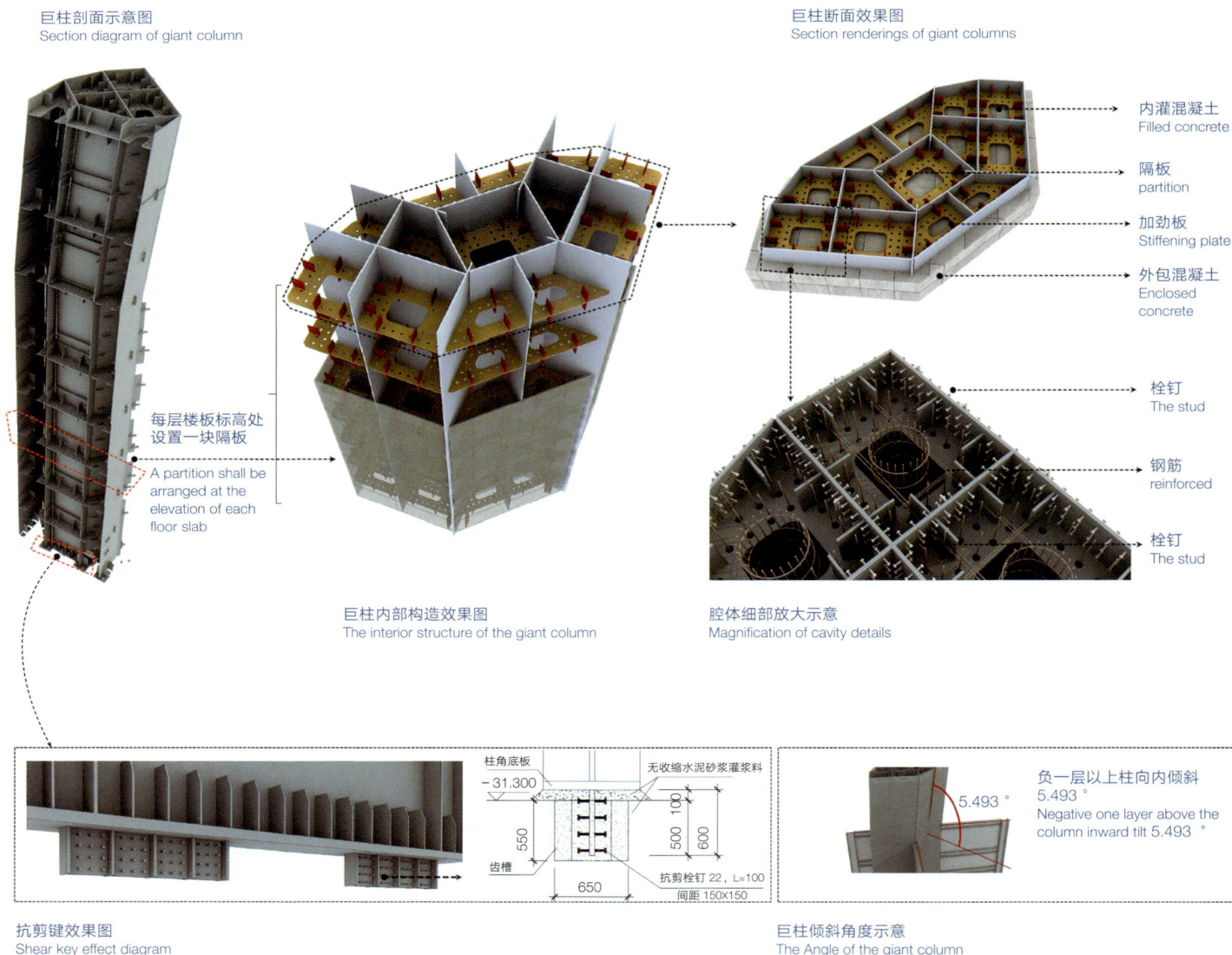

巨柱剖面示意图
Section diagram of giant column

巨柱断面效果图
Section renderings of giant columns

内灌混凝土
Filled concrete

隔板
partition

加劲板
Stiffening plate

外包混凝土
Enclosed concrete

每层楼板标高处设置一块隔板
A partition shall be arranged at the elevation of each floor slab

栓钉
The stud

钢筋
reinforced

栓钉
The stud

巨柱内部构造效果图
The interior structure of the giant column

腔体细部放大示意
Magnification of cavity details

抗剪键效果图
Shear key effect diagram

柱角底板
-31.300
无收缩水泥砂浆灌浆料
550
500
100
600
齿槽
650
抗剪栓钉 22，L=100
间距 150x150

巨柱倾斜角度示意
The Angle of the giant column

5.493°
负一层以上柱向内倾斜
5.493°
Negative one layer above the column inward tilt 5.493°

5.7 大跨度双曲悬挑雨篷高空原位安装技术
High Altitude In-Situ Installation Technology Of Large-Span Hyperbolic Cantilever Awning

雨篷为大跨度空间双曲箱型悬挑结构，高度方向跨5个结构层，向外最大悬挑跨度为14.1m。下部最低点距离地面高度约4.1m，顶部标高为29.75m。主要由顶横梁、边梁、主挑梁、横竖龙骨及支撑组成，悬挑雨篷通过支撑与巨柱、桁架下弦及结构边梁相连组成结构稳定体系。

受运输条件限制，雨篷分段运至现场，采用部分散件组拼、高空整体吊装的方式进行安装，由巨柱本体部位对称向中间扩展，直至合拢。雨篷主构件分段按照1根主挑梁+1根竖龙骨+4根横梁+2根支撑牛腿，两榀间的横龙骨现场拼装；分段最大重量为单榀7.2t，拼装后总重16.5t。雨篷安装采用标准化胎架。主要由标准节、顶部钢板及底部柱脚组成。胎架最大高度14m，共32组，单组标准化胎架可承受最大载荷300t。

施工前通过有限元软件对施工全过程进行模拟分析，根据现场实际工况对雨篷安装进行优化。安装中，根据计算的结构下挠值，对雨篷在施工中进行预调值起拱，确保安装精度。

The canopy is a double-curved box-type cantilever structure with a large span, spanning 5 structural layers in the height direction, and the maximum overhang span is 14.1m. The lowest point of the lower part is about 4.1m from the ground and the top level is 29.75m. It is mainly composed of top beam, side beam, main picking beam, horizontal and vertical keel and support. The cantilevered canopy is connected with the giant column, the truss lower string and the structural side beam to form a structural stability system.

Limited by the transportation conditions, the canopy is transported to the site in sections, and some parts are assembled and assembled in a high-altitude overall hoisting manner. The main body of the giant column is symmetrically extended to the middle until it is closed. The main components of the canopy are segmented according to 1 main picking beam +1 vertical keel + 4 beam + 2 supporting the ox leg, and the horizontal keel between the two rafts is assembled on site; the maximum weight of the section is 7.2t, and the total weight is 16.5t after assembly. The canopy is installed using a standardized tire frame. Mainly composed of standard section, top steel plate and bottom column foot. The maximum height of the tire frame is 14m, a total of 32 groups, and a single set of standardized tire frames can withstand a maximum load of 300t.

Before the construction, the finite element software was used to simulate the whole process of the construction, and the installation of the canopy was optimized according to the actual working conditions on site. During the installation, according to the calculated deflection value of the structure, the pre-adjusted value of the canopy during the construction is arched to ensure the installation accuracy.

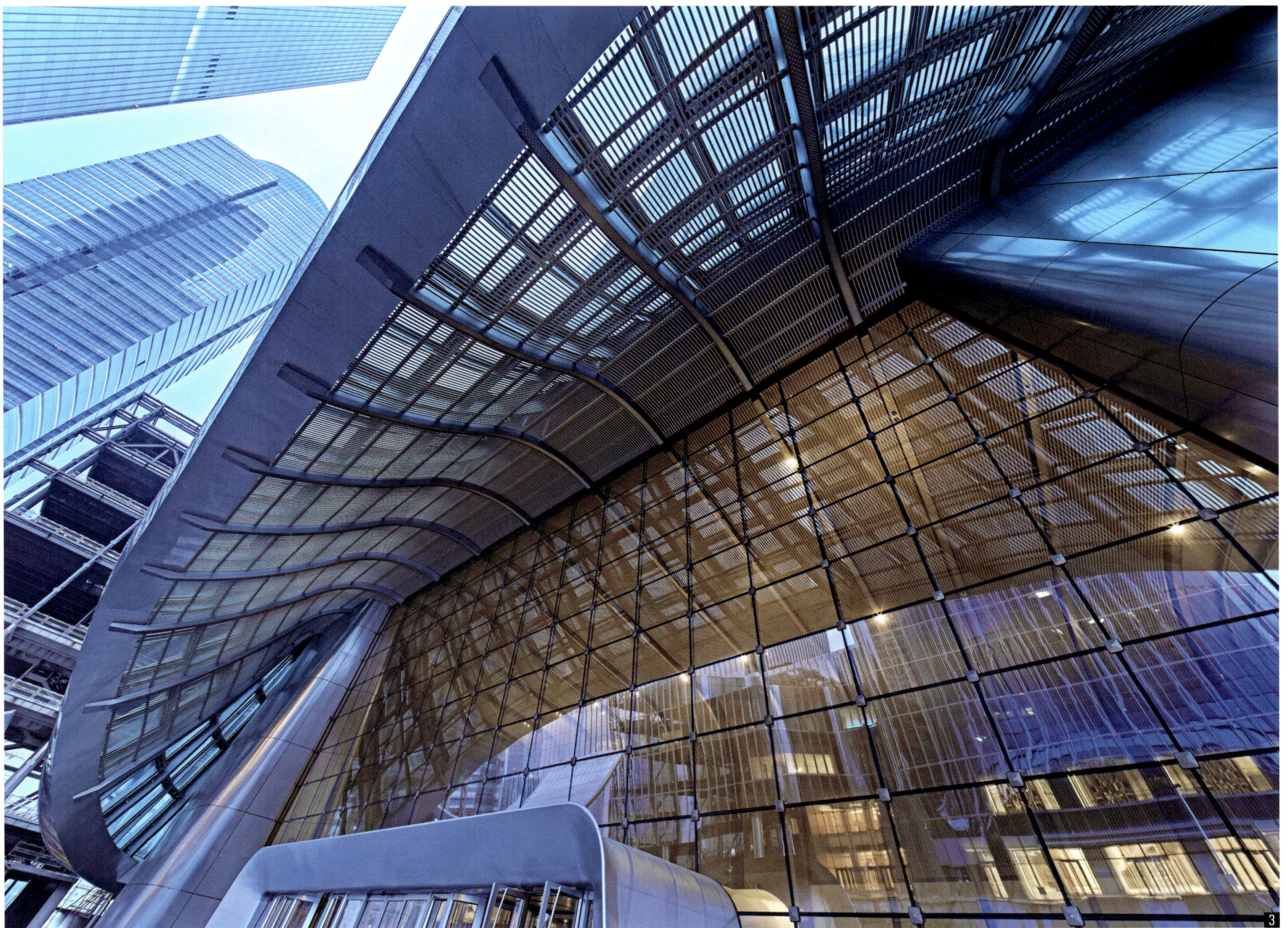

居安思危 警钟常鸣

5.8 高空双曲塔冠"硬支撑无胎架"式安装技术
The "Hard Support Without Frame" Technology For High-Altitude Hyperbolic Tower Crown

大数据 Data

503m
塔冠高空施工安装位置
Installation altitude of tower crown construction

30.3m
塔冠结构高度
The structure height of tower crown

204件
塔冠钢结构构件数量
The number of steel structure components of tower crown

25.3t
塔冠最大构件分段重量
Maximum component segmental weight of the tower crown

12m
塔冠构件最大长度
Maximum length of the tower crown component

50mm
塔冠构件钢板厚度
Steel plate thickness of the tower crown component

图片说明 Comments

1. 塔冠钢结构图
Drawing of the tower crown steel structure

2. 施工安装中的塔冠
Tower crown under construction and installation

3. 塔冠屋面板安装旁站监督
On-spot supervision of the installation of the tower crown roof panel

延伸阅读 Links

P188

塔冠结构位于503m高空的F105楼面梁上，并向上延伸至屋顶，高度为30.3m。塔冠钢结构构件总数量约204件，双轴对称布置。其中1/4塔冠结构共分18段，最大构件分段重量约为25.3t，主构件为18件，补档构件为25件，构件最长为12m，使用最大板厚50mm。

项目在施工前进行安装模拟分析，依据模拟结果，确定合理的施工顺序和施工方法。在塔冠结构安装过程中采用地面散件组拼+塔吊高空原位安装的方式。受塔冠结构、现场场地、工期等诸多因素影响，塔冠施工采用"硬支撑无胎架施工技术"。实现了结构构件的高效、安全安装，缩短了工期和获得了经济效益。

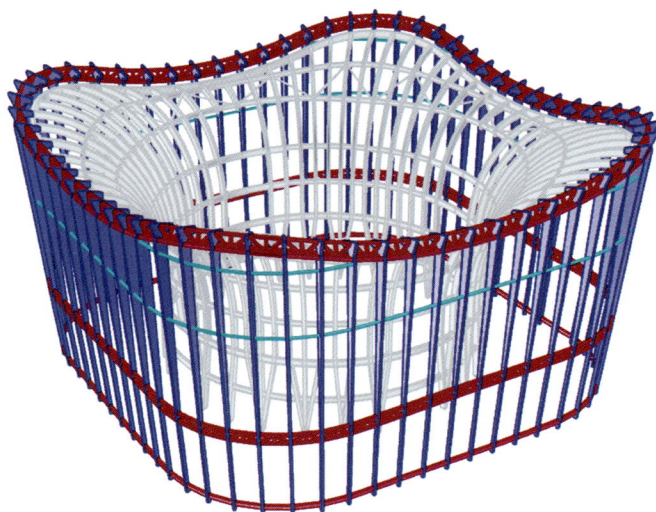

The tower crown structure is located on the floor F105 beam and extends upwards to the roof with a structure height of 30.3m. The total number of the components with the structure of tower crown steel are about 204, which are arranged in biaxially symmetrical manner. The 1/4 tower crown structure is divided into 18 sections, the weight of the section with the maximum component is about 25.3t. There are 18 main components and 25 gear-compensating components. The maximum length of components is 12m, and the maximum plate thickness is 50mm.

The installation simulation analysis is carried out before the construction of the project, and the reasonable construction sequence and method is determined according to the simulation results. In the process of tower crown structure installation, the method of ground loose parts assembly + tower crane in situ at high altitude is adopted. Due to many factors, such as tower crown structure, site and construction period, the "hard support without frame" technology is adopted in tower crown construction. The efficient and safe installation of the structural components is realized with enabling construction period and sound economic benefits.

2 台 M900D 塔吊在工作高度约 600m，以"硬支撑无胎架"技术吊装构件安装位于 500 多米高空的双曲塔冠。
Two M900D tower cranes with the working height of about 600m are hoisted and constructed, and the "hard support without frame" type hyperbolic tower crown is installed at the height of 500m.

双曲塔冠高度
The height of the tower crown is 528m
528m

塔吊施工高度
Tower crane construction height

>600m

5.9

塔冠高空幕墙悬臂式移动吊装技术
Cantilever Mobile Hoisting Technology For High-Altitude Curtain Wall Of Tower Crown

大数据 Data

1,808 块
塔冠区域玻璃板块
Glass plates in tower crown area

4 组
屋面异形钢结构上架设悬臂式
移动吊装施工设备
Four sets of cantilever mobile
hoisting construction equipment
erected on the special-shaped
steel structure of the roof

1,200 kg
吊臂前后端吊钩总额定荷载
Total fixed load of the lifting hooks
at the front and rear-end of lifting arm

1,850 mm
吊臂及篮筐前后行程
The fore-and-aft travel of
lifting arm and basket

14 °
吊装施工设备爬坡能力
Gradeability of hoisting
construction equipment

0-40 m
吊装施工设备提升高度范围
The hoisting height range of the
hoisting construction equipment

32 t
4 组吊装施工设备总重量
Total mass weight of 4 sets of
hoisting construction equipment

图片说明 Comments

1. 幕墙项目部进行
F104 单元体吊装
The curtain wall project department
hoists 104-floor units

2. 悬臂式移动吊
工作现场（一）
Cantilever mobile hoist (1)

3. 悬臂式移动吊
工作现场（二）
Cantilever mobile hoist (2)

中信大厦的塔冠拉索幕墙是世界最高的拉索幕墙系统，塔冠 F106 至女儿墙顶玻璃幕墙位置，高度为 503 - 528m。

塔冠区域共 1,808 块玻璃板块，604 套金属装饰线条，双曲造型金属铝单板 4,800m²。施工中借助 4 组悬臂式移动吊吊装施工设备。此设备在双曲造型钢结构上架设安装，大大克服了各种不利的施工环境，同时具备自行走、爬坡、进出调节、材料吊运、人员施工等多项功能。

1. 双轨道间距为 1,200mm；
2. 设备前端吊臂有两个吊钩吊装机构，前面吊钩吊施工篮筐，额定载荷为 200kg（2 人），后端辅助吊钩可装电动葫芦，额定荷载为 1,000kg，主要吊装玻璃板块；
3. 吊臂及篮筐前后行程为 1,850mm；
4. 设备爬坡能力为 14°；
5. 设备提升高度为 40m，可在 0 - 40m 的相对高度范围内吊装和作业；
6. 设备自行走电机功率为 8kW；
7. 每台设备自重 8t；
8. 每台设备由两个单独的大车组成，间距可调整，同时吊篮借用擦窗机轨道，实现了和塔冠外倾造型的完美贴合，大大增加了施工速度和安全性能。

通过楼层悬臂塔吊、自制桁架吊、吊装环形轨道等设备，实现了"楼层悬臂塔吊吊运单元体板块到卸料平台—单元板块转运进楼层—单元板块楼层内水平运输—环形轨道吊装单元板块"的吊装安装模式，该调运体系具有运输稳定、吊速快、覆盖性好、节省人力等优势，有效地保障了垂直运输和安装效率。

Tower crown cable curtain wall system — the world's tallest cable curtain wall, has the tower crown L106 extending to the position of parapet top glass curtain wall with the height of 503 — 528m.

There are 1,808 glass plates, 604 sets of metal decorative lines, hyperbolic metal aluminum veneer of 4,800m² in the tower crown area. Four sets of cantilever mobile hoisting construction equipment are adopted in construction. The equipment is installed on the hyperbolic steel structure, which greatly prevails over all kinds of unfavorable construction environment, and creates multiple functions, such as self-walking, slope climbing, in-out adjustment, material hoisting and personnel construction.

1. Dual track spacing: 1,200mm.
2. There are two lifting hook mechanisms on the lifting arm in the front end of the equipment. The front lifting hook lifts the construction basket with a rated load of 200kg (2 persons), and the rear auxiliary lifting hook can be equipped with an electric hoist with a rated load of 1,000kg, mainly used to hoist the glass plate.
3. The fore-and-aft travel of lifting arm and basket: 1,850mm;
4. Gradeability of the equipment: 14°;
5. The lifting height of the equipment is 40m, which can be hoisted and operated in the relative height range of 0 — 40m;
6. Power of the self-propelled electric motor of the equipment is 8kW;
7. Each equipment weighs 8t.
8. Each equipment is composed of two separate carts with adjustable spacing. Meanwhile, the basket achieves the perfect fit with the tilt modelling of the tower crown by dint of the track of the Window Scrubber, significantly increasing the construction speed and security performance.

With the help of equipment such as floor cantilever tower crane, self-made truss crane and hoisting ring track, the hoisting and installing mode of "floor cantilever tower crane lifting unit plate to discharge platform — unit plate transferring into floor — unit plate horizontal transportation on floor — ring track lifting unit plate" is realized. The hoisting system bears advantages such as stable transportation, fast hoisting speed, good coverage and manpower saving , effectively ensuring the vertical transportation and installation efficiency.

5.10 BIM 在超高层建筑建造中的应用
Application Of BIM In The Construction Of Super High-Rise Buildings

大数据 *Data*

2,773 个
合计 BIM 设计模型
The overall integrated
model of CITIC Tower

12,500 余个
设计及施工阶段累计发现问题
More than 12,500 problems found
in the design and construction stage

70%~**80**%
现场变更量较同类
超高层建筑降低
The number of on-site
changes is 70-80% less
than that of similar super
high-rise buildings

7,000 ㎡
增加使用面积
Increase the usable area by
more than 7,000㎡

10%
建造垃圾 LEED 金级评定标准
比率
LEED gold evaluation standard
ratio of construction waste

延伸阅读 *Links*

P094

中信大厦项目 BIM 模型构件管理体系采用与建筑体系相同的划分方式：划分为 7 大类，包括一级系统 31 个，二级系统 46 个，并根据 26 个分区原则进行科学组合。项目模型按区段拆分为 10 大区（ZB-Z8），以及 F8 设备层、F8 避难层及幕墙等。合计 BIM 设计模型 2,773 个。

在中信和业投资有限公司的总体规划下，施工阶段继承设计阶段的 BIM 模型，并进行模型深化，替换部分设计模型，增加必要过程信息，最终将模型深度提升至竣工模型标准。最终的竣工模型包含了运维所需的所有信息，并最终用于基于 BIM 模型的可视化运维系统。

为配合钢结构、幕墙、机电、装饰等专业的需求，在部分区域采用统一专业软件进行建模。最终所有专业模型需要导入轻量化平台进行综合应用。

各参建方利用 BIM 技术在可视、协调、模拟方面的优势，有效地提高了设计质量和效率。在设计及施工阶段累计发现 12,500 余个问题，大量减少了可能发生的拆改和返工。据统计，现场变更数量较同类超高层建筑降低了 70%~80%（被动变更占比更低）。据测算，结合 BIM 对建筑空间进行的优化，为大厦增加了超过 7,000㎡ 的使用面积；优化了超过 20 个大型设备用房的机电排布，使得物业运维更加便捷；大量的构件实现场外加工或预制生产，有效减少了现场扬尘及污染，产生建筑垃圾仅为 LEED 金级评定标准的 10%。

中信和业主导"全生命周期"采用 BIM 建造技术				
阶段	设计阶段	施工阶段 I	施工阶段 II	运维管理
成果	设计模型	施工模型	竣工模型	运维模型
应用	建立 7 大类，97 项模型文件	审核各专业深化设计模型 165 批次，优化和调整各专业深化设计问题 2,422 项	组织对设备层 / 避难层的机电深化设计 BIM 成果进行审核、优化 42 批次，共优化和调整 1,408 问题问题	数字化资产库
	41 家单位协同搭建模型平台	组织 52 次 BIM 综合协调，共审核设计问题共 2,129 项	三维激光扫描，辅助施工质量控制，目前已完成地上所有楼层的结构三维扫描工作，数据容量达 2.3TB	设备设施管理
	审核初步设计、施工图设计模型 56 批次，优化和调整施工图设计问题 6,526 项	组织施工现场 BIM 巡检 30 次，对机电、电梯及幕墙预埋件、结构预留洞口与 BIM 一致性核查，发现并解决 271 项设计 / 施工问题	开发运维 BIM 轻量化平台	物业管理
	重点难点施工模拟 16 次	基于 BIM 的施工进度计划模拟、施工方案模拟、施工交底等共 33 项	编制二维码系统及监理运维信息数据库	联动运营管理平台
	辅助预制化构建			

BIM CONSTRUCTION TECHNOLOGY IS ADOPTED IN "FULL LIFE CYCLE" BY CITIC HEYE				
Stage	Design stage	Construction stage I	Construction stage II	Operation management
Result	Design model	Construction model	Model of completed project	Operation model
Application	Establish 7 major categories; 97 model files	Review 165 batches of in-depth design models for each specialty, and optimize and adjust 2,422 indepth design problems for each specialty	Organize to review and optimize 42 batches of the BIM results of the electromechanical in-depth design of the equipment floor / refuge floor with 1,048 problems optimized and adjusted in total.	Digital asset bank
	41 companies collaborate to build model platforms	Organize 52 rounds of BIM comprehensive coordination and review 2,129 design problems.	The 3D laser scanning, assisted by construction quality control, of the structures of all floors above ground has been completed, with a data capacity of 2.3TB	Equipment and facilities management
	Review 56 batches of preliminary design and construction drawing design model, and optimize and adjust 6,526 problems of construction drawing design	Organize 30 times of BIM patrol inspections on the construction site, check the consistency of the embedded parts and the reserved openings of the structure of the electromechanics, elevator and curtain wall and BIM, and find and solve 271 design/construction problems	Develop BIM Operation lightweight platform	Property management
	16 times of key and difficult construction simulation	There are 33 BIM based construction schedule simulation, construction scheme simulation and construction disclosure.	Compile QR code system and supervisor Operation information database	Linkage operation management platform
	Auxiliary prefabrication			

中信大厦设计阶段 BIM 管理流程
BIM management flow of CITIC Tower in design stage

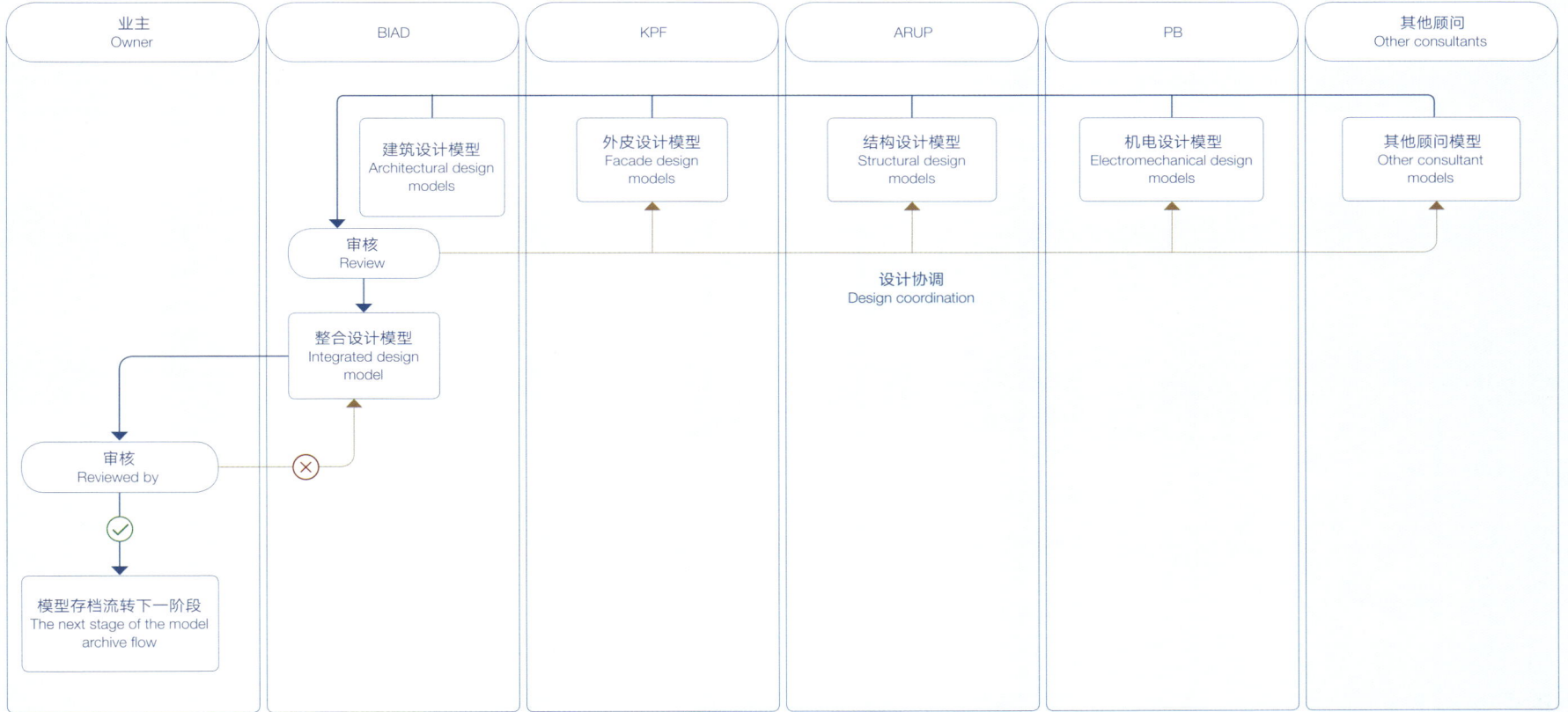

业主 Owner	BIAD	KPF	ARUP	PB	其他顾问 Other consultants
	建筑设计模型 Architectural design models	外皮设计模型 Facade design models	结构设计模型 Structural design models	机电设计模型 Electromechanical design models	其他顾问模型 Other consultant models

审核
Review

设计协调
Design coordination

整合设计模型
Integrated design model

审核
Reviewed by

模型存档流转下一阶段
The next stage of the model archive flow

BIM 模型成果审核优化流程
Design BIM management process

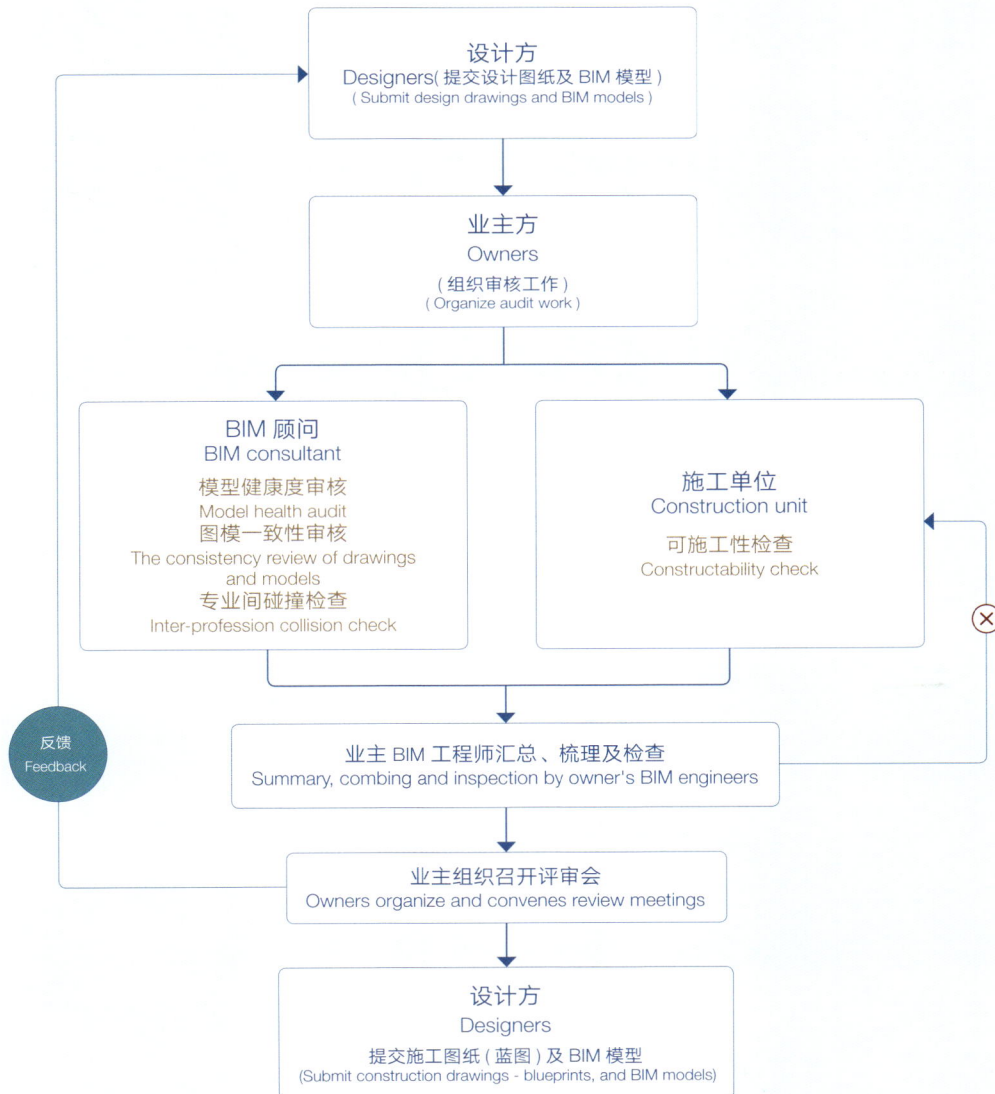

设计方
Designers(提交设计图纸及 BIM 模型)
(Submit design drawings and BIM models)

业主方
Owners
(组织审核工作)
(Organize audit work)

BIM 顾问
BIM consultant
模型健康度审核
Model health audit
图模一致性审核
The consistency review of drawings and models
专业间碰撞检查
Inter-profession collision check

施工单位
Construction unit
可施工性检查
Constructability check

反馈
Feedback

业主 BIM 工程师汇总、梳理及检查
Summary, combing and inspection by owner's BIM engineers

业主组织召开评审会
Owners organize and convenes review meetings

设计方
Designers
提交施工图纸 (蓝图) 及 BIM 模型
(Submit construction drawings - blueprints, and BIM models)

科技与创新
Technology & Innovation

图片说明 Comments

1. 实际施工对比 BIM 图
Comparison of actual construction
with the BIM diagram

2. BIM 施工模拟图
BIM construction simulation drawing

3. BIM 深化设计
F010-F021 核心筒钢板墙
BIM in-depth design F010-F021
core tube steel plate wall

4. 首层大堂深化设计 BIM 模型
BIM in-depth design of
glass curtain wall

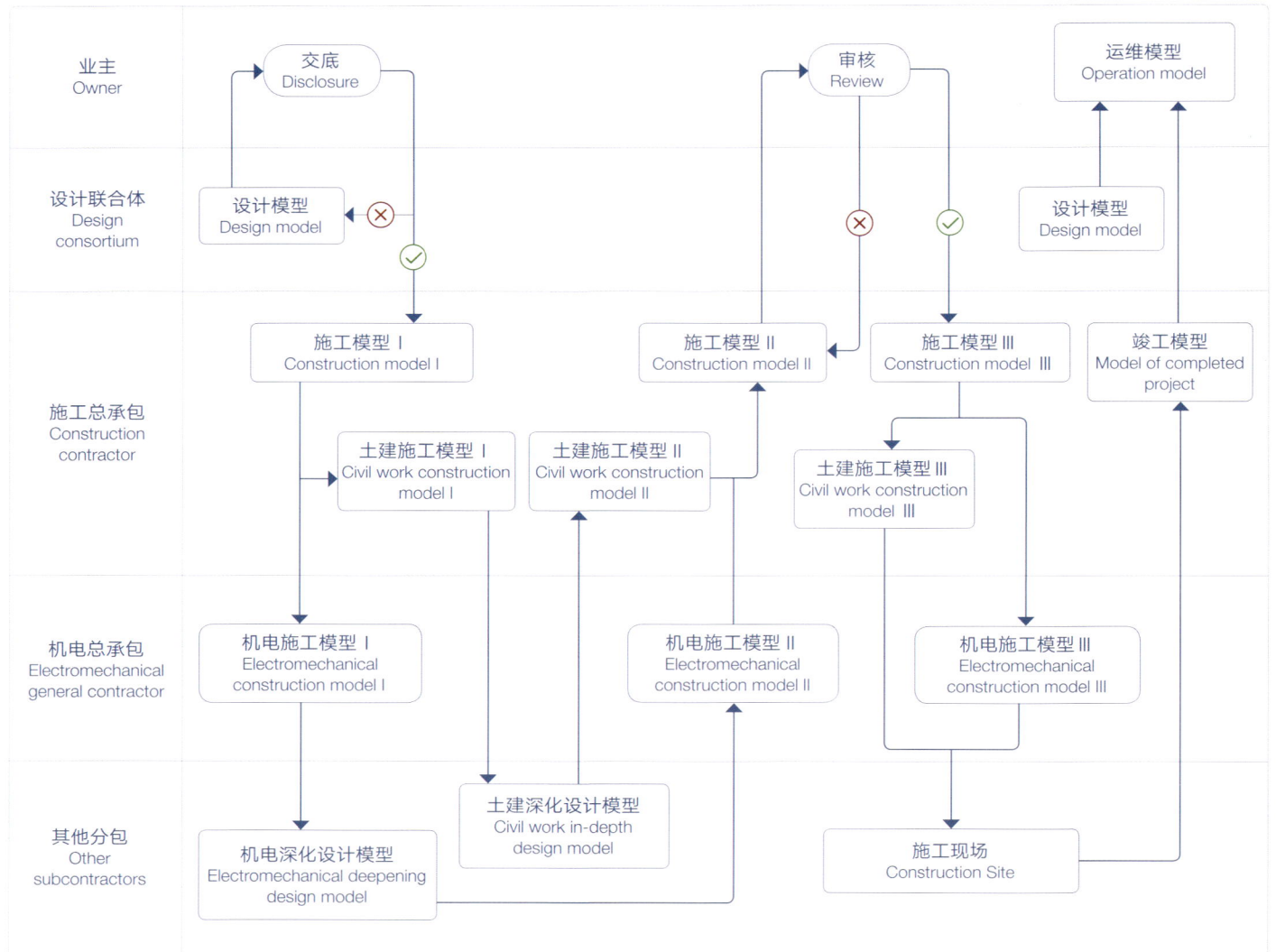

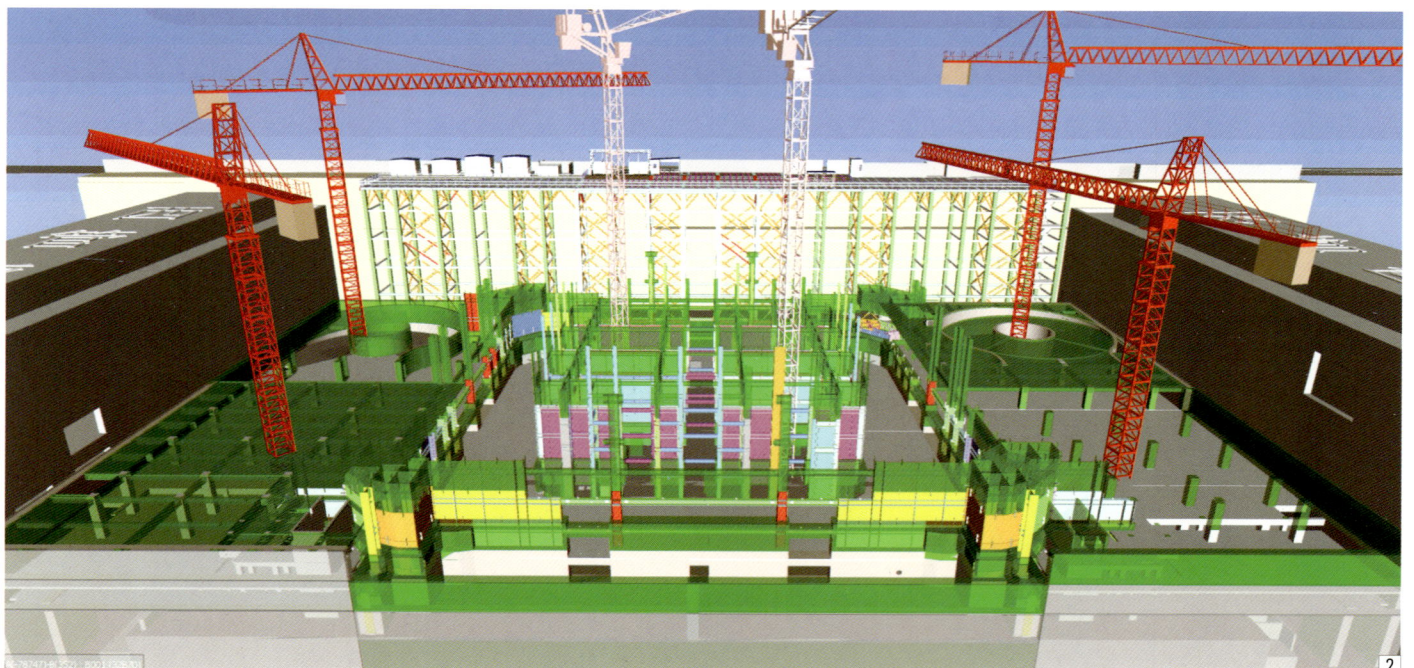

业主 Owner — 交底 Disclosure — 审核 Review — 运维模型 Operation model

设计联合体 Design consortium — 设计模型 Design model — 设计模型 Design model

施工总承包 Construction contractor — 施工模型 I Construction model I — 施工模型 II Construction model II — 施工模型 III Construction model III — 竣工模型 Model of completed project

土建施工模型 I Civil work construction model I — 土建施工模型 II Civil work construction model II — 土建施工模型 III Civil work construction model III

机电总承包 Electromechanical general contractor — 机电施工模型 I Electromechanical construction model I — 机电施工模型 II Electromechanical construction model II — 机电施工模型 III Electromechanical construction model III

其他分包 Other subcontractors — 机电深化设计模型 Electromechanical deepening design model — 土建深化设计模型 Civil work in-depth design model — 施工现场 Construction Site

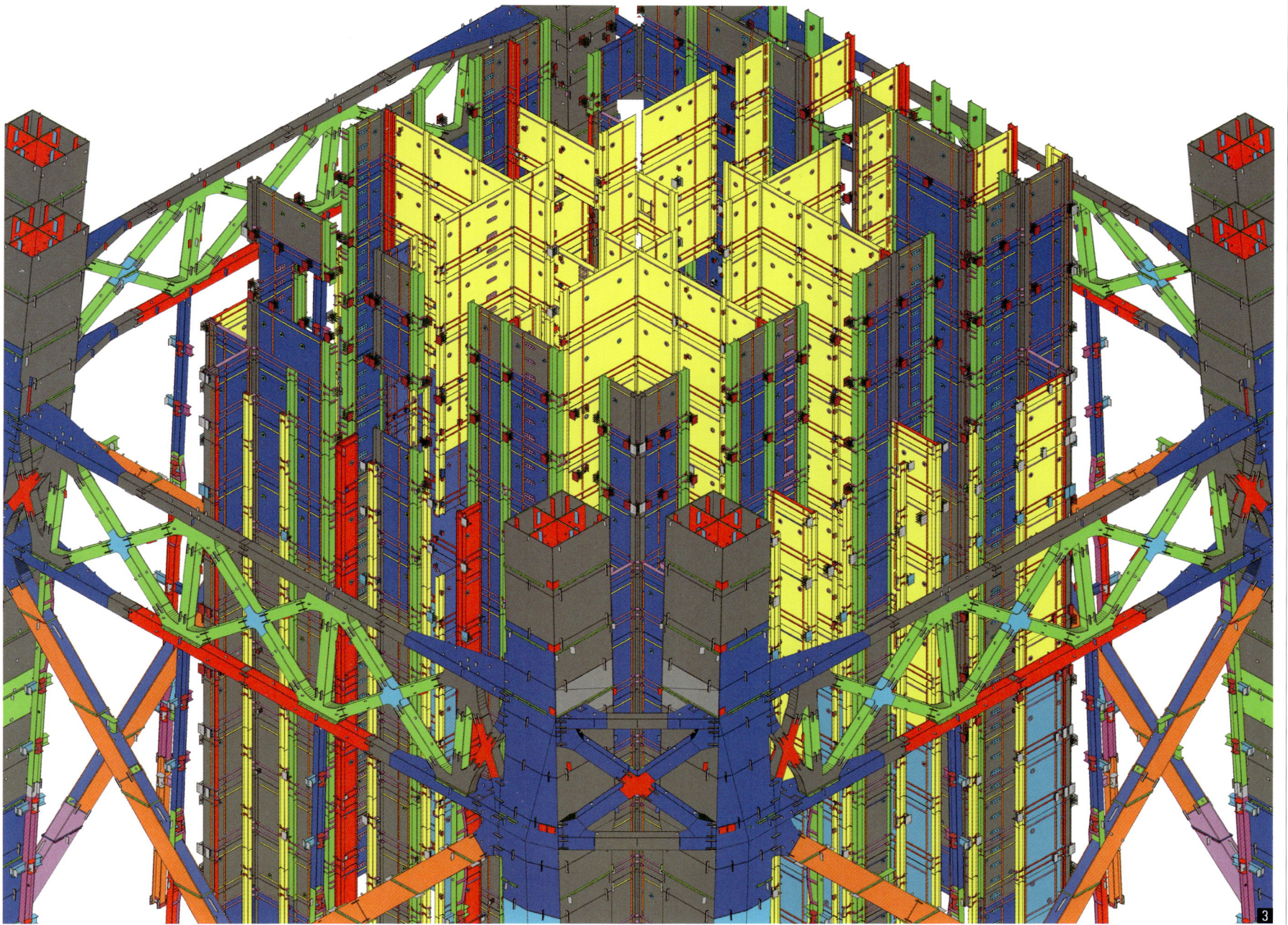

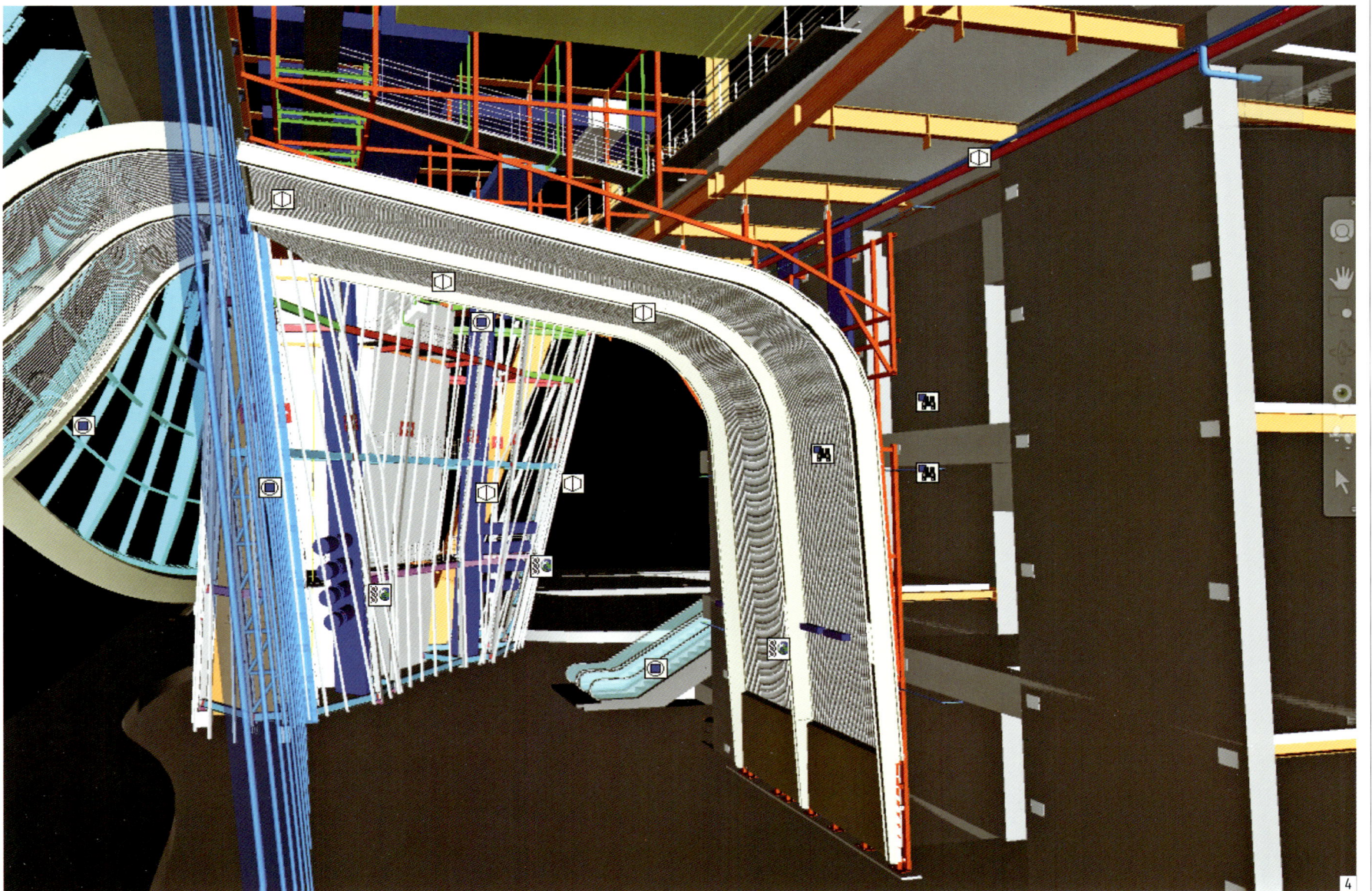

科技与创新
Technology & Innovation

The same division as that of the building system is adopted in the component management system of CITIC Tower BIM model: Systems are divided into 7 categories, including 31 primary systems and 46 secondary systems, which are combined scientifically in line with 26 subarea principles. The project model is divided into 10 zones (ZB-Z8), 8 equipment layers, 8 refuge floors and curtain walls. 14 underground models and 83 aboveground models are established, totaling 97 models.

Under the holistic planning of CITIC Heye Investment Co., Ltd., the BIM model adopted in the design stage is continued to be used in the construction stage, which is upgraded to replace part of the design model, with necessary process information added. Ultimately, the mode is upgraded to meet the standard of model of completed project. The final model of completed project contains all the information needed for operation and maintenance, and is finally adopted in the visual Operation system based on the BIM model.

In order to meet the needs of steel structure, curtain wall, electromechanics, decoration and other specialties, the application area of BIM in the construction of super high-rise buildings is modeled by unified professional software. Finally, all professional models need to be imported into the lightweight platform for comprehensive application.

The various participants effectively improve the design quality and efficiency by using the advantages of BIM technology in visualization, coordination and simulation. More than 12,500 problems are found in the design and construction stage, which greatly reduces the potential dismantling, modification and rework. According to the preliminary statistics, the number of changed sites is 70%~80% less than that of the similar super high-rise buildings (the proportion of passive changes is even lower). The estimated calculation shows that, the optimization of the building space by BIM provides more than 7,000m² of usable areas for the building; optimizes the electromechanics layout of more than 20 large equipment rooms, making the Operation of the property more convenient; allows a large quantity of components to achieve off-site processing or precast production and effectively reduce on-site dust and pollution with construction waste generated of only 10% by the LEED gold evaluation standard.

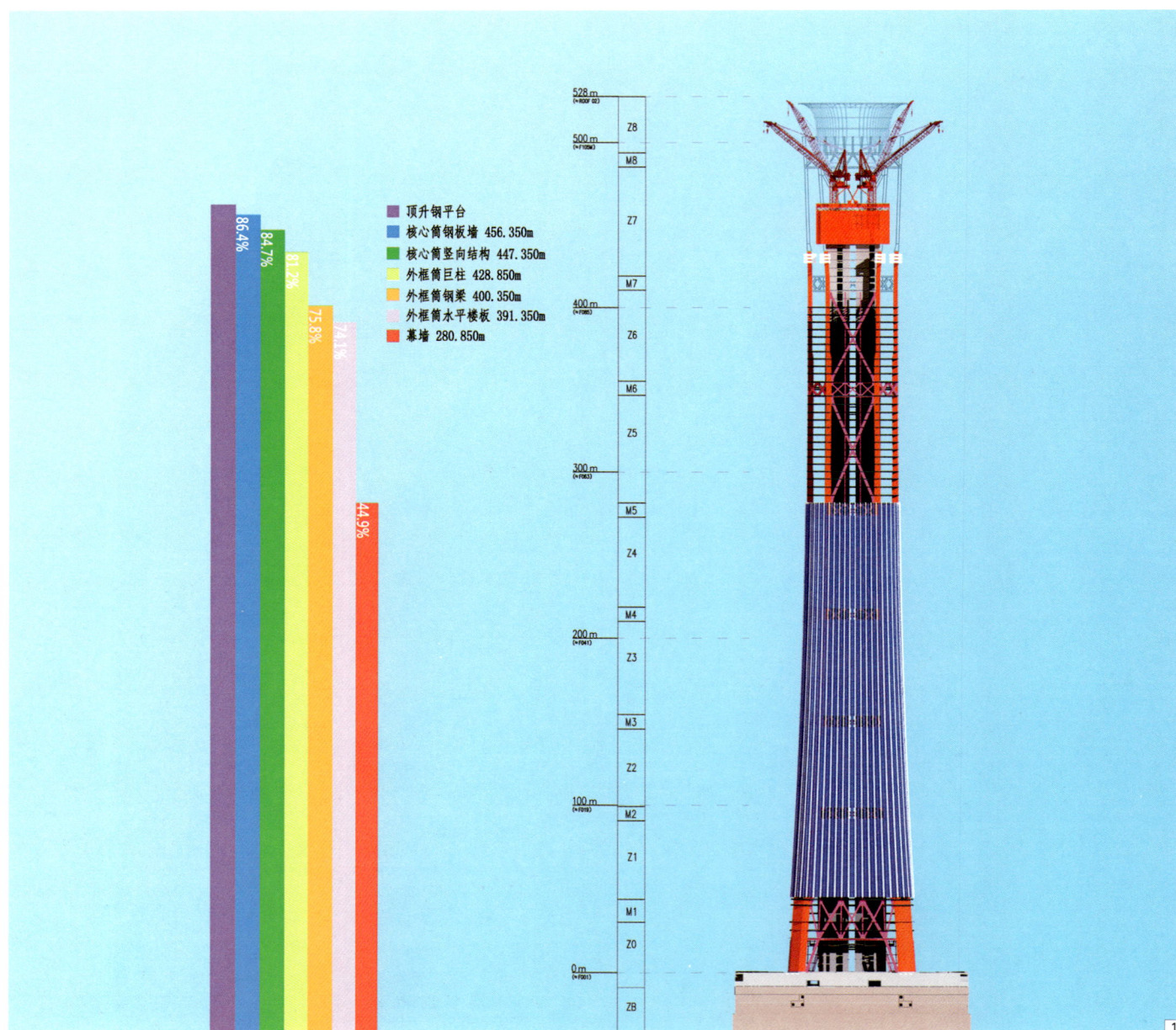

顶升钢平台
核心筒钢板墙 456.350m
核心筒竖向结构 447.350m
外框筒巨柱 428.850m
外框筒钢梁 400.350m
外框筒水平楼板 391.350m
幕墙 280.850m

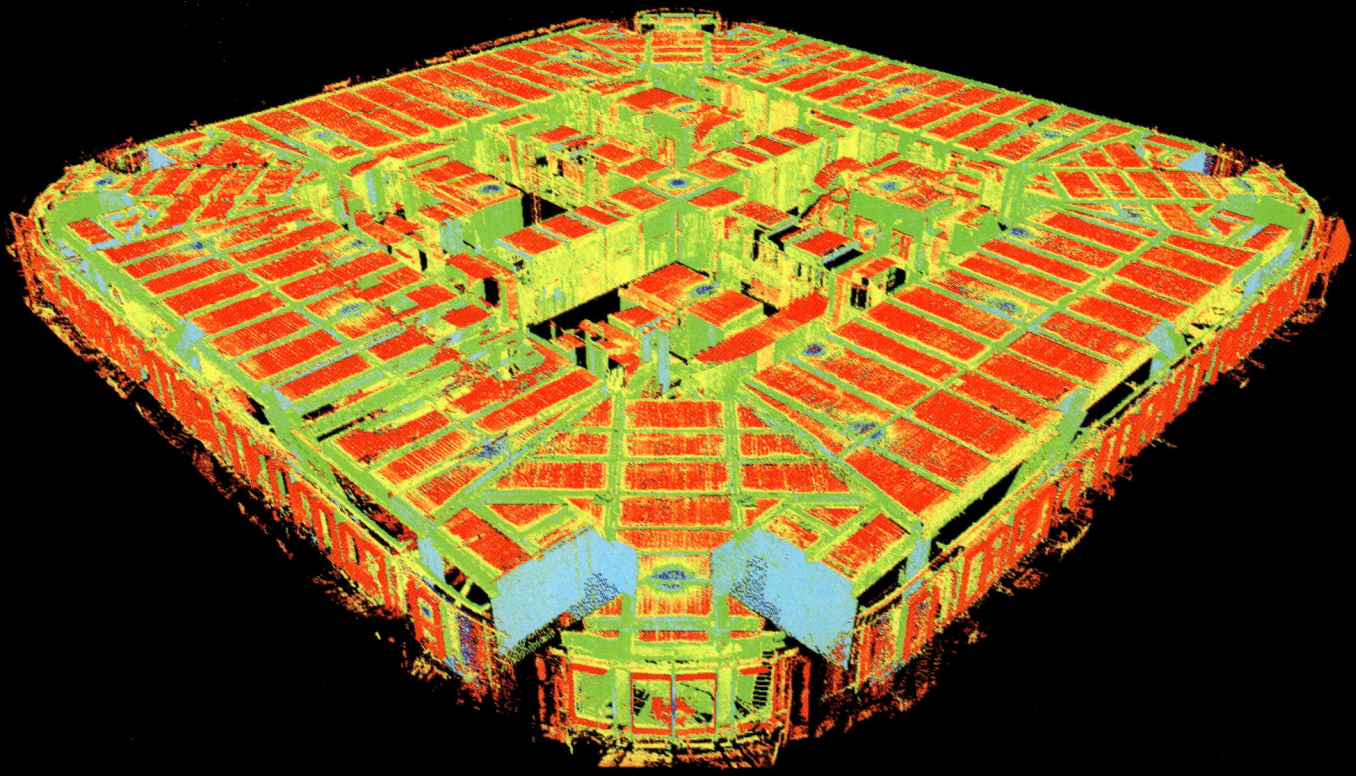

2

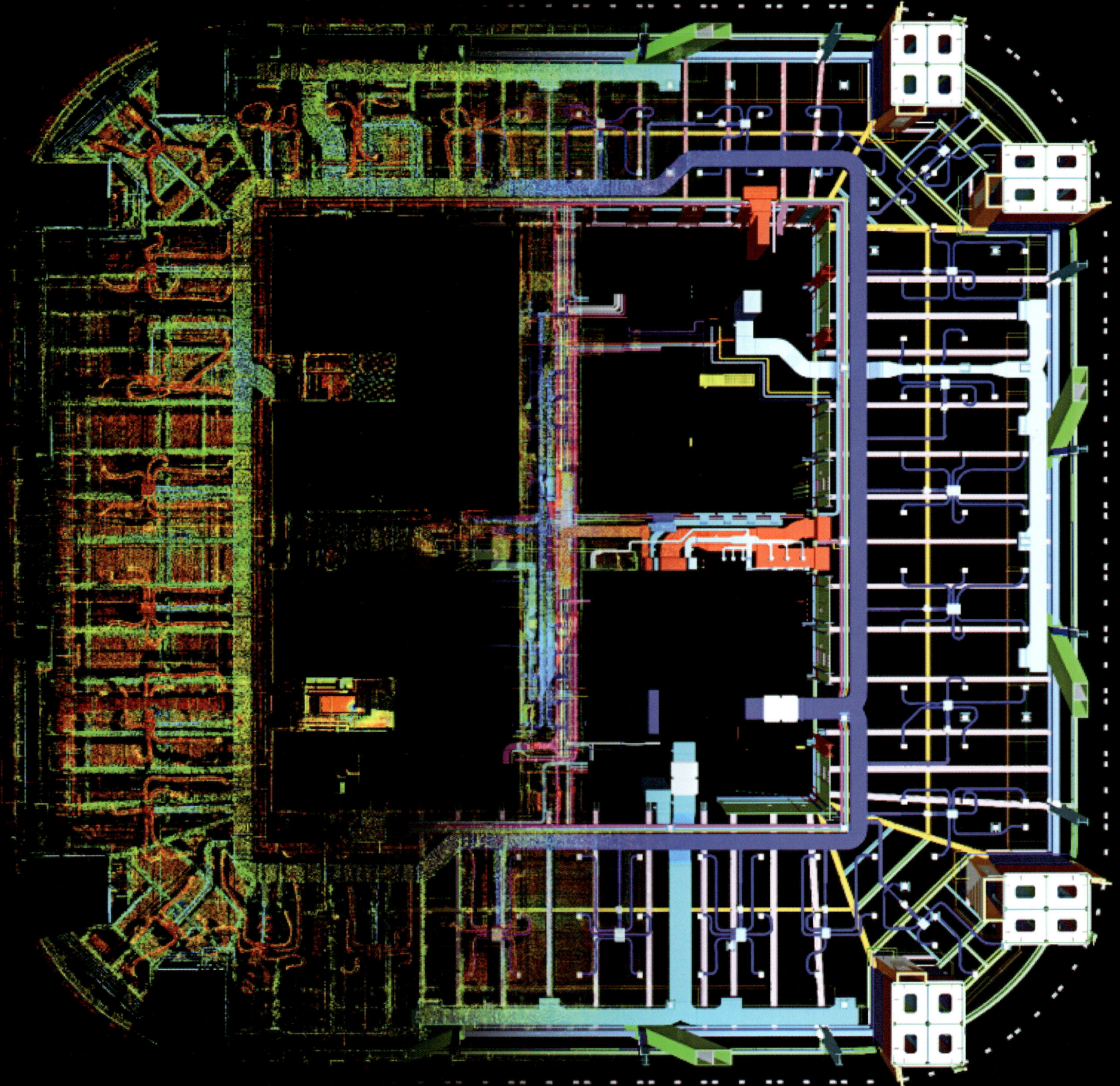

3

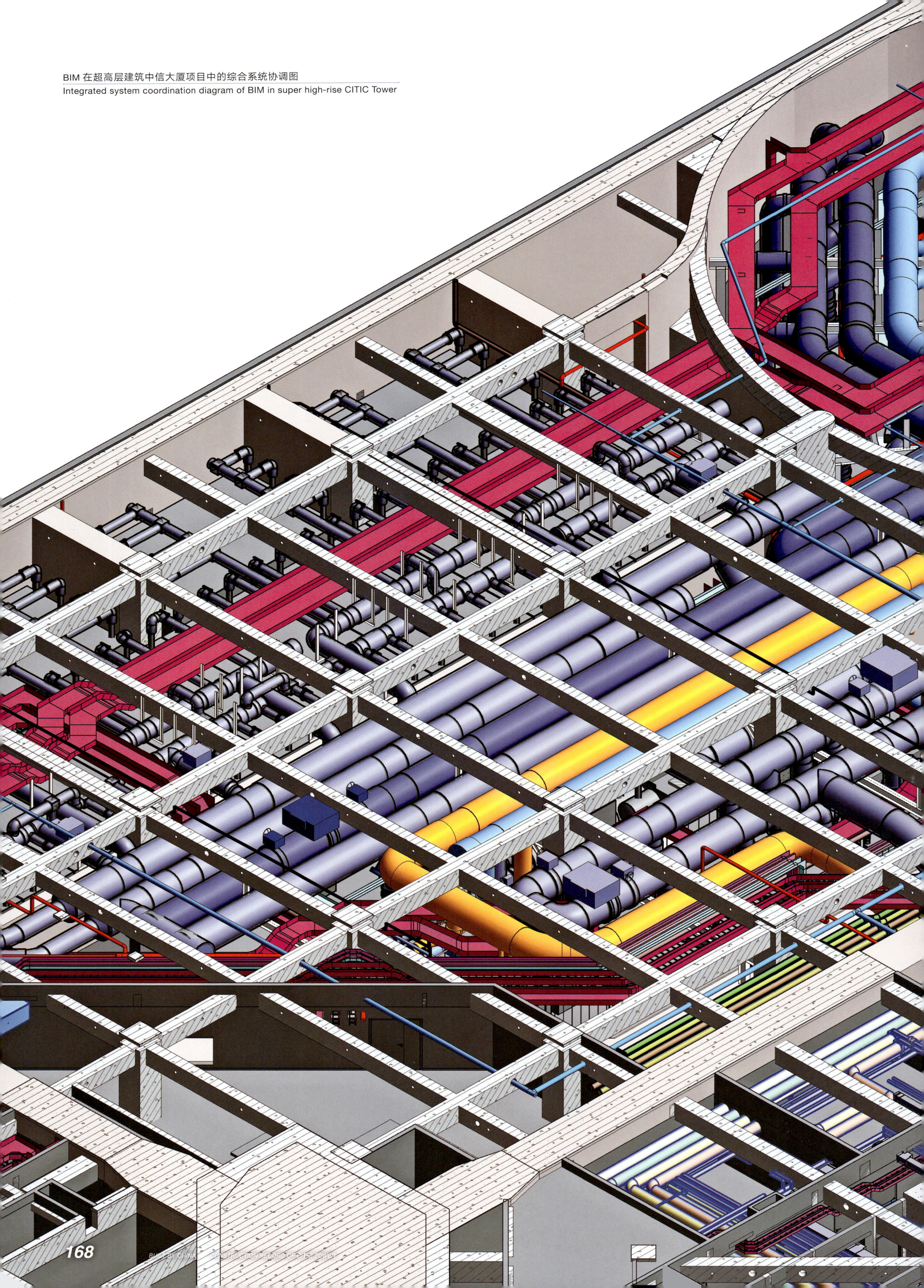

BIM 在超高层建筑中信大厦项目中的综合系统协调图
Integrated system coordination diagram of BIM in super high-rise CITIC Tower

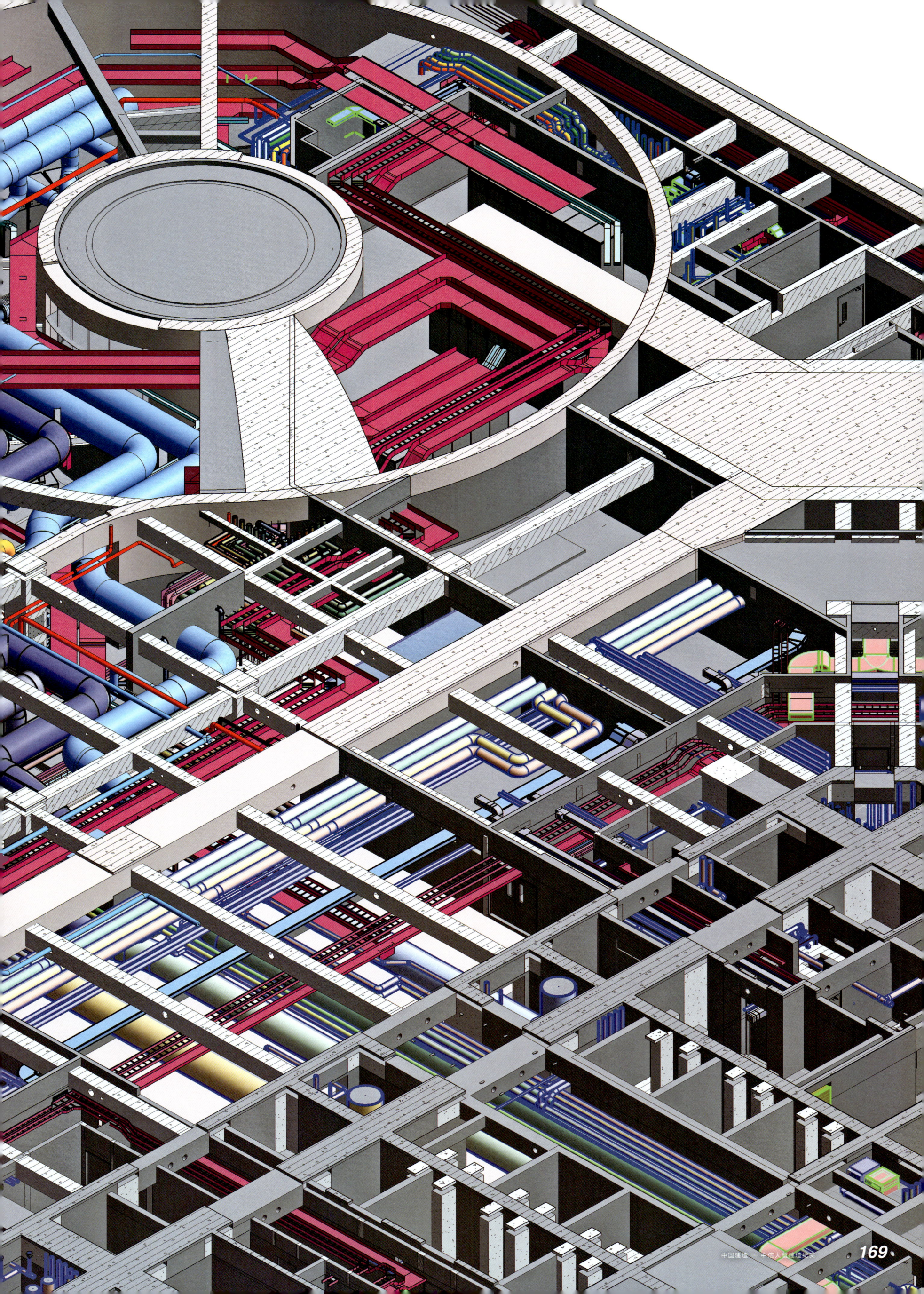

5.11 超高层建筑 10KV 高压垂吊电缆敷设技术
Laying Technology Of 10kV High-Voltage Hanging Cable In Super High-Rise Building

大数据 *Data*

24 根
高压垂吊电缆
The number of high-
voltage hanging cables

13,936 m
高压垂吊电缆总长
Total length of the high-
voltage hanging cable

9,308 m
高压垂吊电缆垂直段长度
Length of vertical section of
high-voltage hanging cable

5.7 t
单根高压垂吊电缆重量
Weight of a single high-
voltage hanging cable

图片说明 *Comments*

1. 高压垂吊电缆剖面图
Sectional drawing of the high-
voltage hanging cable

2. 高压垂吊电缆敷设示意图
Schematic diagram of the laying
of high-voltage hanging cable

3. 高压垂吊电缆敷设
Laying of high-voltage
hanging cable

4. 高压垂吊电缆固定用
"一板双吊"承载板
"One plate for double hanging
purpose" of bearing plate for
fixing high-voltage hanging cable

5. 固定用承载板连接
吊具本体拉力试验
Tension test of lifting sling
body connecting with
bearing plate for fixation

6. 电缆成品燃烧耐火特性试验
Fire resistance test of finished cable

延伸阅读 *Links*

P086. P128

中信大厦采用 24 根高压垂吊电缆共 13,936m，其中垂直段长 9,308m。此电缆由下水平段、垂直段和上水平段组成，垂直段采用扇形塑料包覆的钢丝绳与 3 根电缆芯绞合，是针对超高层建筑设计的专用电缆。单根钢丝绳可满足电缆整体自重 2 倍，专用吊具经过 3 倍自重的拉力试验及长寿命热循环拉力评估。电缆整体制作，无中间接头，导体不受力，供电更加安全。电缆占用空间小，整体一次性吊装、敷设效率高，且方便安全，并拥有 7 项专利技术。

电缆单根总长 820m，垂直段长 482m，创世界超高层建筑新纪录。首创"一板双吊"技术，电缆重 5.7t。

24 high-voltage hanging cables are adopted in CITIC Tower, totaling 13,936m, of which the vertical section is 9,308m. The cable is composed of lower horizontal section, vertical section and upper horizontal section. The twisted three cable cores and steel rope wrapped by fan-shaped plastics are adopted in the vertical section of the cable, which are specially designed for super high-rise buildings. A single steel rope can meet twice the weight of the cable as a whole, the special lifting sling passes the test of the tension of 3 times of the weight of itself and long-life thermal cycle tensile evaluation. Cables are produced integrally with no intermediate joints and provide safer power supply if the conductor is unstressed. The cable bears the advantages of small footprint, integral one-time hoisting, high laying efficiency, convenience and safety, and possesses 7 patented technologies.

A single cable boasts the total length of 820m and the vertical section of 482m, reaching a record high of super high-rise building in the world. Innovate "one plate for double hanging purpose" with 5.7t of cable weight.

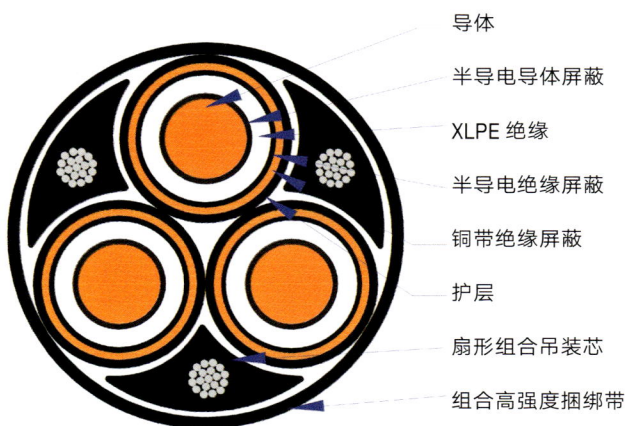

导体
半导电导体屏蔽
XLPE 绝缘
半导电绝缘屏蔽
铜带绝缘屏蔽
护层
扇形组合吊装芯
组合高强度捆绑带

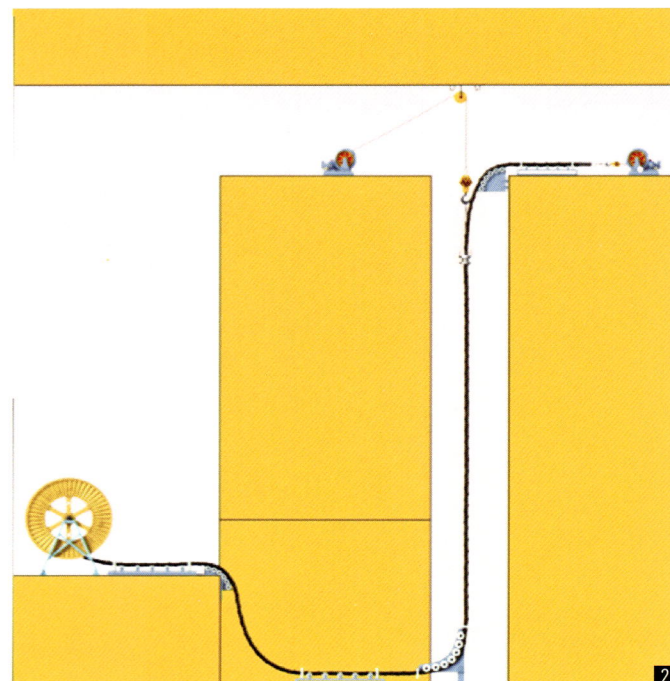

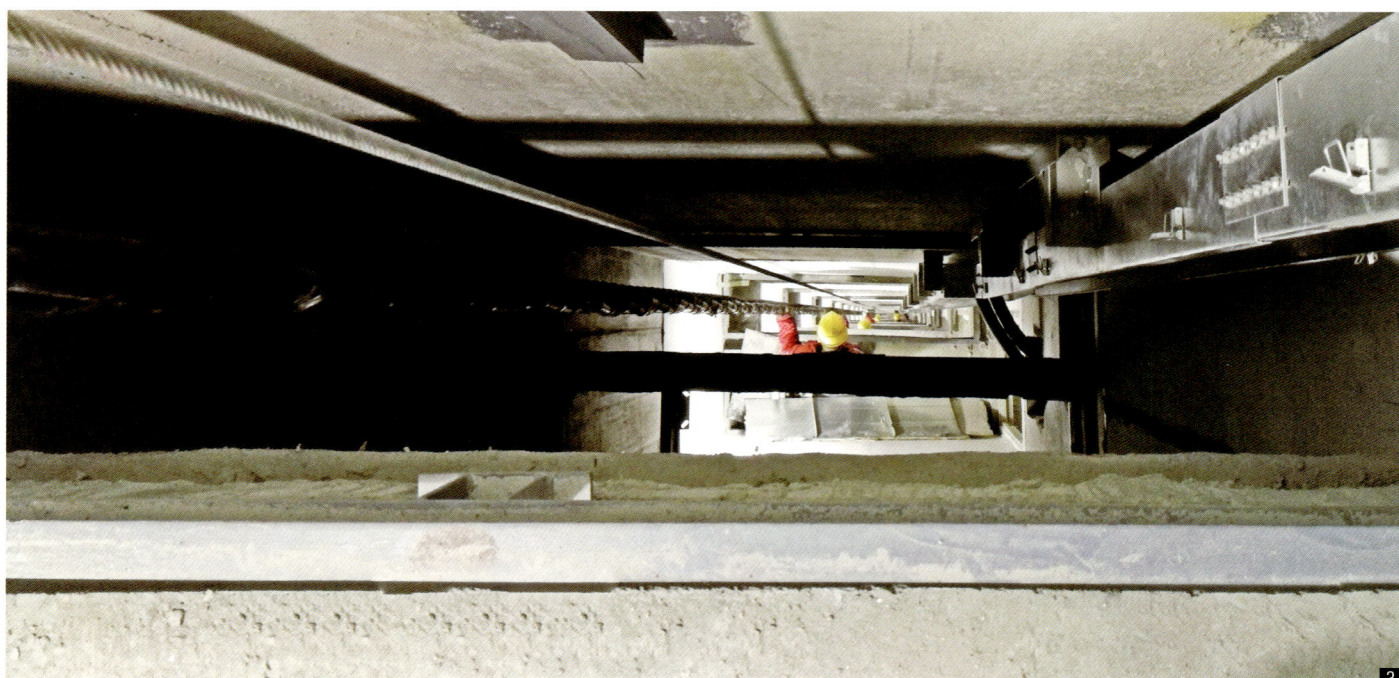

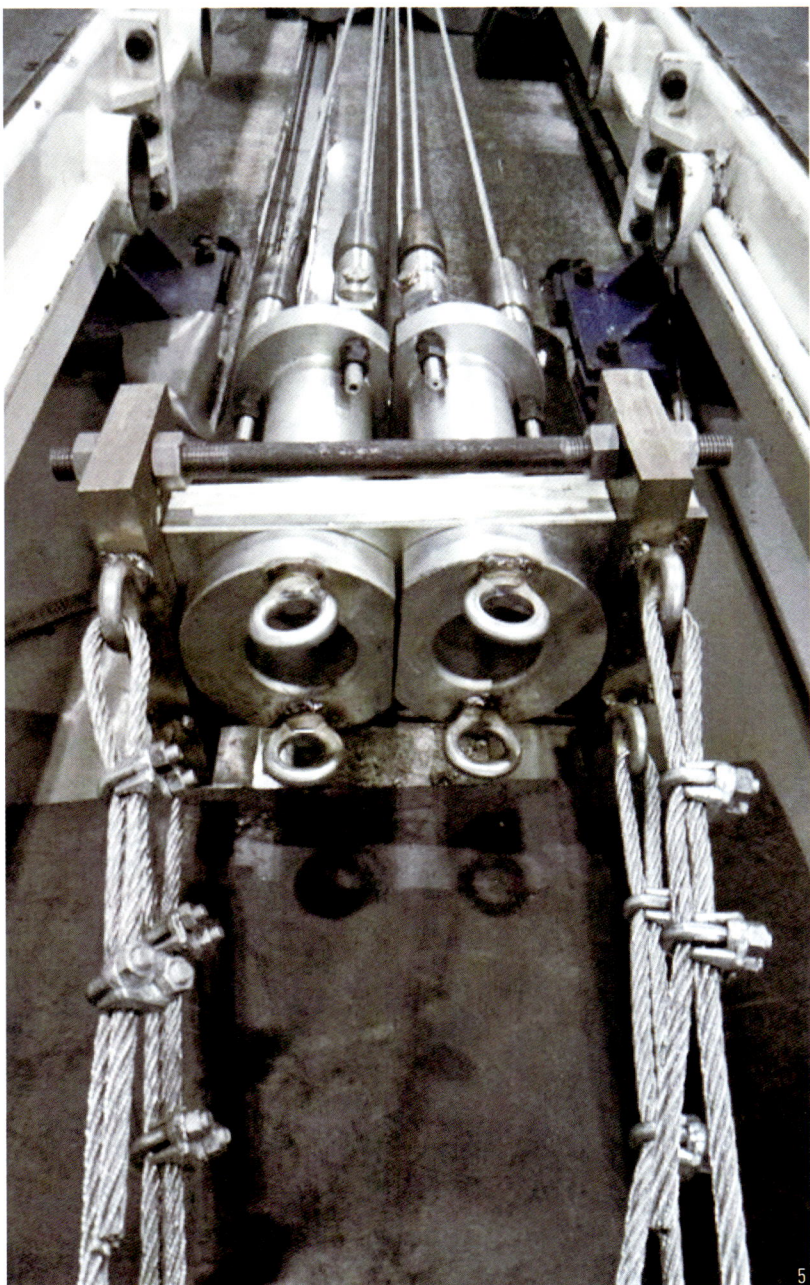

大数据 Data

-35m
超高层建筑最深制冷机房
Super high-rise deepest
refrigeration machine room

660v
降压变频
Step-down frequency conversion

35,000rth
总蓄冷量
Total ice storage

18-20mm
结冰层厚度
Icing layer thickness

>30年
使用寿命
Service life

图片说明 Comments

1. 冷源系统流程图
Flow diagram of cold source system

2. 蓄冰机房 BIM 图
BIM diagram of ice storage
machine room

3. 蓄冰盘管
Ice storage coil

4. 永磁同步离心机房
Permanent magnet synchronous
centrifuge room

延伸阅读 Links

P179. P211. P234

5.12　冰蓄冷及永磁同步变频技术
Ice Storage And Permanent Magnet Synchronous Frequency Conversion Technology

制冷机房位于 B7（-35m），创世界超高层建筑最深制冷机房纪录。冰蓄冷系统采用 4 台高效永磁同步变频离心式双工况机组和 4 台双级高效永磁同步变频离心机组。该机组应用了高速永磁同步电机直驱叶轮、双级压缩补气循环、降压 660V 变频等创新技术，制冷更加节能高效。永磁同步变频离心式基载冷机制冷工况 COP 为 6.46，IPLV 为 9.57，经鉴定达到国际领先水平。

总蓄冷量为 35,000RTH，采用 56 台纳米导热复合冰盘管。结冰层厚度为 18-20mm，结冰、融冰性好，系统效率更高。耐腐蚀，使用寿命达 30 年以上。复合材料重量轻，韧性好，可减小蓄冰槽底部的承重，安装维护方便。

混凝土蓄冰槽在国内首次采用了双层聚脲防水处理，确保安全不渗漏。在机房一侧做了保温加强措施，可有效减少冷损失和避免墙体结露。

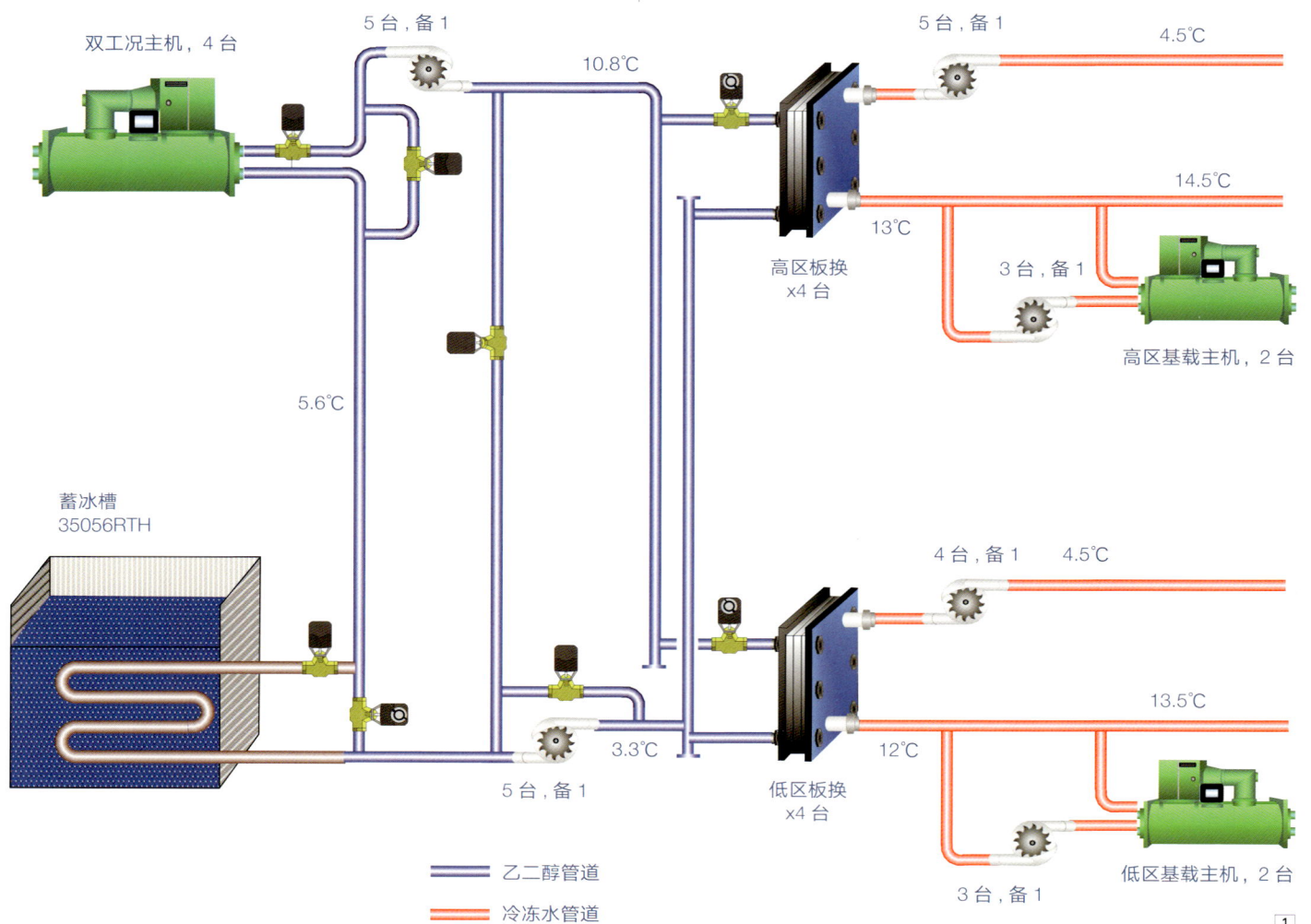

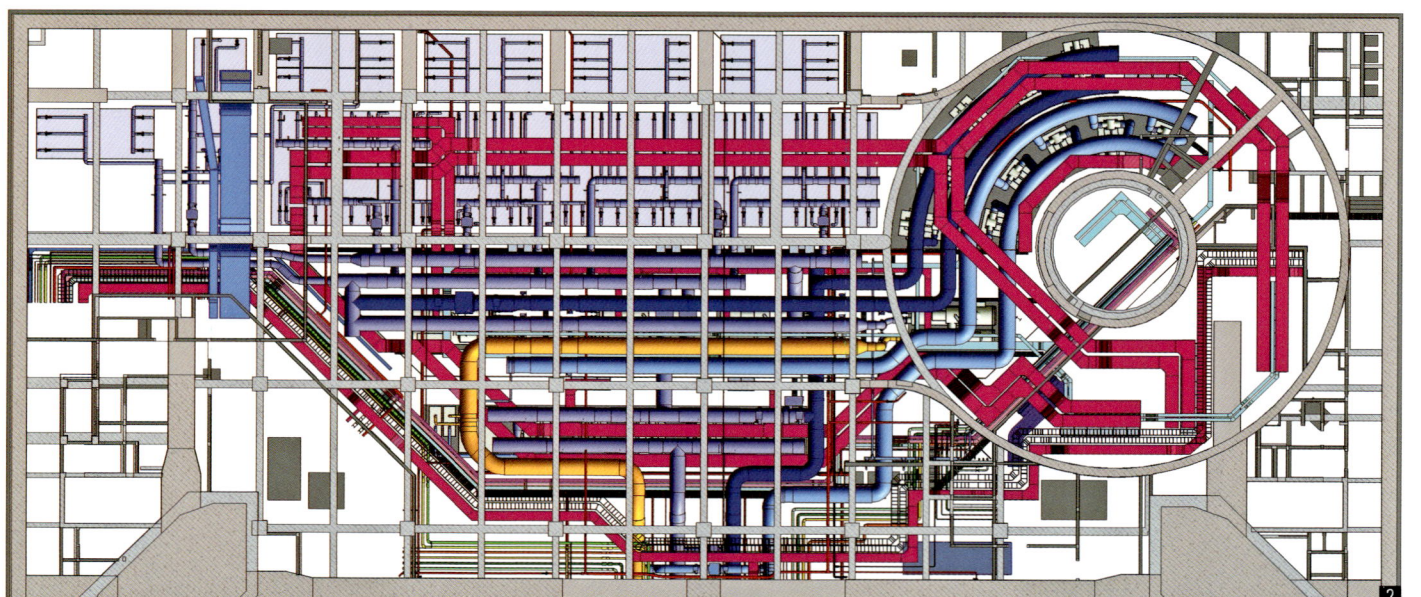

The refrigeration machine room is located on the underground 7th floor at -35m, reaching the record high of the deepest refrigeration machine room in super high-rise buildings in the world. Four high-efficiency permanent magnet synchronous frequency conversion centrifugal duplex status units and four two-stage high-efficiency permanent magnet synchronous frequency conversion centrifugal units are adopted for the ice storage system. Innovative technologies such as the high-speed permanent magnet synchronous electric motor for direct-drive impeller, two-stage compression air supply cycle and step-down 660V frequency conversion are applied in the unit for more energy-saving and efficient refrigeration. The refrigeration condition COP and IPLV of the permanent magnet synchronous frequency conversion centrifugal base load cooler are 6.46 and 9.57 respectively, which are identified to reach the international leading level.

The total ice storage capacity is 35,000RTH with 56 Runpaq nanothermally conductive composite ice-coil adopted. The thickness of the icing layer is thinner, which is 18-20mm, the ice formation and melting property are better, and the system efficiency is higher. Corrosion resistance; service life is more than 30 years. The composite material has the advantages of light weight, good toughness, reduced load bearing at the bottom of the ice storage tank and more convenient installation and maintenance.

The double-layer polyurea waterproofing treatment is adopted for the first time in China to ensure the security and no leakage of concrete ice storage tank. At the same time, measures are taken at one side of the machine room to strengthen heat preservation, effectively reducing the cold loss and avoiding condensation on the wall.

5.13 安置在超高层建筑室内的引风式冷却塔
The Indoor Air-Induced Cooling Tower Is Arranged In The Super High-Rise Building

中信大厦没有裙房，冷却塔只能设置在大厦内的 F5。经清华大学 CFD 模拟大厦室外周边风环境，采用南、北侧布置冷却塔，东西侧进风，南北侧排风的方式，效果最好，且对室外周边影响最小。

采用 10 台引风式双层冷却塔（其中 4 台用于租户冷却）。引风双层横流设计，风阻小，更加节能。区别与传统的鼓风式冷却塔，引风式双层冷却塔结构紧凑，满足设备层安装高度；塔体内负压，有效降低飘水率，提高冷却塔的热力性能，有效防止细菌滋生，运行环境更加清洁。多叶高静压、直接驱动、采用变频电机的轴流风机运行更加节能，噪声低。800mm 深集水盘，无涡流，防溢满。

通过增大风量、多台冷却塔风量组合控制和热水盘管加热器的措施来有效除雾。冷却塔出风口经过静压箱再加装阵列式消声器，满足白天噪声限值 65dB(A) 要求。

As there is no podium in CITIC Tower, so the cooling tower can only be set on the 5th floor of the building. According to the CFD simulation of the outdoor ambient wind environment of the building by Tsinghua University, the cooling towers arranged at the south and north sides enjoy the best effect and the minimum impact on the outdoor ambient, with the east-west side for air intake while the north-south side for air exhaustion.

10 Ryowo air-induced double-layer cooling towers (4 of which are for tenant cooling) are used, with double-layer cross-flow design, small wind resistance, higher energy-efficiency. Different from the traditional air-blown cooling tower, the air-induced double-layer cooling tower is compact in structure, catering to the installation height of equipment floor. Negative pressure in the tower can effectively reduce the floating rate, improve the thermal performance of the cooling tower, effectively prevent the growth of bacteria, and make the operating environment cleaner. Axial flow fan with multi-blade high static pressure, direct drive and frequency conversion electric motor is more energy-saving with lower noise. The water collection plate is 800mm in depth with no eddy current and anti-overflow function.

By means of increasing air flow rate, combining control of air flow rate of several cooling towers and adopting hot water coil heaters, the mist can be removed effectively. The wind out of the air outlet of the cooling tower firstly passes through the static pressure tank and then the array muffler, so that the noise occurred is within the daytime noise limit of 65dB (A).

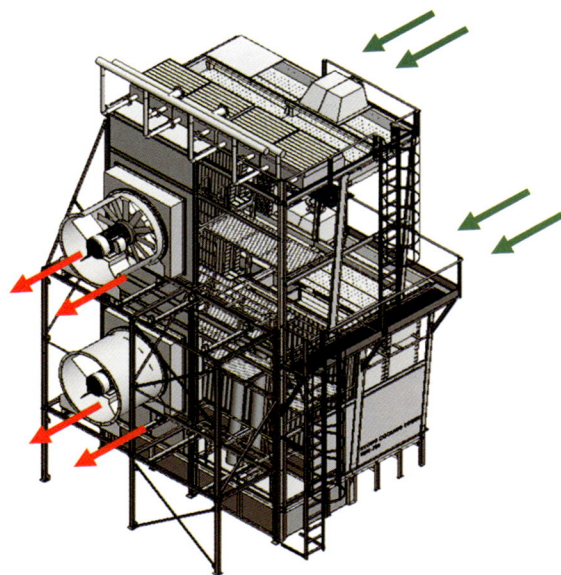

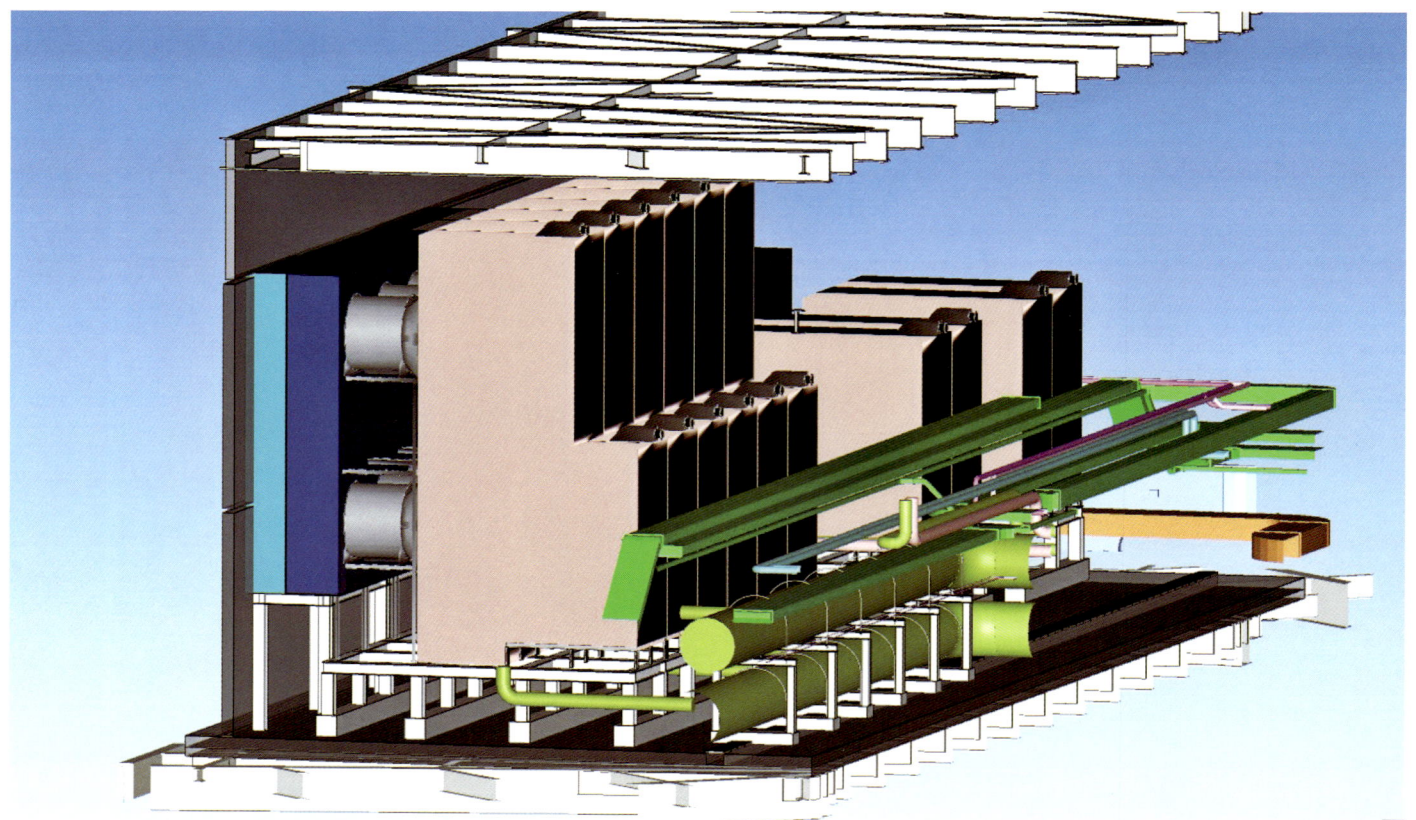

5.14 高速电梯安装无脚手架组合技术
Scaffold-Less Assembly Technology For High-Speed Elevator Installation

大数据 *Data*

100 部
井道高度超过 100m 的电梯共计 20 部，超过 60m 的电梯共计 80 部
In which, 20 lifts with shaft height over 100 m and 80 lifts with shaft height over 60 m

530 m
中信大厦电梯最高到达高度
The height of the elevator reaches 530m, setting a Beijing record

图片说明 *Comments*

1. 2:1 提拉克轿厢方法
2:1 Tirak elevator car method

2. 1:1 提拉克轿厢方法
1:1 Tirak elevator car method

3. 无脚手架安装方法
Scaffold-less installation method

延伸阅读 *Links*

P208. P218. P248

中信大厦有 100 台直梯，井道高度超过 60m 的共计 80 部，还有 20 部是超过 100m 以上。若采用传统的脚手架安装方法，过程中将会消耗大量的时间和资源。运用提拉克（Tirak）无脚手架施工方法来解决这个问题，对中信大厦不同方案的电梯采用如下两个方案：

2:1 提拉克轿厢方法，用于井道高度约 200m 以下电梯
2:1 提拉克轿厢方法利用在机房吊点和 2:1 绕绳比进行提升，包括轿厢和轿顶上的工作平台，可以从下往上把导轨安装上去，一遍遍爬升，工人只需在轿顶上工作，相对轻松，并且装置本身设有安全钳和限速器，同时，头顶保护也提供了可靠安全保障。

1:1 提拉克平台方法，用于井道高度超过 200m 的电梯
1:1 提拉克平台方法适用于更高的井道，通力电梯公司使用一个自行设计的工作平台，装置本身设有安全钳和限速器，利用机房吊点，提拉克可提升平台，让工人从下往上去安装导轨和井道设备，此施工方法本身没有高度限制，所以我们在所有穿梭电梯安装中都采用这种方法，中信大厦电梯最高到达 530m，创下北京最高纪录。

无脚手架安装方法，可减少施工人员在井道内攀爬，不需要上下运输大量工具物料，节约了安装和拆卸脚手架的时间，有效地缩短了大楼整体安装时间。

井道分段安装措施
中信大厦穿梭电梯行程超过 500m，为了节约整体工期，施工时采用特殊的安装方法，把井道分为 3 个部分，F1 - F60、F61 - F86、F87 - F105M 机房，这样让电梯安装人员能提前到现场施工，无需等待整个电梯井道完成，这样把土建施工与电梯安装在工序上重叠起来，有效地节约了整体工期。

在每个分段，中信大厦都在井道内建造了硬隔离、防坠斜台、防水保护装置，这些措施确保下面的电梯井道施工安全及不受天气影响；硬隔离提供基本保护和无脚手架安装设备的空间；防坠斜台能够抵挡高空坠物，并把它们反弹出井道；防水保护措施则保证了雨雪天气下井道的工作环境，分段安装让电梯安装时间更有弹性。

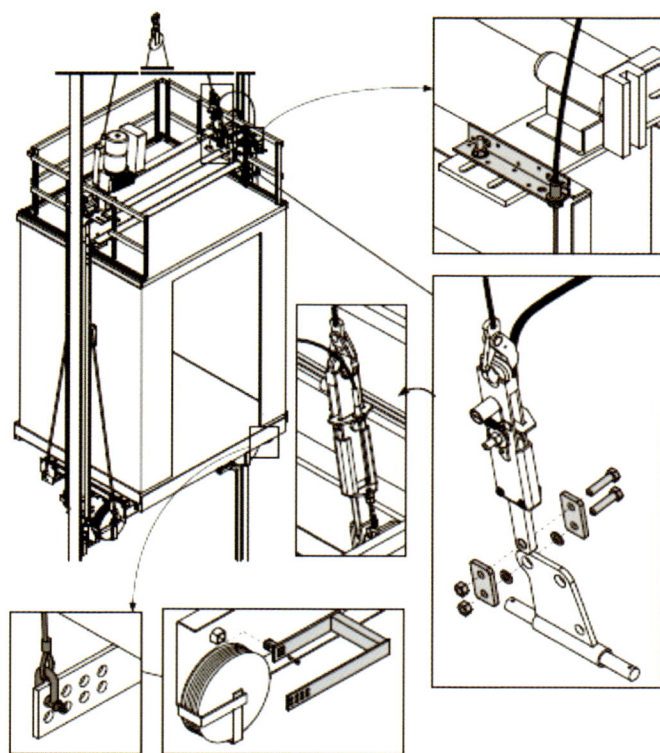

1

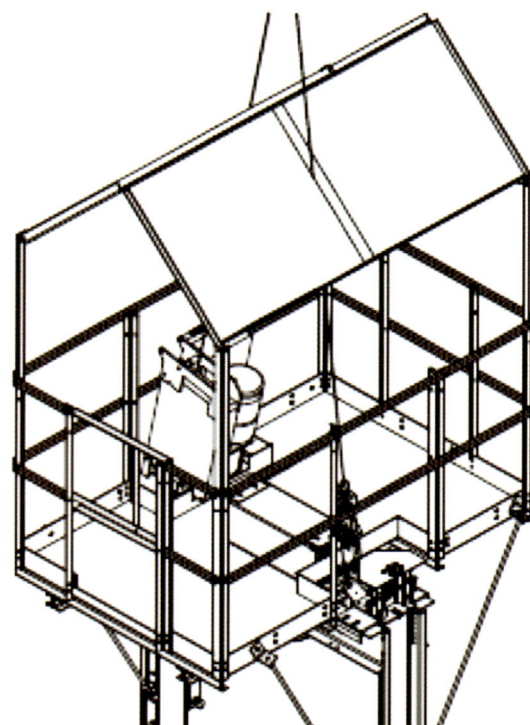

2

There are 100 straight ladders in CITIC Tower, with a total of 80 shafts with over 60m in height and 20 shafts with over 100m in height. If the traditional scaffolding installation method is adopted, plenty of time and resources will be consumed during the process. Tirak scaffold-less construction method can be adopted to resolve this problem. The following two schemes are adopted for the elevators of different schemes in CITIC Tower:

2:1 Tirak elevator car method for elevators below about 200m of shaft

2: 1 The lifting point in the machine room and 2:1 winding rope ratio is used for Tirak elevator car lifting. The working platforms on the elevator car and the elevator car roof are involved in the method. The guide rails can be mounted from bottom to top through multiple times of climbing, while the work is relatively easy and safe as the workers only need to work on the elevator car roof, protected by safety tongs and the speed limiter of the device, as well as overhead reliable safety guarantee.

1:1 Tirak platform method for elevators with shaft exceeding 200m.

1:1 Tirak platform is suitable for higher shafts. A self-designed working platform is adopted by Kone. The device itself is equipped with safety tongs and speed limiters. The platform can be lifted by dint of the lifting point of machine room through Tilac. The workers are allowed to install guide rails and shaft equipment from bottom to top. This construction method has no height restriction, so it can be adopted in all shuttle elevators. The height of elevators in CITIC Tower reaches 530m, setting a Beijing record.

The method for scaffold-less installing can reduce the climb of the construction personnel in the shaft, without need to transport a large amount of tool materials up and down, save the time for installing and disassembling the scaffolding, and effectively shortens the overall installation time of the building.

Installation measures of shaft in sections

The travel of shuttle elevators in CITIC Tower exceeds 500m. In order to save the whole construction period, a special installation method is adopted, which divides the shaft into three parts: 1-60th floor, 61-86th floor and 87-105th floor machine room. In this way, the elevator installer can launch field construction ahead of time without waiting for the completion of the whole elevator shaft. The overlapping of the civil work construction with the elevator installation on the working procedure effectively saves the whole construction period.

In each section, CITIC Tower have built hardware-based isolation, anti-fall tilt platform and waterproof protection in the shaft. These measures ensure the safe and weather-proof construction of the elevator shaft below. Hardware-based isolation provides basic protection and scaffold-less installation space; Anti-fall tilt platform can resist falling objects from high altitude and bounce them out of the shaft; Water-proof protection ensures the normal working environment of underground shaft in rainy and snowy weather, and the installation of the elevator in sections offers more elastic installation time.

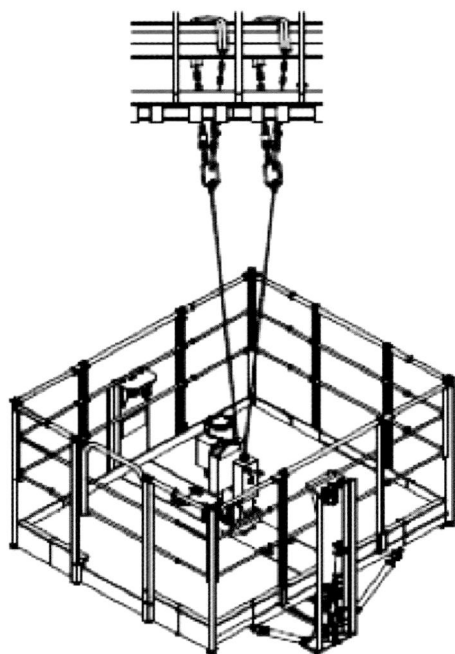

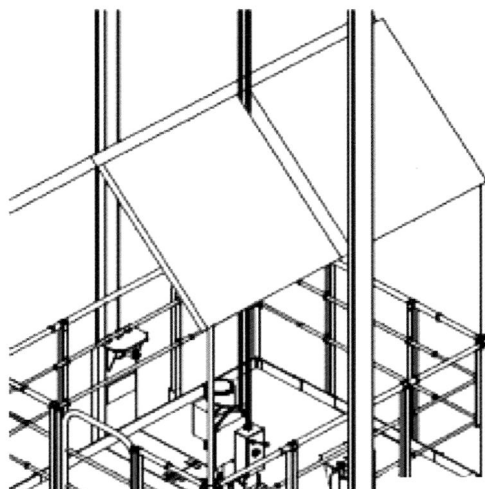

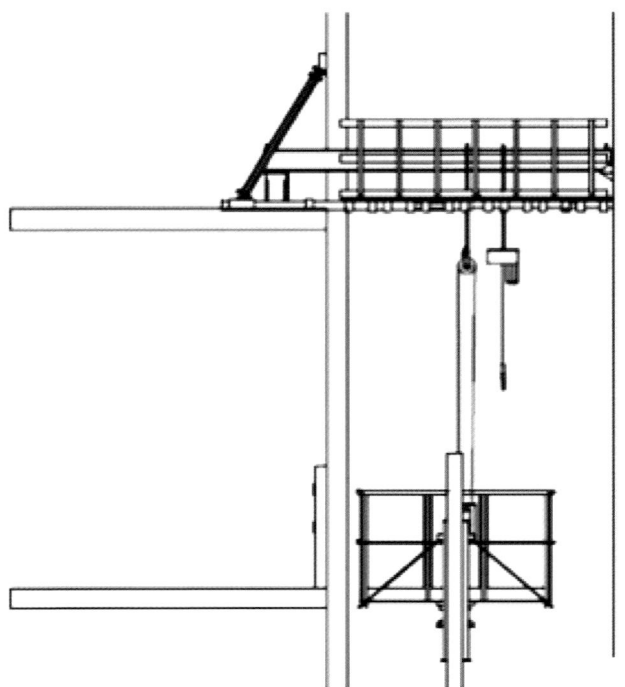

5.15 一体化集成空调机组技术
Integrated Air-Conditioning Unit Technology

图片说明 Comments

1. 集成空调机组原理图
Principle diagram of integrated
air-conditioning unit

2. 集成空调机组
Integrated air-conditioning unit

延伸阅读 Links

P179. P206. P238

空调系统采用 339 台集成空调机组。风机段与消声器集成设计，上下叠放，整体消声。20,000m³/h 的空调机组噪声为 52.1dB(A)，创国际先进。机组内设旁通阀，过渡季运行更节能。整体漏风率 0.32%，远低于国家标准 0.5%。

设置 5 道空气过滤器，采用"新风端设 G4 板式初效过滤，空调机组送风端设 G4 板式初效过滤 + 双驱静电中效过滤 +F7 中效袋式过滤"的过滤方式，PM2.5 过滤效率可达 99.8%，使室内 PM2.5 可控制在 50μg/m³ 内。

339 integrated air-conditioning units are adopted in the air-conditioning system. Fan section and muffler are stacked and designed integrally to realize the overall sound suppression. The noise of air-conditioning unit with an efficiency of 20,000m³/h is 52.1dB(A), which is internationally advanced. The unit is equipped with the bypass valve to save energy in transition season. The overall air leakage rate is 0.32%, far lower than the national standard of 0.5%.

5 air filters are set up and the filtering method of "G4 plate primary filtering at the fresh air end, G4 plate primary filtering at fresh air end of air-conditioning unit + double-drive electrostatic medium filtering + F7 medium bag filtering" is adopted. The filtering efficiency of PM2.5 can reach 99.8%, controlling the indoor PM2.5 within 50μg/m³.

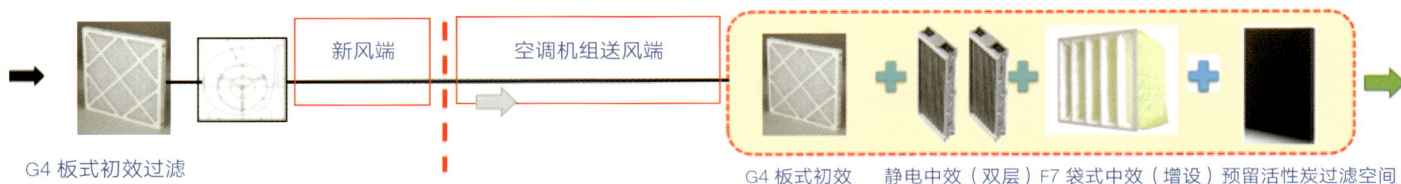

新风端　　空调机组送风端

G4 板式初效过滤　　　　　　　　　　　　　　　　　　G4 板式初效　静电中效（双层）F7 袋式中效（增设）预留活性炭过滤空间

1

2

中信大厦采用"冰蓄冷 + 大温差 + 低温送风 + 变风量 + PLC"的空调系统集成创新技术：

冰蓄冷合理利用峰谷电价，降低运行费用。国际先进的永磁同步变频冰蓄冷机组，COP 高达 6.46，IPLV 高达 9.57。空调水通过三级换热，由冰槽供冷 3.3℃，经板换依次换热成 4.5℃、5.7℃ 和 6.9℃ 的空调冷水，且采用 10℃大温差供回水，能源利用率高，可减少空调机组送风量。

2,500 台第 2 代叶轮测速可变多孔阀片的节能型 VAV-BOX，叶轮式风速传感器避免了测速孔堵塞，测量数据稳定且精度高（1 ~ 12m/s），可感知微弱风速；可变式多孔叶片使气流稳定，整流效果好，具有线性控制性能；风管无需变径，压力损失小（可减少约 100Pa），运行更节能。

1,970 个风量自平衡一体化送风单元（Flexible Air Supply Unit，简称 FASU），是一种辅助流量分配末端装置，工厂化生产，各支路风量平衡性好，安装方便，节省施工成本，节省工期，也有利于办公区二次装修的空调末端配合调整。

采用防结露、吊顶贴附射流型空调风口，可保证最佳的气流扩散，避免冷风直吹，舒适性更高；且风口压损低（比常规射流型低温风口减少 80%），更加节能。3,600 余台创新研发的大温差窗边风机盘管，运行噪声低，结构超薄，占用空间少。

2,777 套 PLC 控制器，工业 4.0 技术，可按大厦不同负荷需求自主编程，节能性好，可靠性高，可实现数据统计及运算，创国内超高层建筑暖通监控系统 PLC 控制先例。

总风量 + 变静压的变风量节能控制；过渡季节新风比运行，最大新风比可达 70%；空调热回收系统；环形空调风管分区温度控制；大空间分层空调系统；冷却塔直接供冷系统；根据 CO_2 浓度进行新风量调节等多种技术集成和运行策略，实现空调系统舒适节能运行，提升大厦品质。

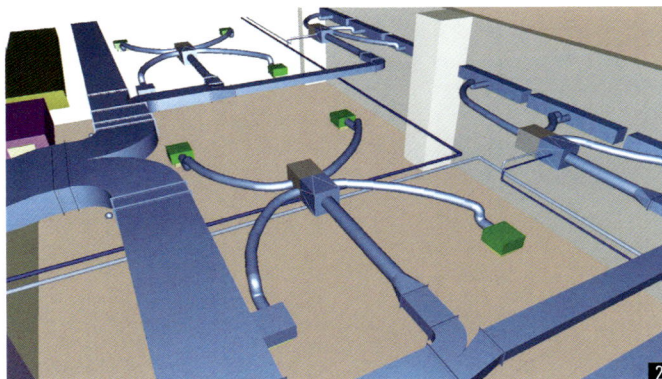

大数据 Data

2,500 台
节能型 VAV-BOX
The number of energy-saving VAV-BOX

1,970 个
风量自平衡一体化送风单元
Air supply system FASU

2,777 套
PLC 控制器
The number of PLC controller

70 %
最大新风比
Maximum fresh air ratio

图片说明 Comments

1. 第二代 VAV
The second generation VAV

2. FASU BIM 设计图
FASU BIM design drawing

3. FASU 系统
FASU system

延伸阅读 Links

P172. P207

传统 VAV-Box：单板叶片 + 无整流效果

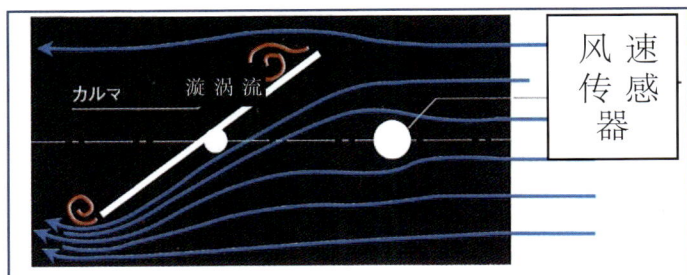

第二代 VAV-Box：多孔叶片 + 整流效果

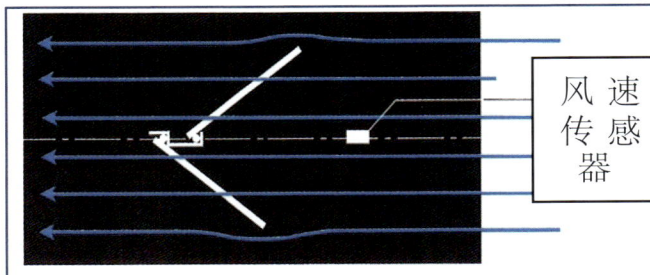

空调水三级换热参数表

空调水系统	供冷温度℃	送风温度	区域
B7 冰槽供冷	3.3	/	/
B7 冷冻水换热	4.5/13.5 4.5/14.5	≥ 9.5	ZB-Z1 Z2
M3 换热站	5.7/15.7	≥ 10.7	Z3-Z5
M6 换热站	6.9/16.9	≥ 11.9	Z6-Z8

The integrated innovation technology of "ice storage + large temperature difference + low temperature air supply + variable air volume + PLC" is adopted in the air-conditioning system of CITIC Tower:

Ice storage makes rational use of peak-valley electricity prices to reduce operating costs. The COP and IPLV of the internationally advanced permanent magnet synchronous frequency conversion ice storage unit are respectively as high as 6.46 and 9.57. The air-conditioning water is cooled by the ice storage tank at 3.3℃ and then changed into the air-conditioning cold water at 4.5℃, 5.7 ℃ and 6.9 ℃ through the plate heat exchanger. The large temperature difference of 10 ℃ is adopted to supply and return water with high energy utilization ratio, reducing the air supply volume of the air-conditioning unit.

The impeller type wind speed sensor of 2,500 energy-saving VAV-BOX with variable multi-hole valve plate of the second generation impeller for speed measurement can avoid clogging the velocity measurement hole with stable and accurate measurement data (1-12m/s). It is able to sense the weak wind speed; Variable multi-hole blade makes the air flow stable, well-rectified, and has the linear control performance; The air duct does not need to change in diameter, leading to small the pressure loss (can reduce about 100 Pa), and more energy-saving operation.

1,970 Flexible Air Supply Unit (FASU) is a kind of auxiliary flow distribution terminal device. FASU unit is manufactured. The air flow of each branch is well balanced and easy to install, which saves construction cost and construction period, and is also beneficial to the coordination and adjustment of the air conditioning terminal of the secondary decoration in the office area.

Adopting anti-condensation and adhering jet type air outlet on the ceiling can ensure the best air diffusion, avoid the direct blowing of cold air, and improve the comfort of the air conditioner. And the pressure loss of the outlet is lower (80% less than that of the conventional jet type outlet), so that the outlet is more energy-saving. More than 3,600 sets of fan coil with ultra-thin window edge and large temperature difference have been innovated and developed, which have low operation noise and ultra-thin structure, and small footprint.

2,777 sets of PLC controllers and Industry 4.0 technology can be independently programmed according to different load requirements of the building, with good energy conservation and high reliability. Meanwhile, data statistics and calculation can be realized, setting a precedent for PLC control of super high-rise HVAC monitoring system in China.

Energy saving control of variable air volume with total air volume + variable static pressure; The maximum fresh air ratio can reach 70% in transition season. Air conditioning heat recovery system; Temperature control of air duct subarea in ring air conditioning; Stratified air-conditioning system with large space; Direct cooling system in cooling tower; Various technical integration and operation strategies, such as the fresh air volume adjustment according to CO_2 concentration can achieve comfortable and energy-saving operation of the air-conditioning system to improve the building quality.

原设计窗台系统（500mm）
Originally designed window sill system (500mm)

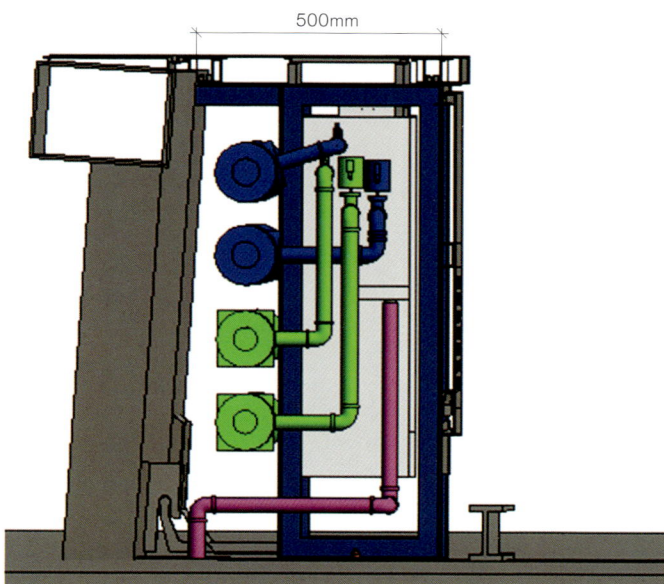

超薄型一体化窗台系统（288mm）
Ultra-thin integrated window sill system (288mm)

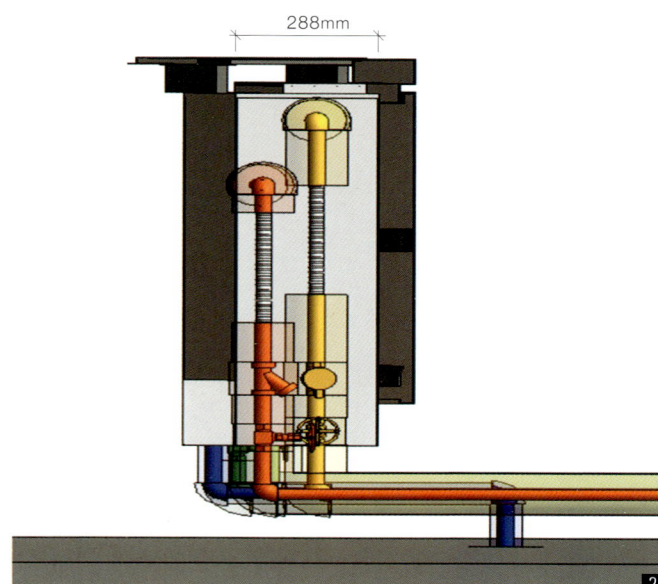

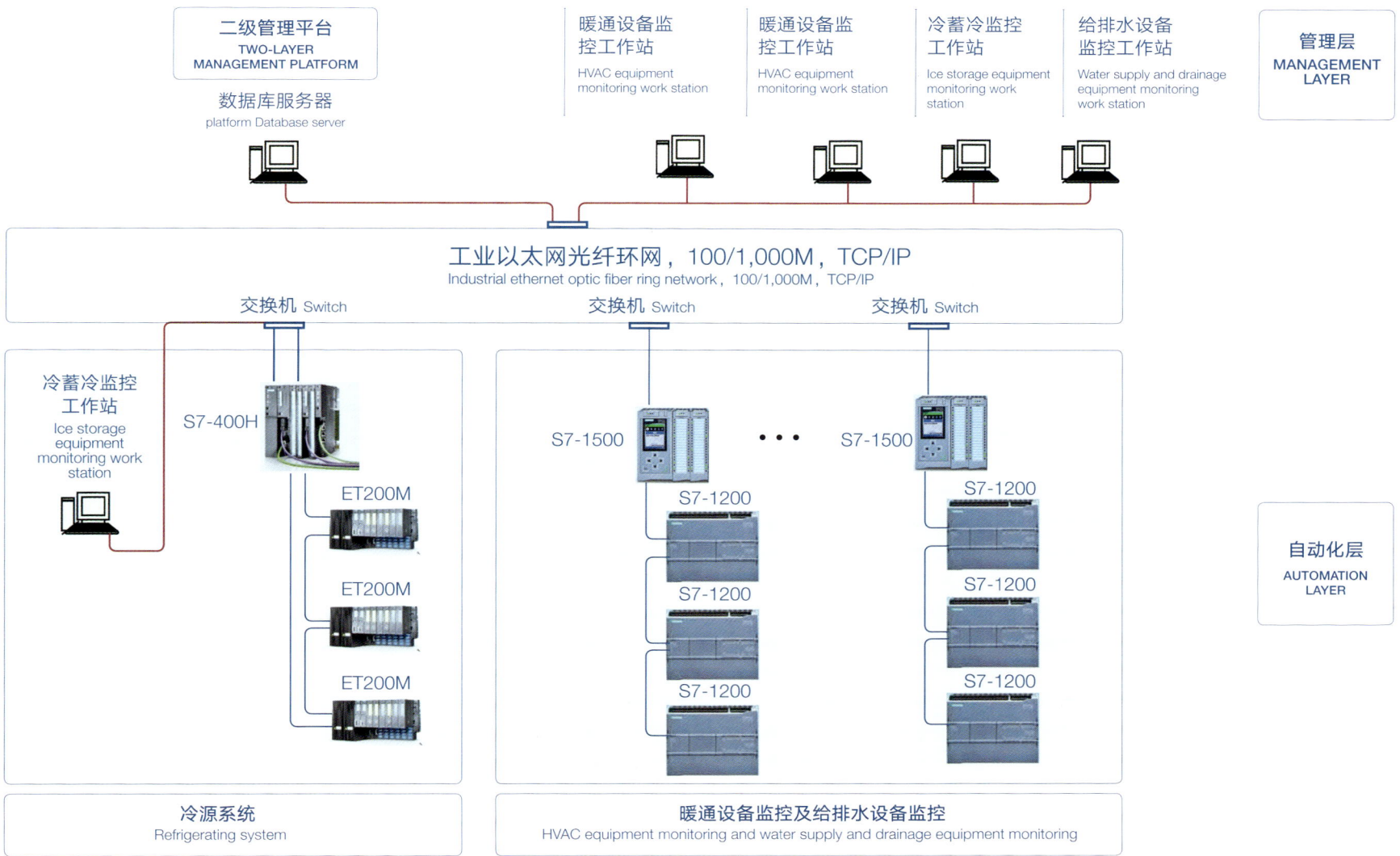

PLC 控制原理图 / PLC control schematic

| 二级管理平台 TWO-LAYER MANAGEMENT PLATFORM | 暖通设备监控工作站 HVAC equipment monitoring work station | 暖通设备监控工作站 HVAC equipment monitoring work station | 冷蓄冷监控工作站 Ice storage equipment monitoring work station | 给排水设备监控工作站 Water supply and drainage equipment monitoring work station | 管理层 MANAGEMENT LAYER |

数据库服务器 platform Database server

工业以太网光纤环网，100/1,000M，TCP/IP
Industrial ethernet optic fiber ring network，100/1,000M，TCP/IP

交换机 Switch　　交换机 Switch　　交换机 Switch

冷蓄冷监控工作站 Ice storage equipment monitoring work station

S7-400H　ET200M　ET200M　ET200M

S7-1500 ··· S7-1500
S7-1200　S7-1200
S7-1200　S7-1200
S7-1200　S7-1200

自动化层 AUTOMATION LAYER

| 冷源系统 Refrigerating system | 暖通设备监控及给排水设备监控 HVAC equipment monitoring and water supply and drainage equipment monitoring |

PLC 智能空调系统 / PLC intelligent air conditioning system

中国中信集团公司 CITIC Group　中信大厦 CITIC Tower

RUNPAQ 源牌　runpaq　2019/3/11 18:46:23

系统总览　实时报表　ZB区域　Z0区域　Z1区域　Z2区域　Z3区域　Z4区域　Z5区域　Z6区域　Z7区域　Z8区域　报警记录

1区VAV

编号	运行模式	运行工况	低阀位故障	高阀位故障	阀位状态	室内温度(℃)	夏季温度(℃)	冬季温度(℃)	实际风量(cmh)	需求风量(cmh)	过冷过热	风量范围	风量保持	温度范围	温度保持
3	占用	制热	正常	正常	适中	21.2	22.5	20.5	348	339		正常	无保持	正常	无保持
4	占用	制热	正常	正常	适中	22.1	22.5	20.5	329	316		正常	无保持	正常	无保持
5	占用	制热	正常	正常	适中	21.5	22.5	20.5	349	302		正常	无保持	正常	无保持
6	占用	制热	正常	正常	适中	21.9	22.5	20.5	324	302		正常	无保持	正常	无保持
7	占用	制热	正常	正常	适中	22.1	22.5	20.5	331	350		正常	无保持	正常	无保持
8	占用	制热	正常	正常	适中	21.8	22.5	20.5	356	300		正常	无保持	正常	无保持
9	占用	制热	正常	正常	适中	22.1	22.5	20.5	342	321		正常	无保持	正常	无保持
10	占用	制热	正常	正常	高阀位	20.9	22.5	20.5	592	577		正常	无保持	正常	无保持
11	占用	制热	正常	正常	高阀位	20.3	22.5	20.5	830	1157		正常	无保持	正常	无保持
12	占用	制热	正常	正常	适中	24.0	23.5	21.5	396	364		正常	无保持	正常	无保持
13	占用	制热	正常	正常	适中	22.0	22.5	20.5	384	329		正常	无保持	正常	无保持
14	占用	制热	正常	正常	适中	22.6	22.5	20.5	359	330		正常	无保持	正常	无保持
15	占用	制热	正常	正常	适中	22.5	22.5	20.5	374	328		正常	无保持	正常	无保持
16	占用	制热	正常	正常	适中	22.4	22.5	20.5	377	333		正常	无保持	正常	无保持
17	占用	制热	正常	正常	适中	21.3	22.5	20.5	378	345		正常	无保持	正常	无保持
18	占用	制热	正常	正常	适中	22.2	22.5	20.5	377	360		正常	无保持	正常	无保持
19	占用	制热	正常	正常	适中	22.9	22.5	20.5	298	225		正常	无保持	正常	无保持
20	占用	制热	正常	正常	适中	22.9	22.5	20.5	372	315		正常	无保持	正常	无保持
21	占用	制热	正常	正常	适中	21.0	22.5	20.5	373	354		正常	无保持	正常	无保持
22	占用	制热	正常	正常	适中	22.5	22.5	20.5	336	304		正常	无保持	正常	无保持
23	占用	制热	正常	正常	适中	22.2	22.5	20.5	336	290		正常	无保持	正常	无保持
24	占用	制热	正常	正常	适中	22.3	22.5	20.5	346	316		正常	无保持	正常	无保持
25	占用	制热	正常	正常	适中	22.5	22.5	20.5	213	229		正常	无保持	正常	无保持
26	占用	制热	正常	正常	高阀位	21.2	22.5	20.5	59	106		正常	无保持	正常	无保持
27	占用	制热	正常	正常	低阀位	21.2	22.5	20.5	108	106		正常	无保持	正常	无保持
28	占用	制热			低阀位	21.2	22.5	20.5	104	91		正常	无保持	正常	无保持

故障智能诊断

室温全部在舒适度区间

VAV运行工况自动判断　　　系统平衡

4F　VAV设备　FCU设备　数据记录　趋势记录　楼层区域示意图

1区　2区　Z8　Z7　Z6　Z5　Z4　Z3　Z2　Z1　Z0　ZB

5.17 动态 UPS 飞轮储能技术
Dynamic UPS Flywheel Energy Storage Technology

中信大厦在 ZB-Z2 区采用了动态 UPS 技术来保障智能化系统重要的大容量服务器和存储设备的运行，替代传统的静态蓄电池 UPS 系统。无需后备电池，减少维护量，无需装设专用空调，极大降低了维护成本。更加绿色环保。

动态 UPS 占地面积小，为大厦增加净使用面积约 60m²，总寿命周期内节省费用约 660 万元。

DUPS technology is adopted in ZB-Z2 zone of CITIC Tower to ensure the important large-capacity server and storage equipment of the intelligent system, replacing the traditional UPS system with static battery. No need for backup battery reduces maintenance while no need to install special air conditioning, greatly cutting maintenance costs, and making it greener and more environment-friendly.

DUPS has small footprint, increasing the net usable area of the building by about 60m², and saving about 6.6 million yuan within the total life cycle.

动态 UPS 飞轮构造图
Dynamic UPS flywheel structural diagram

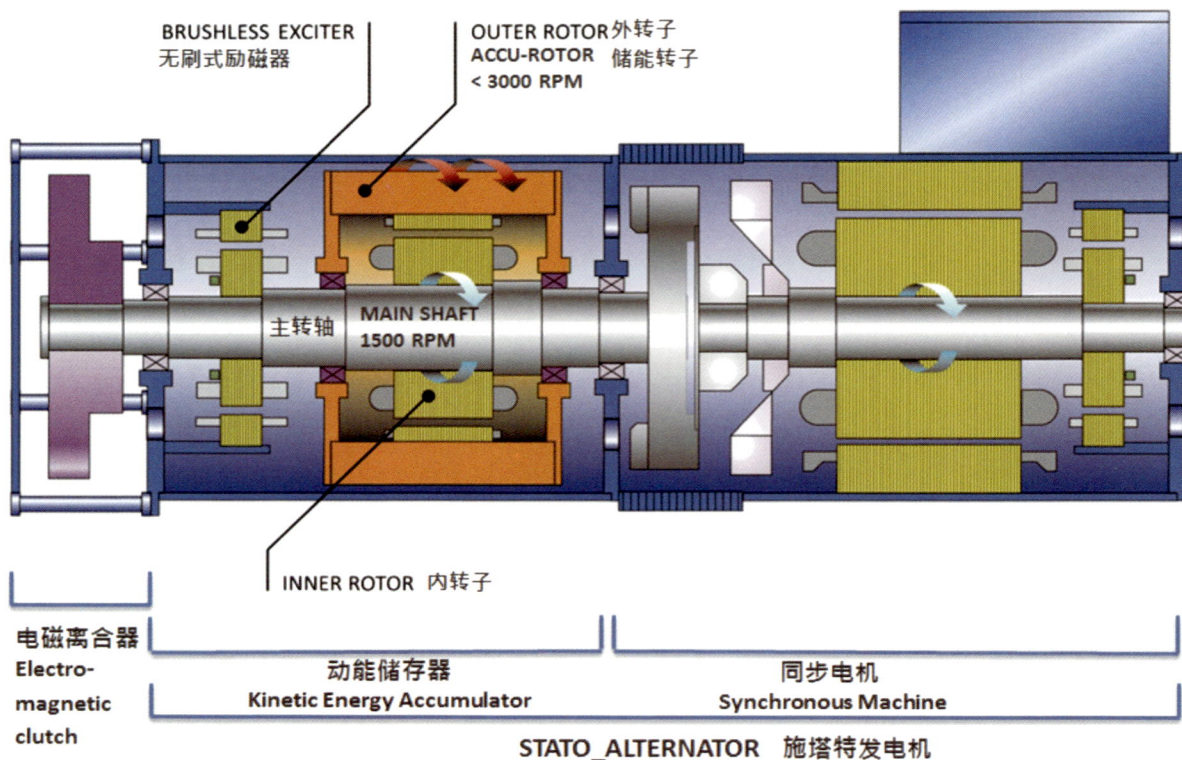

正常 → 应急 → 正常模式转换下飞轮及柴发转速变化示意图
Normal → emergency → schematic diagram of flywheel and diesel generator speed under normal switch mode

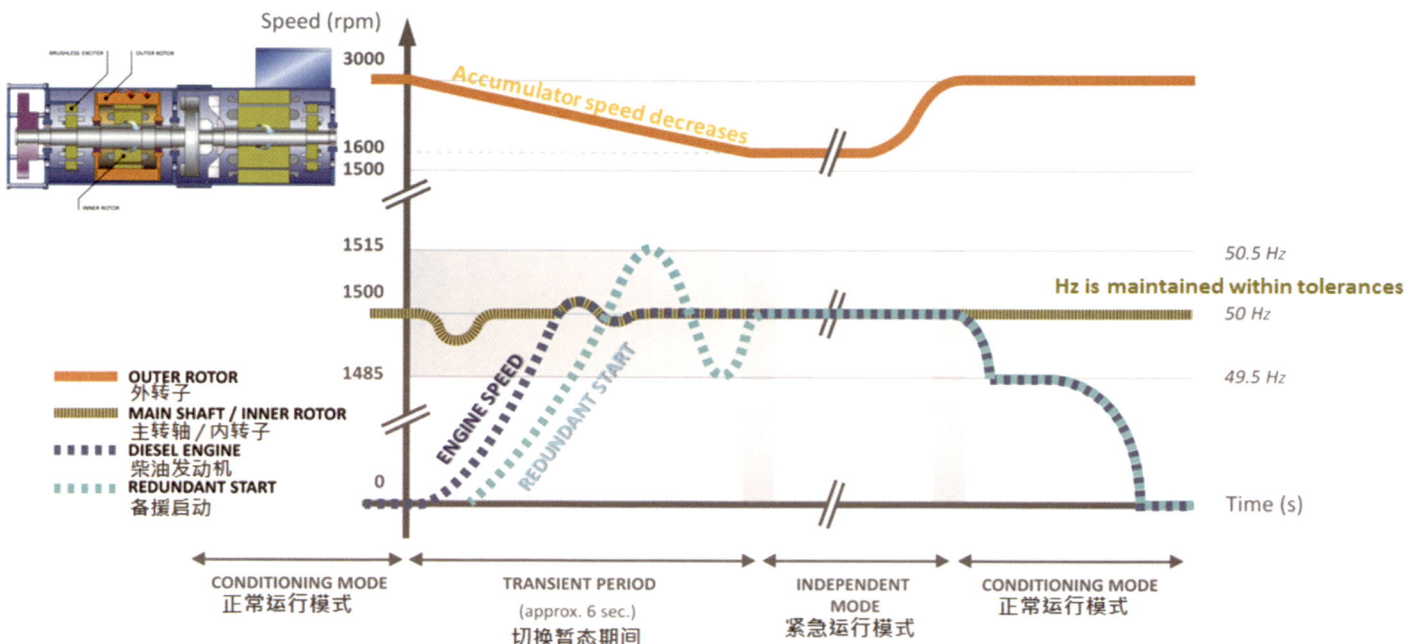

动态 UPS 飞轮示意图
Schematic diagram of dynamic UPS flywheel

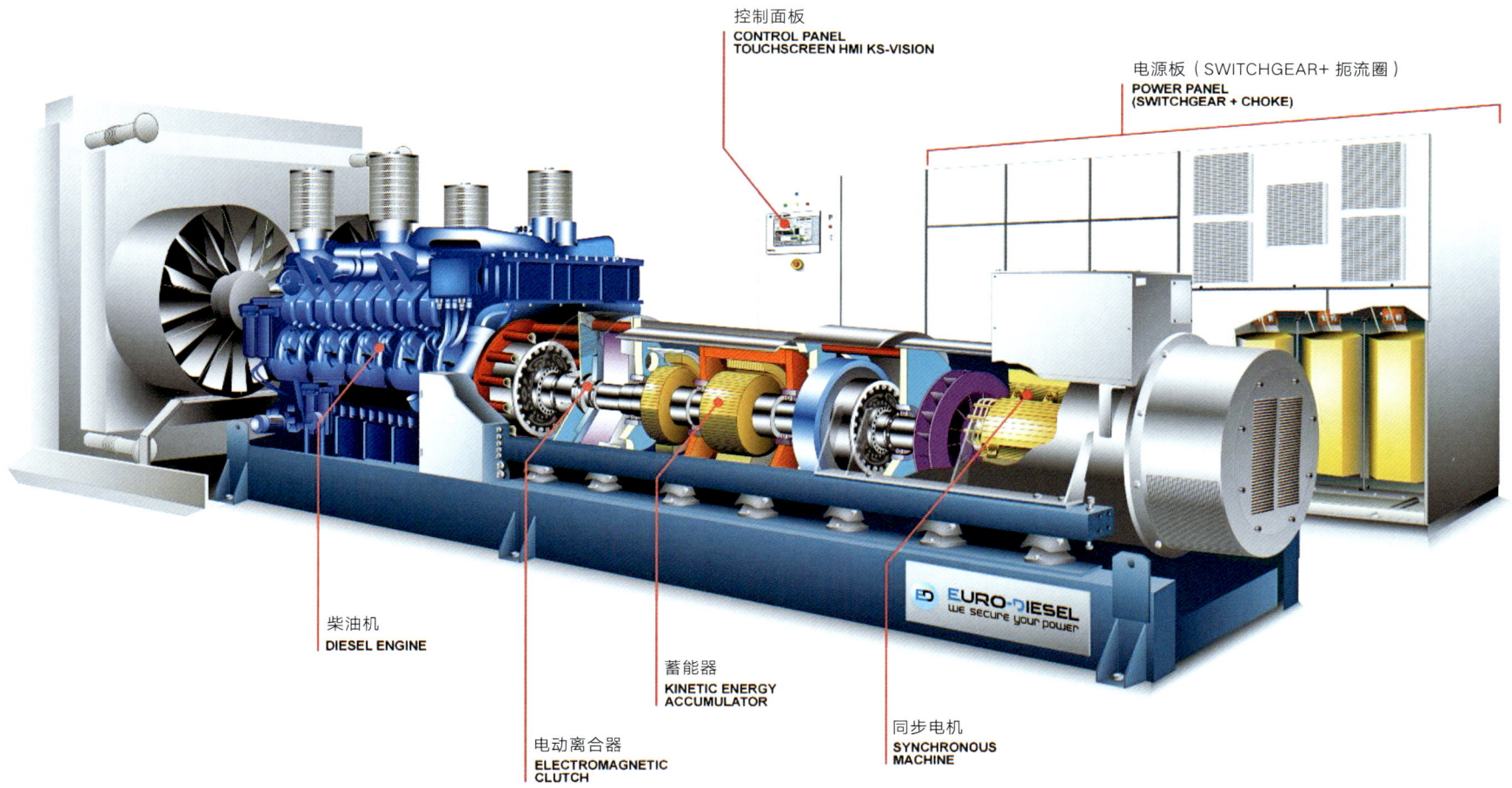

控制面板
CONTROL PANEL
TOUCHSCREEN HMI KS-VISION

电源板（SWITCHGEAR+ 扼流圈）
POWER PANEL
(SWITCHGEAR + CHOKE)

柴油机
DIESEL ENGINE

蓄能器
KINETIC ENERGY
ACCUMULATOR

电动离合器
ELECTROMAGNETIC
CLUTCH

同步电机
SYNCHRONOUS
MACHINE

EURO·DIESEL
we secure your power

5.18 夜景照明节能创新技术
Innovative Technology Of Energy Saving For Nightscape Lighting

大数据 Data

>100,000m
LED 线性灯总长度
Total length of LED linear lamp

1.4m
LED 最窄反光面间距
LED narrow reflector spacing

AC220v
LED 供电电压
LED supply voltage

350kW
总运行功率比设计降低值
Total operating power is 350 kW lower than the designed power

10,000m
节约电缆
Save cable

33%
全年节能约为原设计的比率
Annual energy saving rate is about 33% of the original design

图片说明 Comments

1. LED 灯安装示意图
Installation schematic diagram of LED lamp

2. 定制吊篮和可拆卸擦窗机立体作业
The stereoscopic operation of custom basket and dismountable Window Scrubber

3. 蜘蛛人高空灯具安装
Installation of spider-man high altitude lamps

4. 夜景照明效果设计方案
Design scheme of nightscape lighting effect

延伸阅读 Links

P202. P244

夜景照明采用 72,208 套 LED 灯具,正面全彩灯具可实现 1,670 万种颜色。LED 线性灯总长度超过 100,000m,是国内 LED 线性灯数量最大、总长度最长、控制点数最多的超高层建筑。设计上,巧妙利用幕墙单元板块使用灯线分离可拆卸式装饰条系统,单元板块全部预埋线管,工厂提前做好预开孔,现场只装灯具,大大缩减了现场夜景照明施工时间。幕墙结构小挺正面安装 RGB 光源,两侧安装白色光源,利用大挺作为反光基面,最窄处仅 1.4m,得益于较小的间距,为大厦夜景照明带来令人震撼的视觉效果。

通过将原设计的 DC24V 供电优化为 AC220V 供电方式,简化供电回路划分,降低线路的工作电流,减少供电损耗,降低缆线规格,创国内超高层建筑夜景照明 LED 灯具 AC220V 供电先例。通过精确设计灯具亮度,取消格栅、提高灯具的发光效率等措施,优化后的夜景照明系统总运行功率比设计降低了 350kW,节约电缆 10,000m,线路损耗也大幅度降低,达到节能环保的要求,全年节能约为原设计的 33%。

中信大厦独特的外立面造型,施工难度大。采用了幕墙板块预装、传统蜘蛛人、低区无配重悬挂定制吊篮、临时可拆卸擦窗机和正式擦窗机等多种措施相结合的立体作业,保质、保量、安全地完成了高空灯具安装。

幕墙板块 | Curtain wall plate
白光 LED | White light LED
装饰鳍板 | Decoration plate
白光 LED | White light LED
RGB LED

图片说明 *Comments*

1．定制吊篮和可拆卸擦窗
机立体作业现场（一）
The stereoscopic operation of
custom basket and dismountable
Window Scrubber 1

2．定制吊篮和可拆卸擦
窗机立体作业现场（二）
The stereoscopic operation of
custom basket and dismountable
Window Scrubber 2

72,208 sets LED lamps are used for nightscape lighting, and front full-color lamps can create 16.7 million colors. The LED linear lamp with the total length of over 100,000m is the longest with the largest number of LED linear lamps, which controls the super high-rise with the most points in China. In the design, the detachable decorative strip system with light and line separation is cleverly used in the curtain wall unit plate, and all the unit plates are pre-embedded with line pipes. The factory makes pre-opening with only lamps and lanterns are installed on the site, which greatly reduces the construction time of night scene lighting on the site. RGB light source is installed on the front side of the small tappet, and white light sources are installed on both sides of the small tappet. The large tappet with the narrowest part of 1.4m is adopted as the reflective base surface. Thanks to the small spacing, nightscape lighting of the building brings about shocking visual effect.

By optimizing the original DC24V power supply to AC220V power supply, the division of power supply circuit is simplified, the working current of power line is reduced, the power supply loss and the cable specification are reduced, and the precedent of AC220V power supply for super high-rise nightscape lighting LED lamps in China is set. Accurately design the luminance of the lamp; by eliminating the grating and improving the luminous efficiency of the lamp, the total operating power of the optimized nightscape lighting system is reduced by 350 kW compared with the designed one. The cable is saved by 10, 000 m, and the line loss is also greatly reduced, which meets the requirements of energy conservation and environmental protection. The energy conservation of the whole year is about 33% of the original design.

The construction of the unique facade modelling of CITIC Tower is difficult. Multiple measures, such as curtain wall plate pre-installation, traditional spider-man, low- zone non-counterweight hanging customized basket, temporary detachable Window Scrubber and formal Window Scrubber are adopted to ensure the completion of 3D operation with high quality, sufficient quantity and security.

基坑监测及大厦沉降监测
Foundation Pit Monitoring And Building Settlement Monitoring

中信大厦基础沉降和应力监测分为主体沉降观测和基础应力监测两个大项目,具体监测项目包括建筑物主体沉降监测、建筑物周边场地沉降观测、基底孔隙水压力监测、基底桩间土压力监测、桩顶反力监测、筏板基础钢筋应力监测等。

中信大厦基坑第三方监测按照北侧22m深度处可划分为两部分,上半部分监测项目主要包括北侧护坡桩桩顶水平位移、桩体深层水平位移、桩体钢筋应力、锚杆轴力,栈桥立柱的水平位移、沉降,地下水位,北侧道路地表及地下管线沉降。深度22m以下部分监测项目在基坑上半部分的基础之上,新增地下连续墙顶水平位移、沉降,地下连续墙深层水平位移,地下连续墙钢筋应力及锚杆轴力,内支撑立柱的水平位移、沉降及应力,目前中信大厦沉降量约11cm。

The foundation settlement and stress monitoring of CITIC Tower are divided into two major projects, namely, the main settlement monitoring and the foundation stress monitoring. The specific monitoring projects include the main settlement monitoring of the building, the settlement monitoring of the site around the building, the pore water pressure monitoring of the foundation, the soil pressure monitoring between the foundation piles, the pile-top counterforce monitoring and the rebar stress monitoring of the raft foundation.

The third-party monitoring of the foundation pit of CITIC Tower can be divided into two parts according to the depth of 22m on the north side. The upper part of the monitoring project mainly includes the horizontal displacement of pile top, horizontal displacement of pile deep layers, stress of rebar and axial force of anchor rod of the pile body, horizontal displacement and settlement of trestle column, groundwater level and settlement of surface and underground pipelines on the north side of road. On the basis of the upper part of foundation pit, some monitoring projects below 22m in depth are added with horizontal displacement and settlement of the roof of the diaphragm wall, horizontal displacement of the deep layer of the diaphragm wall, rebar stress and axial force of the anchor rod of the diaphragm wall and horizontal displacement, settlement and stress of the internal support column. At present, the settlement of the building is about 11cm.

大数据 *Data*

11cm
目前沉降量
Current settlement

图片说明 *Comments*

1. 中信大厦底板下的摩擦桩
Friction piles under the bottom plate of the building

[1]

沉降观测示意图
Settlement test diagram

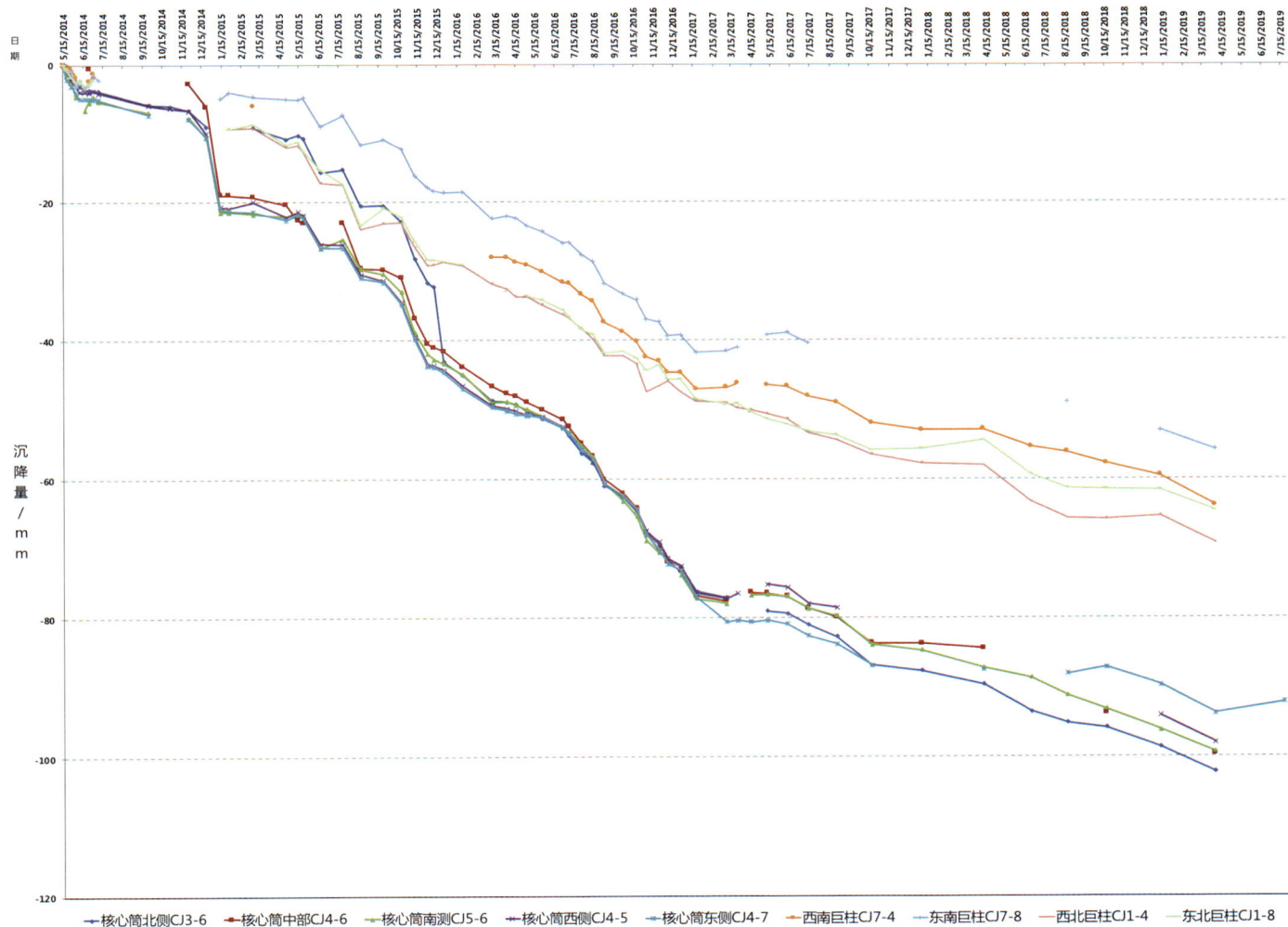

核心筒北侧CJ3-6　核心筒中部CJ4-6　核心筒南测CJ5-6　核心筒西侧CJ4-5　核心筒东侧CJ4-7　西南巨柱CJ7-4　东南巨柱CJ7-8　西北巨柱CJ1-4　东北巨柱CJ1-8

5.20 工厂预制化建造技术
Factory Prefabrication Construction Technology

中信大厦全力推进工厂预制化步伐，实现节能环保、绿色建造。现场钢筋全部采用工厂预制加工，现场安装；钢构件全部使用 BIM 完成深化设计，通过定制的钢结构全生命周期信息化管理平台对构件的下料、运输、安装进行全过程追踪管理。优化排版取料顺序，实时更新材料精确位置，减少材料浪费，显著提高加工速度。

中信大厦地下室疏散楼梯采用预制混凝土楼梯，梯段预制，休息平台现浇。现浇平台板与梯段相接处预留安装企口，企口处应按照图纸预留安装埋件。地下室共 146 部预制楼梯，单部最重为 3.11t。

中信大厦积极应用工厂预制化建造技术，包括全部风管和桥架的制作，14,600m 窗台板和风机盘管一体化系统。F7 至F102 核心内筒 4 个大型管道井的空调水、给排水和消防水DN70 至 DN600 管道共 13,365m 采用了预制立管技术。预制立管每节 9m，共 222 节，最大管组含 12 根管道重 4t。利用塔吊和核心内筒行车吊两次吊装，共 576 吊次，施工难度大。该技术创超高层建筑预制立管在核心内筒施工的先河。220个空调机房水管及阀部件；全部给水泵及阀部件；办公区 LED集成灯盘等实现了 100% 的工厂预制装配化。

幕墙专业制作高精度模型进行生产加工。BIM 模型及信息通过 ERP 系统管理。生产过程中，可通过装配明细表查询物料编码、属性编码及加工图号等信息。构件运至现场后，读取编码信息即可运至指定的楼层，高效准确地完成单元板块的安装。装饰专业在精装修区域以预制加工手段完成异形构件的制作。通过 BIM 实现对复杂构件的参数化设计，并导入数控机床加工，在工厂完成组装后实现现场整体吊装。

大量预制化构件的使用，不仅实现了节能、节材、节地的目标，同时也提高了工程整体的施工质量、加快了实施进度。据统计，中信大厦施工中较常规工程少产生 90% 的建筑垃圾，施工用水和施工用电消耗仅为常规工程的 20%，有效解决了现场零场地施工组织的难题。

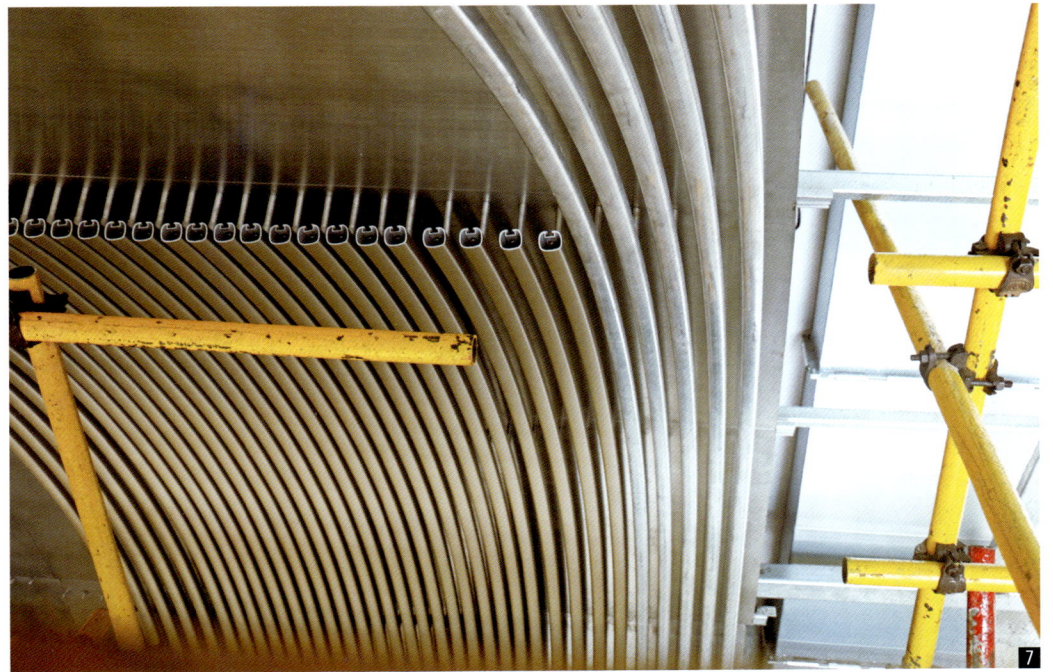

科技与创新
Technology & Innovation

The pace of factory prefabrication is fully promoted by CITIC Tower to achieve energy conservation, environmental protection and green construction. All bars on site shall be prefabricated and processed by the factory and installed on site; BIM is used to complete the in-depth design of steel components, and the whole process of blanking, transportation and installation of components is tracked and managed through the customized steel structure full life cycle information management platform. Optimizing the layout order and updating the exact position of the material in real time, can reduce the waste of the material, and improve the processing speed significantly.

The precast concrete staircase, staircase and cast-in-situ rest platform shall be adopted for the evacuation staircase in the basement of CITIC Tower. Installation groove shall be reserved at the junction of cast-in-situ platform plate and ladder section. The embedded parts shall be reserved and installed at the opening according to the drawings. There are 146 prefabricated staircases in the basement, with a maximum single staircase weight of 3.11t.

CITIC Tower actively applies factory prefabrication construction technology. Production of all air ducts and bridge frames; The integrated system of 14,600m window sill plate and fan coil; A total of 13,365m of air conditioning water, water supply and drainage and firefighting water DN70 to DN600 pipelines for the four large pipe wells of the core inner tube on the F7 to F102 are equipped with prefabricated stand pipe technology. The prefabricated stand pipe is 9m in length for each section, with a total of 222 sections. The maximum pipe group consists of 12 pipes, weighing 4t. The tower crane and the core inner tube crane are used for hoisting twice, with a total of 576 hoisting times, which is difficult in construction. This technology makes its debut in the core inner tube construction of super high-rise prefabricated stand pipe. 220 water pipes and valve parts in air conditioner room; All feed pumps and valve parts; The LED integrated lamp panel in the office area realizes 100% factory prefabrication and assembly.

High-precision model of the curtain wall is made for professional production and processing. BIM model and information are managed by ERP system. During the production process, the material code, attribute code and processing drawing number are queried through the assembly list. After the components are transported to the site, the coded information is read and transported to the designated floor, in line with which, the unit plates are installed efficiently and accurately.

The specially shaped components are produced with prefabrication processing means by the decoration specialty in the refined decoration zone. The parameterized design of complex components is realized with BIM, and the components are imported into CNC machine tools for machining and then assembled in the factory, after which the whole hoist on site can be realized

The use of a quantity of prefabricated components not only achieves the goal of saving energy, materials and land, but also improves the overall construction quality of the project and speeds up the implementation progress. According to statistics, the project produces 90% less construction rubbish than the conventional project, and the consumption of construction water and construction electricity is only 20% of the conventional project, which effectively solves the problem of zero-site construction organization.

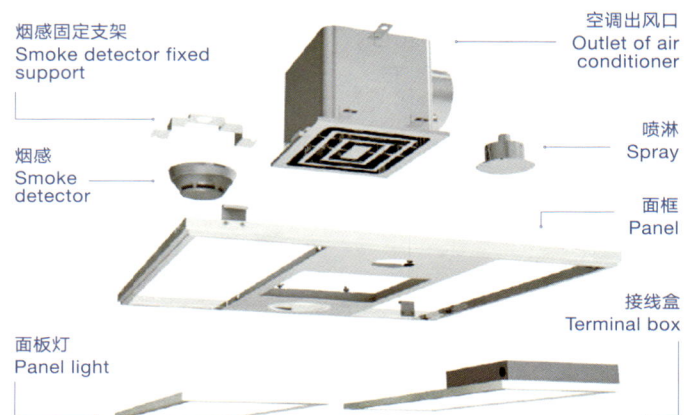

烟感固定支架
Smoke detector fixed support

空调出风口
Outlet of air conditioner

烟感
Smoke detector

喷淋
Spray

面框
Panel

接线盒
Terminal box

面板灯
Panel light

1

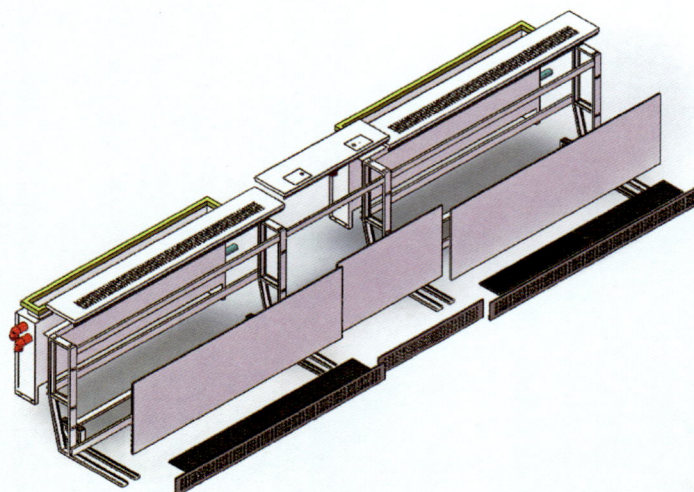

2

5.21 智能建筑云平台创新技术
Innovative Technology Of Smart Building Cloud Platform

大数据 Data

45 个
子系统
The number of integrated systems

650,505 个
信息点
The number of information points

75,000 m
光纤总长度
Total length of optical fiber

99.99 %
自主容错技术使系统可靠性达
Reliability of the system with autonomous fault tolerance technology

99,211 个
监控数据点
The number of monitoring data points

图片说明 Comments

1. 对接信息点饼图
Pie chart of docking information points

2. 中信大厦物联网架构
CITIC Tower IoT structure diagram

延伸阅读 Links

P088. P133

中信大厦构建了智能建筑云平台（SBC-Cloud），汇集了云计算、大数据、物联网、流式处理和 BIM 等创新信息技术，集成 45 个子系统共 650,505 个信息点，运用虚拟技术分出约 30 台虚拟机，光纤总长度约 75,000m。采用 8 台华为高性能服务器，独特的自主容错技术使系统可靠性高达 99.99％。

一级平台包括基于 BIM 的综合监控平台、智能建筑云平台（SBC-Cloud）、物联网配置系统、OA 办公系统、应急指挥系统。二级平台包括基于 BIM 的物业及设施管理系统（PM+FM）、建筑设备管理平台（BMS）、综合安防管理平台（SMS）、信息设施管理平台（ITSI）。物联网将所有的设备在线路层面互联互通，打破信息孤岛，实现信息资源的共享和协同管理。

智能监控系统（Smart monitoring system）是真正将框架、塑壳、微断以及智能网关模块等完美整合在一起的一体化方案，是传统强电和弱电的有机结合，使低压系统中的配电设备信息互通，监控数据共 99,211 个点。基于信息技术 IT 和运营

技术 OT 的融合，用数字化远程管理平台"千里眼"助力运行维护人员提高效率，管理用电设备，提供维修计划，从配电柜拓展到服务代维业务。

CITIC Tower has built the Smart Building Cloud Platform (SBC-Cloud), which integrates innovative information technologies, such as cloud computing, big data, Internet of Things, streaming and BIM. It integrates 45 subsystems with 650,505 information points, and adopts virtual technology to separate about 30 virtual machines with a total fiber length of about 75,000m. The reliability of the system is as high as 99.99% by adopting 8 Huawei high-performance servers and unique self-fault-tolerant technology.

The first-level platform includes BIM-based Integrated Monitoring Platform, Smart Building Cloud Platform (SBC-Cloud), Internet of Things Configuration System, OA Office System and Emergency Command System. The secondary platform includes BIM-based Property and Facility Management System (PM + FM), Building Equipment Management System (BMS), Integrated Security Management System (SMS) and Information Technology System Infrastructure (ITSI). The Internet of Things interconnects all devices at the line level, breaks the information isolated island, and realizes the sharing and collaborative management of information resources.

Smart Panel is a perfect integration of frame, plastic shell, micro circuit breaker and intelligent gateway module. It is the organic combination of traditional strong and weak current, which makes the information of distribution equipment in low-voltage system intercommunicate. The monitoring data total 99,211 points. Based on the integration of information technology (IT) and operation technology (OT), the digital remote management platform "Clairvoyance" is used to empower operation and maintenance personnel to improve efficiency, manage electrical equipment, provide maintenance plans, and expand service from distribution cabinets to business maintenance services.

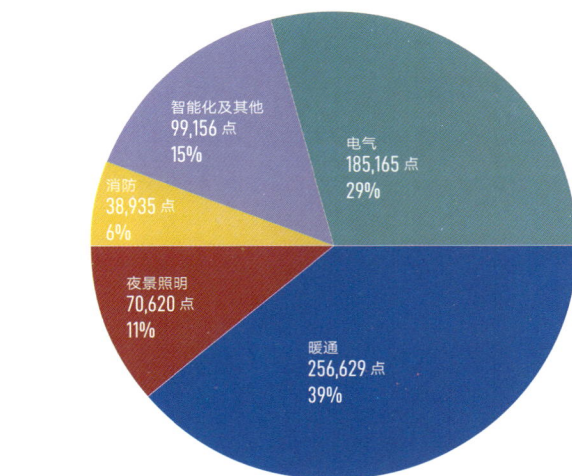

对接总点数 650,505 点

● 电气 ● 暖通 ● 夜景照明 ● 消防 ● 智能化及其他

[1]

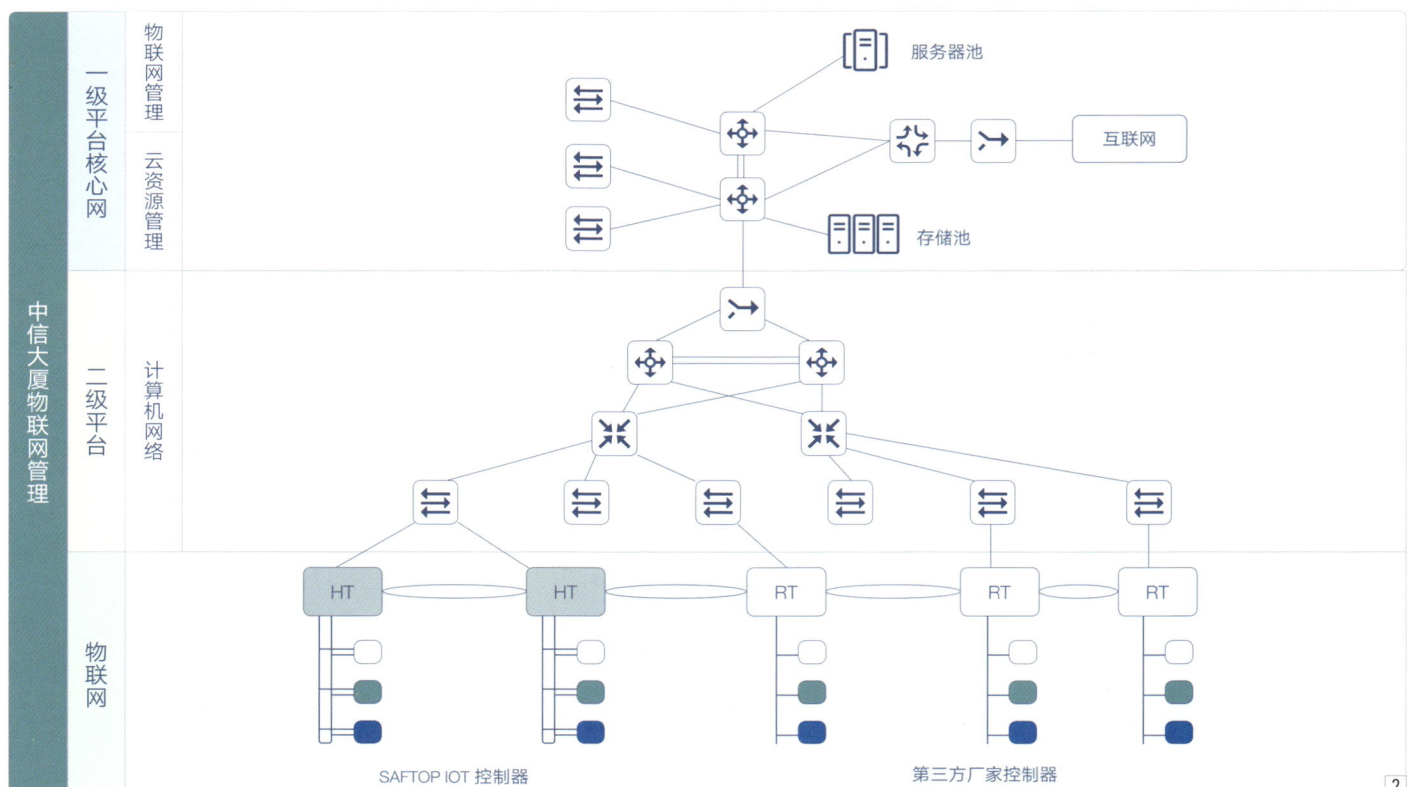

SAFTOP IOT 控制器　　　第三方厂家控制器

[2]

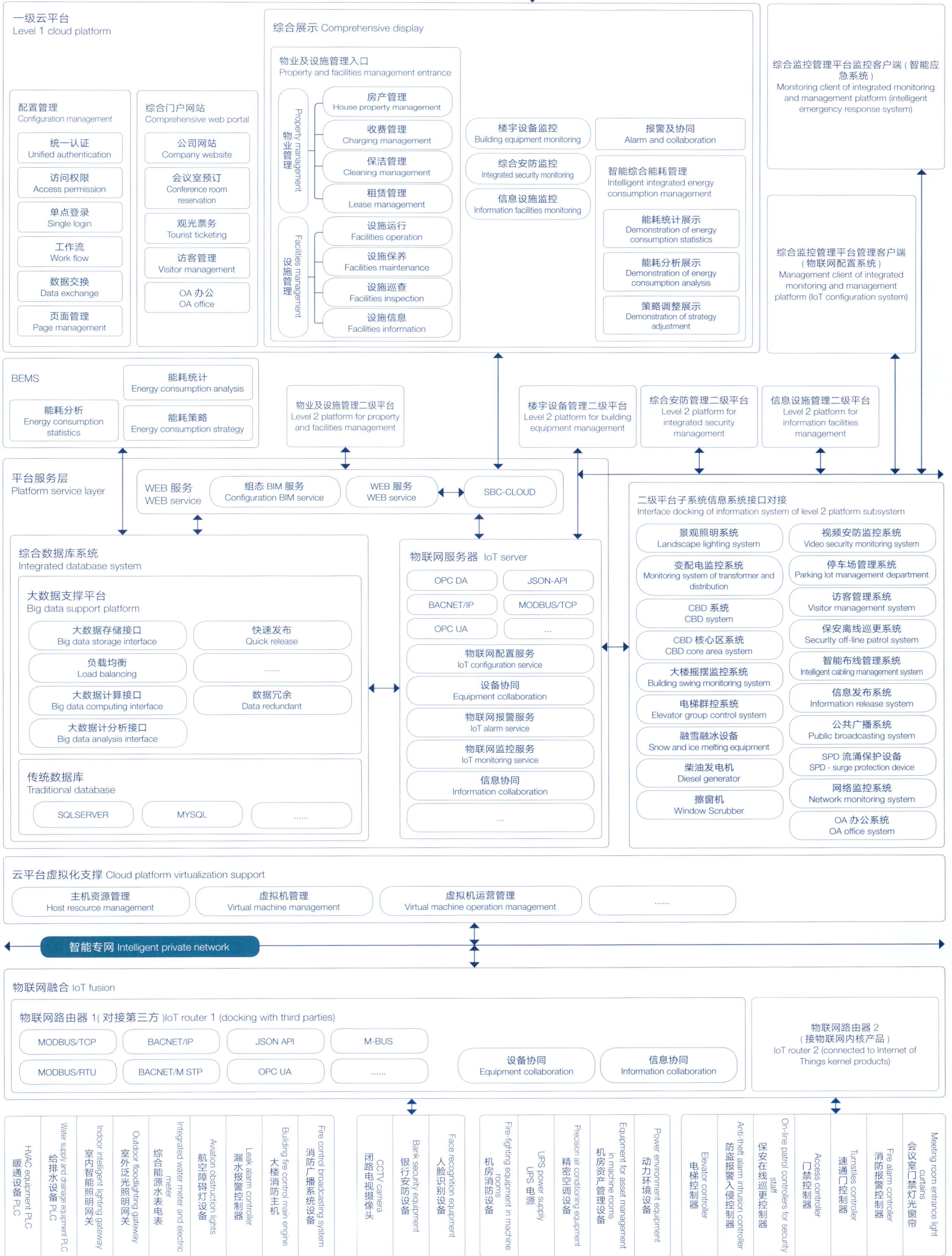

中信大厦智能云平台层级图
Layer diagram in smart cloud platform of CITIC Tower

浏览器，移动端，Android 应用，IOS 应用
Browsers, Mobile, Android Applications, IOS Applications

一级云平台 Level 1 cloud platform

配置管理 Configuration management
- 统一认证 Unified authentication
- 访问权限 Access permission
- 单点登录 Single sign login
- 工作流 Work flow
- 数据交换 Data exchange
- 页面管理 Page management

综合门户网站 Comprehensive web portal
- 公司网站 Company website
- 会议室预订 Conference room reservation
- 观光票务 Tourist ticketing
- 访客管理 Visitor management
- OA 办公 OA office

综合展示 Comprehensive display

物业及设施管理入口 Property and facilities management entrance

物业管理 Property management
- 房产管理 House property management
- 收费管理 Charging management
- 保洁管理 Cleaning management
- 租赁管理 Lease management

设施管理 Facilities management
- 设施运行 Facilities operation
- 设施保养 Facilities maintenance
- 设施巡查 Facilities inspection
- 设施信息 Facilities information

- 楼宇设备监控 Building equipment monitoring
- 综合安防监控 Integrated security monitoring
- 信息设施监控 Information facilities monitoring

- 报警及协同 Alarm and collaboration

智能综合能耗管理 Intelligent integrated energy consumption management
- 能耗统计展示 Demonstration of energy consumption statistics
- 能耗分析展示 Demonstration of energy consumption analysis
- 策略调整展示 Demonstration of strategy adjustment

综合监控管理平台监控客户端（智能应急系统）
Monitoring client of integrated monitoring and management platform (intelligent emergency response system)

综合监控管理平台管理客户端（物联网配置系统）
Management client of integrated monitoring and management platform (IoT configuration system)

BEMS
- 能耗分析 Energy consumption statistics
- 能耗统计 Energy consumption analysis
- 能耗策略 Energy consumption strategy

物业及设施管理二级平台
Level 2 platform for property and facilities management

楼宇设备管理二级平台
Level 2 platform for building equipment management

综合安防管理二级平台
Level 2 platform for integrated security management

信息设施管理二级平台
Level 2 platform for information facilities management

平台服务层 Platform service layer

- WEB 服务 WEB service
- 组态 BIM 服务 Configuration BIM service
- WEB 服务 WEB service
- SBC-CLOUD

综合数据库系统 Integrated database system

大数据支撑平台 Big data support platform
- 大数据存储接口 Big data storage interface
- 快速发布 Quick release
- 负载均衡 Load balancing
- ......
- 大数据计算接口 Big data computing interface
- 数据冗余 Data redundant
- 大数据计分析接口 Big data analysis interface

传统数据库 Traditional database
- SQLSERVER
- MYSQL
- ......

物联网服务器 IoT server
- OPC DA
- JSON-API
- BACNET/IP
- MODBUS/TCP
- OPC UA
- ...
- 物联网配置服务 IoT configuration service
- 设备协同 Equipment collaboration
- 物联网报警服务 IoT alarm service
- 物联网监控服务 IoT monitoring service
- 信息协同 Information collaboration
- ...

二级平台子系统信息系统接口对接
Interface docking of information system of level 2 platform subsystem
- 景观照明系统 Landscape lighting system
- 视频安防监控系统 Video security monitoring system
- 变配电监控系统 Monitoring system of transformer and distribution
- 停车场管理系统 Parking lot management department
- CBD 系统 CBD system
- 访客管理系统 Visitor management system
- CBD 核心区系统 CBD core area system
- 保安离线巡更系统 Security off-line patrol system
- 大楼摇摆监控系统 Building swing monitoring system
- 智能布线管理系统 Intelligent cabling management system
- 电梯群控系统 Elevator group control system
- 信息发布系统 Information release system
- 融雪融冰设备 Snow and ice melting equipment
- 公共广播系统 Public broadcasting system
- 柴油发电机 Diesel generator
- SPD 流涌保护设备 SPD - surge protection device
- 擦窗机 Window Scrubber
- 网络监控系统 Network monitoring system
- OA 办公系统 OA office system

云平台虚拟化支撑 Cloud platform virtualization support
- 主机资源管理 Host resource management
- 虚拟机管理 Virtual machine management
- 虚拟机运营管理 Virtual machine operation management
- ......

智能专网 Intelligent private network

物联网融合 IoT fusion

物联网路由器 1（对接第三方）IoT router 1 (docking with third parties)
- MODBUS/TCP
- BACNET/IP
- JSON API
- M-BUS
- MODBUS/RTU
- BACNET/M STP
- OPC UA
- ......
- 设备协同 Equipment collaboration
- 信息协同 Information collaboration

物联网路由器 2（接物联网内核产品）
IoT router 2 (connected to Internet of Things kernel products)

下方设备列表：

- HVAC equipment PLC 暖通设备 PLC
- Water supply and drainage equipment PLC 给排水设备 PLC
- Indoor intelligent lighting gateway 室内智能照明网关
- Outdoor floodlighting gateway 室外泛光照明网关
- Integrated water meter and electric meter 综合能源水表电表
- Aviation obstruction lights 航空障碍灯设备
- Leak alarm controller 漏水报警控制器
- Building fire control main engine 大楼消防主机
- Fire control broadcasting system 消防广播系统设备
- CCTV camera 闭路电视摄像头
- Bank security equipment 银行安防设备
- Face recognition equipment 人脸识别设备
- Fire-fighting equipment in machine rooms 机房消防设备
- UPS power supply UPS 电源
- Precision air conditioning equipment in machine rooms 精密空调设备
- Equipment for asset management in machine rooms 机房资产管理设备
- Power environment equipment 动力环境设备
- Elevator controller 电梯控制器
- Anti-theft alarm intrusion controller 防盗报警入侵控制器
- On-line patrol controllers for security staff 保安在线巡更控制器
- Access controller 门禁控制器
- Turnstiles controller 速通门控制器
- Fire alarm controller 消防报警控制器
- Meeting room entrance light curtains 会议室窗灯光窗帘

中国建造 — 中信大厦建造纪实　**193**

5.22 人脸识别技术
Face Recognition Technology

大数据 Data

10,000人
快速比对
Quick comparison of 1,000 people

<1s
单人通行时间
Single occupancy time

>99%
通过率
Passing rate

图片说明 Comments

1. 人脸识别速通门闸机
Face recognition turnstile brake

延伸阅读 Links

P088. P222. P232

中信大厦采用了人脸识别速通门系统和人脸识别门禁系统。首层大堂、地下大堂和地下3层共设置38个速通门通道，采用定制化人脸识别摄像机安装在速通门闸机内部，可满足高达10,000人的快速比对，单人通行时间<1s，通过率>99%。

在VIP大堂及电梯厅、通往屋面室外和地下后勤货运处的通道门旁，安装人脸识别门禁机48台，采用高清双目摄像头，实现人脸和智能IC卡的混合验证。

结合高清视频监控系统进行出入人员权限管理、人像库管理、电子地图上获取人员定位及路径信息、考勤管理、黑名单布控报警、重点人员管控等。也可定义各种联动，如当用户走到某处，摄像头就能识别出人员身份，触发门禁、灯光、电梯、空调等设备，让识别变得智慧有趣。

Face recognition of turnstile system and face recognition access control system are adopted in CITIC Tower. A total of 38 turnstiles are provided in the lobby on the first floor, the underground lobby and the third floor. The customized face recognition camera is installed inside the turnstile, which can conduct the fast comparison of up to 10,000 people with the single-person travel < 1s, and the passing rate > 99%.

In the VIP lobby, elevator hall, and the passageway leading to the roof outdoor and underground logistic freight service, 48 access control machines with a face recognition function are installed, and HD binocular cameras are used to realize the mixed verification of face and smart IC card.

Combined with HD video surveillance system, they can carry out personnel authority management and portrait database management, obtain personnel location and path information on the electronic map, and conduct attendance management, blacklist distribution alarm as well as key personnel management and so on. They can also define a variety of linkages, such as when the user goes somewhere, the camera can identify people, trigger access control, lighting, elevators, air conditioning and other devices, so that identification becomes smart and interesting.

数据机房创新技术
Innovative Technology Of Data Machine Room

B2 的大厦数据机房机柜"背靠背、面对面"摆放，设置空调送回风 + 精密空调 + 冷通道，气流组织合理，精密空调利用率高，制冷效果好。

采用动态 UPS 飞轮储能系统供电，直接输入机房，确保数据机房可靠运行。使用 3 台数字 KVM 管理器对数据机房中的被管理设备进行汇集，在本地配置 3 台液晶套件对本地端进行管理，远程可以集中管控多达 96 台服务器，使用网线从被管理设备的 I/O 口连接到服务器端口上，采用 Over IP 技术将模拟信号转换成 IP 数据包并连接到 IP 网络，可从远程对设备进行管理。

综合布线采用网格式电缆桥架和框架固线器解决方案，呈现完美的布线艺术。

The cabinet of the data machine room on the B2 floor is placed "back to back, face to face", with air-conditioner returning air + precision air-conditioner + cold channel, reasonable air distribution, realizing high use efficiency of precision air-conditioner and good refrigeration effect.

Dynamic UPS flywheel energy storage system is used to supply power and input directly into the machine room to ensure the reliable operation of the data machine room. 3 digital KVM managers are used in assembling the managed devices in the data machine room, 3 liquid crystal suites are configured locally to manage the local side, and up to 96 servers can be managed centrally from a remote location. Network cables are used to connect the I/O ports of the managed devices to the server ports, and Over IP technology is used to convert analog signals into IP packets and connect them to the IP network, so that the devices can be managed from a remote location.

The comprehensive wiring uses a network cable tray and frame fixer, presenting the perfect wiring technique.

大数据 Data

3 台
数字 KVM 管理器
The number of digital KVM managers

96 台
远程可集中管控服务器数量
The number of remote centrally controlled servers

图片说明 Comments

1. 动态 UPS 飞轮储能系统供电示意图
Power supply schematic diagram of dynamic UPS flywheel energy storage system

2. 数据机房
Data machine room

3. 数据机房气流组织图
Data machine room air distribution diagram

4. 数据机房布线艺术
Wiring technique of data machine room

延伸阅读 Links

P088. P182

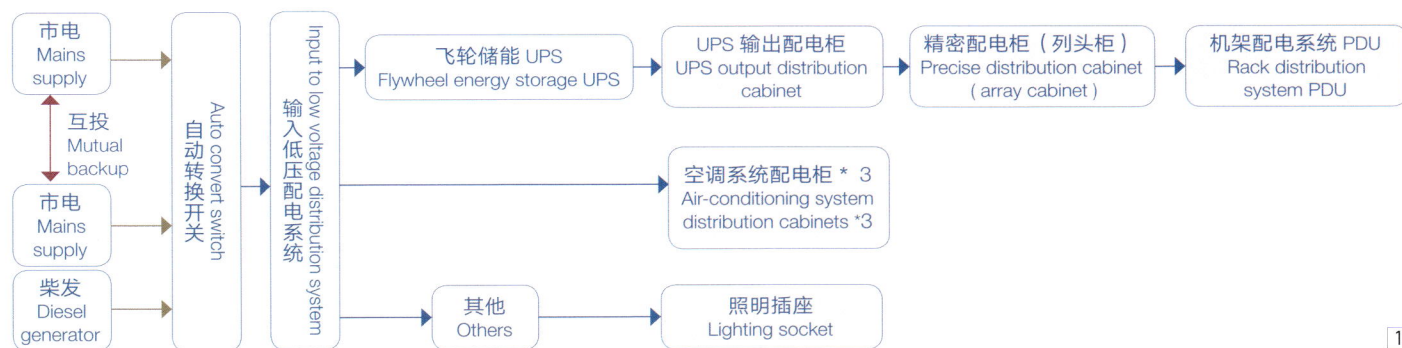

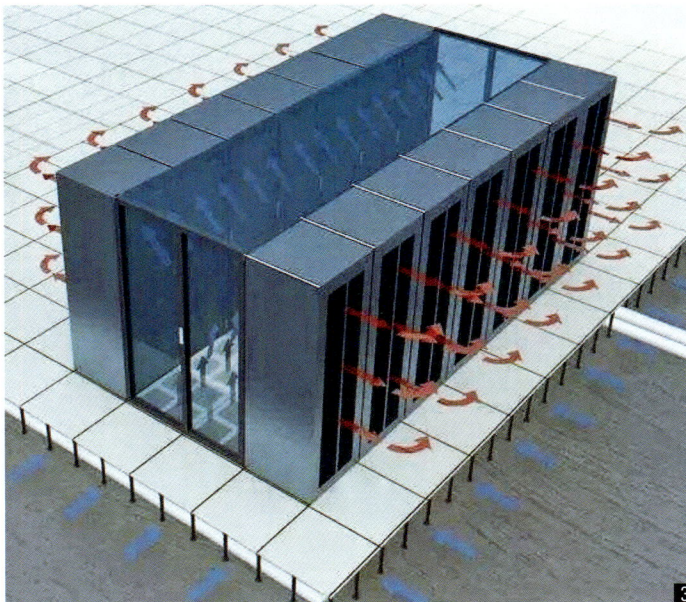

5.24　三维激光扫描点云模型信息技术
3D Laser Scanning Point Cloud Model Information Technology

中信大厦在国内超高层建造过程中首次全面应用三维激光扫描技术辅助项目精细化管理。双总包将点云数据作为公共资源提供给施工管理人员及专业分包单位。将具有实景还原特点的点云与深化设计、质量管理、测量管理、分包协调管理相结合，让所有工程建设参与方准确获取工程实体信息，打通了设计图纸与施工现场之间的信息断链，提高整体工作效率，减少错误与冲突，为总承包管理提供了新的技术解决方案。工程施工过程中，双总包完成了151次的楼层三维激光扫描工作，累计点云数据容量超过2.3TB，可真实准确地还原结构体态原貌。

For the first time, CITIC Tower has used 3D laser scanning technology to assist the project management in the process of super high-rise building construction in China. The general construction contractor provides point cloud data as a public resource for construction managers and professional subcontractors. Through the integration of point cloud with characteristics of scene restoration, in-depth design, quality management, measurement management and subcontractor coordination management, all the project construction participants can accurately obtain the project entity information, open the information chain between the design drawings and the construction site, improve the overall efficiency, reduce errors and conflicts, and provide a new technical solution for the general contractor management. During the construction of the project, the general construction contractor simply completed 121 times of 3D laser scanning for the floors, and the cumulative point cloud data capacity exceeded 2.3TB, which can truly and accurately restore the original appearance of the structure.

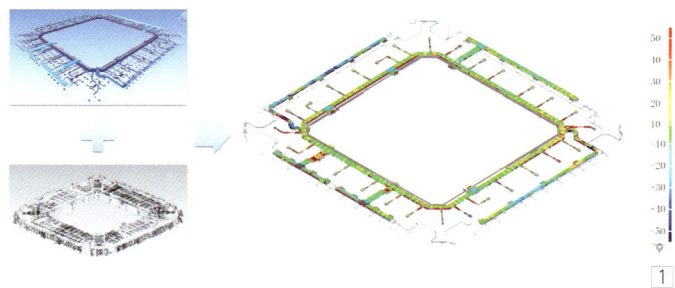

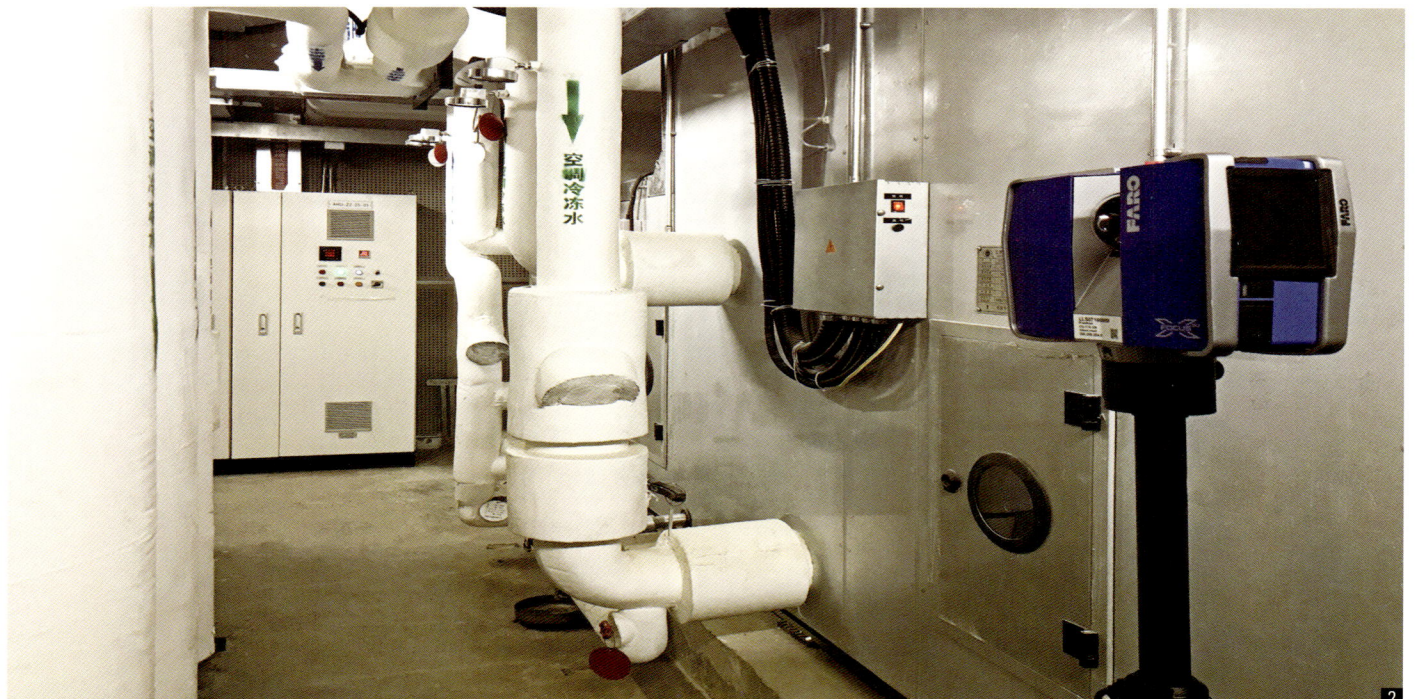

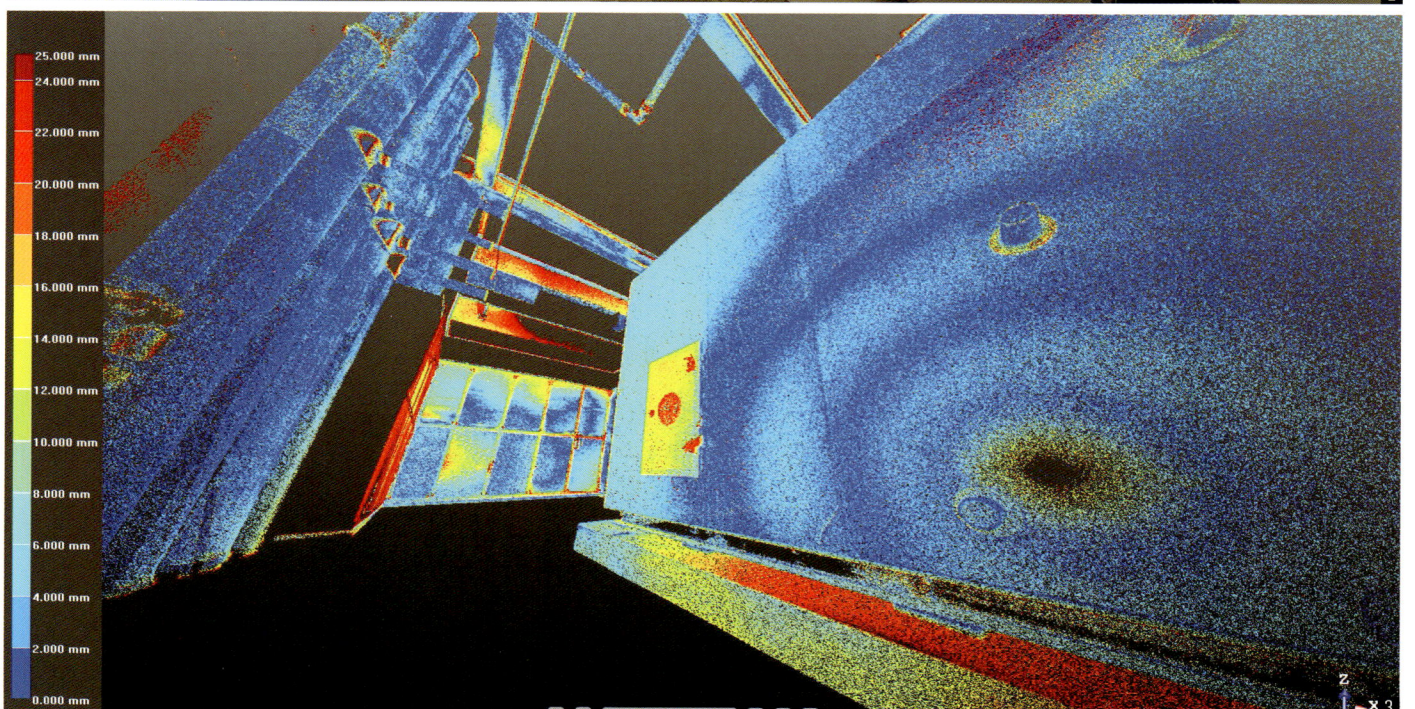

智慧运维云平台技术
Cloud Platform Technology For Intelligent Operation

将 BIM 技术与建筑智能化集成管理系统（IBMS）、物业管理（PM）和设备设施管理（FM）有机结合，构建智慧运维二级云平台（SOC）。

四大功能：以二维码为载体建立大数据库；动态仿真三维化监控；报警快速空间定位；系统联动监控。

采集约 650,505 个数据信息点，对接模型文件约 1,380 个，数据量约 300GB，轻量化后的模型大小仅为 30GB，为原模型的 10%。

BY combining BIM technology with intelligent building management system (IBMS), property management (PM) and facilities management (FM), the intelligent secondary operation cloud platform (SOC) is constructed.

Four main functions: Establish a big database with QR code as the carrier; 3D monitor of dynamic simulation; Alarm fast spatial positioning; System linkage monitoring.

About 650,000 data points of information have been collected, and about 1,380 model files have been delivered, with the data volume of 300GB. The size of the lightweight model was only 30GB, equal to 10% of the original model.

大数据 *Data*

650,505 个
采集数据信息点
The number of data points collected

1,380 个
对接模型文件
The number of connected model files

300 GB
采集数据信息数据量
Data information acquisition

30 GB
轻量化模型为原模型的 10%
Lightweight model is 10% of the original model

图片说明 *Comments*

1. 信息图（一）
Information graph 1

2. 信息图（二）
Information graph 2

延伸阅读 *Links*

P088

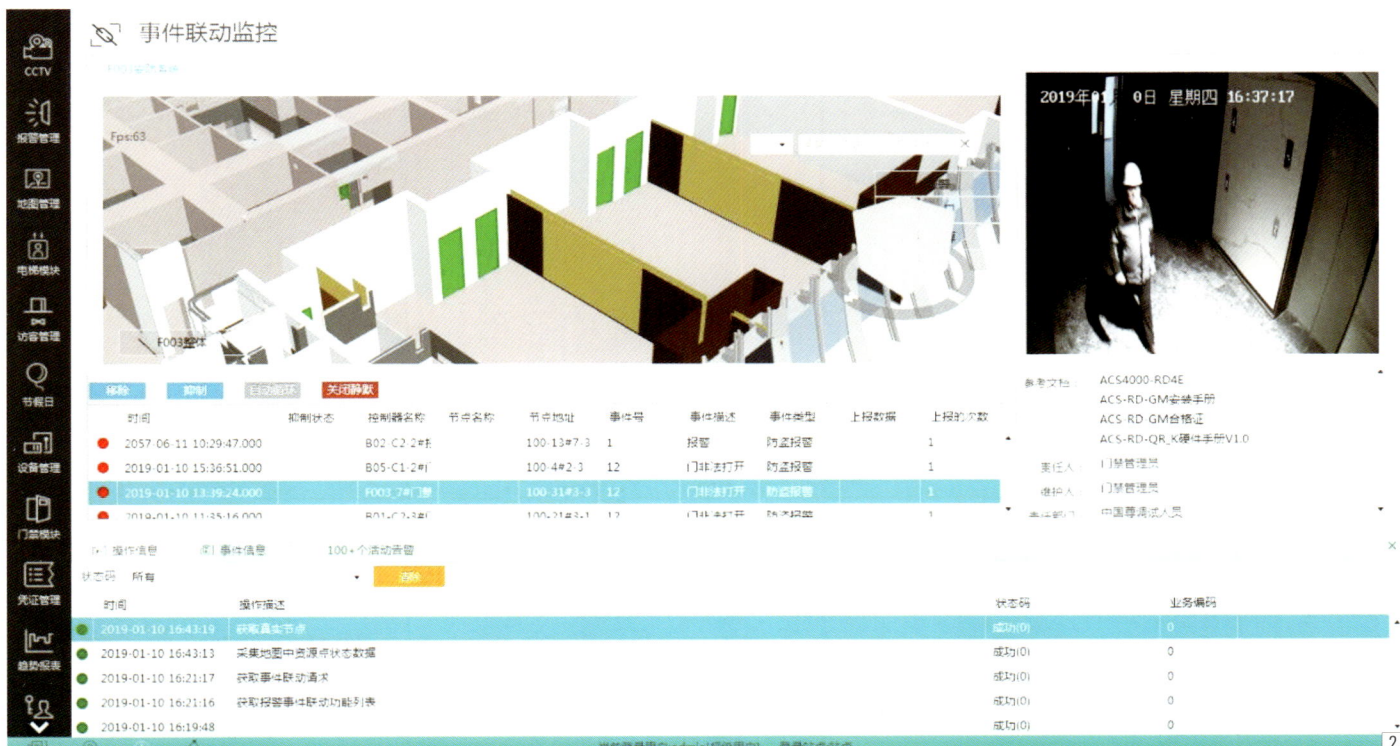

6.1 节能设计优化
Energy-Saving Design Optimization

图片说明 *Comments*

1. 国家三星级绿色建
筑设计标识证书
Certificate of greenbuilding
design label

2. 国际 LEED-CS 金级预认证
Gold pre-certification of
international LEED-CS

延伸阅读 *Links*

P304

设计阶段，中信和业组织专业设计 23 家和 41 家顾问团队，用 9 个月时间对设计成果进行多轮优化，共提出优化建议 1,126 项，被设计采纳 847 项，采纳率为 75%。特别是节能设计优化，效果显著。

大楼总用电量从 78,660kVA 降到 56,560kVA，节省 28%；
暖通空调系统采用了低温大温差冷冻水、温湿度独立控制、变频控制等节能设计策略，较国家标准节能 25.7%；
通过冰蓄冷充分利用北京峰谷电价优势，节费率达 28.6%；
照明系统较国家标准节能 15.8%；
电梯系统全部采用能量回馈系统；
给水、中水、热水水泵优化选型和装机容量，总能耗从 135.1kWh 降到 86.9kWh，节能率为 35.63%；
窗墙比为 0.44 的双中空超白玻璃幕墙系统较国家标准节能 7.7%；
柴油发电机组由 9 台优化为 6 台，节约成本 800 万元；
通过优化核心筒和巨柱边管井排布、优化窗边一体化系统等共为大楼增加使用面积约 7,500m²（相当于 3 层办公楼使用面积）。

已获得 2015-2016 年国家绿色三星建筑设计标识证书；
已获得国际 LEED-CS 金级预认证。

During the design stage, CITIC Heye and more than 30 consultative groups conducted multiple rounds of optimization of the design results over nine months. A total of 1,126 optimization recommendations are made, and 847 of which are adopted by the design project, representing an adoption rate of 75%. In particular, the effect of energy-saving design optimization is significant.

The total electricity consumption of the building is reduced from 78,660kVA to 56,560kVA, saving 28%;
The HVAC air-conditioning system adopts energy-saving design strategies such as chilled water with a large temperature difference from low temperature, independent control of temperature and humidity, frequency conversion control and so on, which is 25.7% more energy-saving than the national standard;
The advantage of Beijing valley peak electricity price is fully utilized through ice storage, and the saving rate is up to 28.6%;
The energy-saving effect of the lighting system better than the national standard: 15.8%;
The elevator system adopts the energy feedback system;
Selection and installed capacity of water supply, reclaimed water and hot water pumps are optimized, reducing the total energy consumption from 135.1kWh to 86.9kWh, with the energy saving rate of 35.63%;
The double hollow ultra-white glass curtain wall system with a window-wall ratio of 0.44 can save energy by 7.7% compared with the national standard;
The diesel generating set is optimized from 9 sets to 6 sets, saving 8 million yuan in cost.
By optimizing the arrangement of core tube and giant column side tube wells and optimizing the integrated system of windows and sides, the usable area of the building is increased by about 7,500m² (equivalent to the usable area of three-storey office building).

Certified with three-star green building design label from 2015 to 2016;
Gold pre-certification of international LEED-CS;

绿色建筑设计标识
GREEN BUILDING DESIGN LABEL
三星级绿色建筑设计标识证书
CERTIFICATE OF GREEN BUILDING DESIGN LABEL

公共建筑   NO.PD30140

建筑名称：北京市朝阳区CBD核心区Z15地块（中国尊大厦）

建筑面积：43.70万 m²

完成单位：中信和业投资有限公司、北京市建筑设计研究院有限公司、清华大学建筑学院、北京清华同衡规划设计研究院有限公司

评 价 指 标	设 计 值	说明
建筑节能率	62.85%	1 此证只证明建筑的规划和设计达到《绿色建筑评价标准》(GB/T 50378-2006 三星级水平)
可再生能源利用率	可再生能源发电提供0.26%建筑用电量	
非传统水源利用率	31.88%	2 "评价指标"值为代表性绿色建筑评价指标值，整体评价查阅《绿色建筑评价标识报告》
住区绿地率	公共建筑不参评	
可再循环建筑材料用量比	23.11%	
室内空气污染物浓度	设计阶段不参评	
物业管理	设计阶段不参评	

有效期限：2015年10月09日-2016年10月08日   签发日期：2015年10月09日

1

LEED

THE PROJECT OF PLOT Z15, CBD, BEIJING

Beijing, China

HAS FULFILLED THE REQUIREMENTS OF

PRE-CERTIFICATION

UNDER LEADERSHIP IN ENERGY AND ENVIRONMENTAL DESIGN (LEED®) FOR CORE & SHELL DEVELOPMENT LEVEL

GOLD

DOCUMENTATION HAS BEEN SUBMITTED FOR THIS PROJECT, WHICH DEMONSTRATES AN INTENT TO DESIGN AND BUILD A HIGH PERFORMANCE LEED GREEN BUILDING.

DECEMBER 2013

2

施工中的环保措施
Environmental Protection Measures In The Process Of Construction

现场采用大量创新技术，实现节能减排、绿色施工的目标。施工总承包成立绿色施工领导小组及绿色施工实施小组，负责现场绿色施工的管理及实施。

编制《基坑降水施工专项方案》，使在施工过程中保护场地四周原有地下水形态，在保证项目施工要求和安全的情况下，最大程度减少抽取地下水。

项目危险品、化学品存放处及污物排放采取隔离措施；例如气笼、油库内部设隔油沙等。

项目在施工现场建立洒水清扫制度，配备洒水设备，在清理项目现场工地及周围路面前均先洒水，防止扬尘对环境影响。在 F003 外框筒楼板边缘设置环状水喷雾设施，对首层环境进行除尘降温，为 CBD 核心区唯一配备水喷雾系统的工程。

项目现场对建筑垃圾进行分类收集、集中堆放，每层设置垃圾堆放区域，利用施工电梯运输至首层垃圾集中堆放区域，夜间由垃圾清运车统一处理。

施工中的返工和拆改较常规工程减少 65% 以上；
较常规工程少产生 90% 的建筑垃圾；
施工用水和施工用电消耗仅为常规工程的 20%。

The site adopts a large number of innovative technologies to achieve the goals of energy conservation, emission reduction and green construction. The general construction contractor shall establish a green construction leading group and a green construction group to be responsible for the management and implementation of green construction on site.

Prepare the Special Scheme for Foundation Pit Dewatering Construction, to protect the original groundwater form around the site during the construction process, and minimize the pumping of groundwater under the condition of ensuring the construction requirements and safety of the project.

Separation measures shall be taken in the discharge of dangerous goods and chemicals in the project, such as the use of air cages and the installation of oil separation sand inside the oil depot. Oil depot closure photos and signs are set up. The project safety department shall be responsible for supervision and management.

The project shall establish a sprinkler cleaning system at the construction site, equipped with sprinkler equipment, and sprinkle water before cleaning the project site and surrounding road surface to prevent dust from affecting the environment. A ring-shaped water spray facility is arranged at the edge of the outer frame tube floor on the F003 to remove dust and cool the environment on the first floor. It is the only object equipped with a water spray system for the CBD core area.

At the project site, the construction waste is collected in category and stacked centrally, and the stacking area is set up on each floor. The construction elevator is used to transport the construction waste to the centralized stacking area on the first floor, and the waste is disposed of uniformly by the trucks at night.

Saving 65% and above rework and demolition in construction compared with those in a conventional project；
90% less construction waste than conventional projects；
Consumption of construction water and construction electricity is only 20% of that of conventional projects.

大数据 Data

65%
施工中返工、拆改较常规工程减少
Saving 65% rework and demolition in construction compared with those in conventional project

90%
较常规工程少产生建筑垃圾量
90% less construction waste than conventional works

20%
较常规工程施工用水、用电消耗量
Saving 20% of the consumption of water and electricity for construction compared with conventional projects

图片说明 Comments

1. 降低扬尘措施：喷淋系统
Measure for reducing dust emission: Spray system

2. 垃圾集中分类
Centralized waste sorting

3. 洗车装置
Car washer

延伸阅读 Links

P304

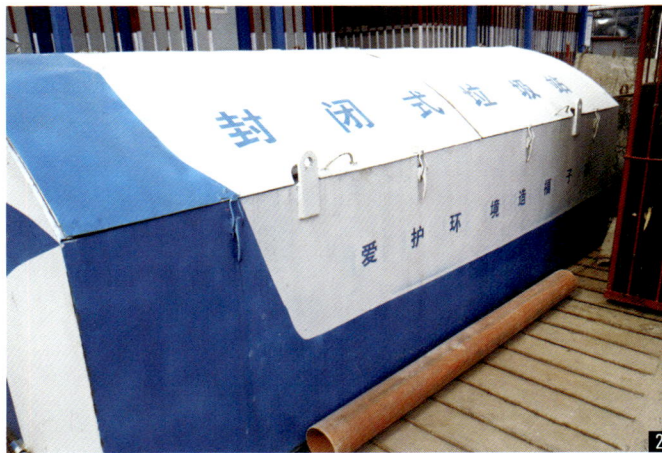

6.3 独具一格的双中空玻璃单元幕墙与擦窗机系统
Unique Double Hollow Glass Unit Curtain Wall & Windows Scrubber System

大数据 Data

0.44
窗墙比
The window-wall ratio

7.7%
较国家标准节能比率
Energy saving rate is improved
by 7.7% compared with
the national standard

4 级
幕墙气密性能
Curtain wall airtightness
performance

3 级
幕墙水密性能
Curtain wall watertightness
performance

6-8 级
热工性能
Thermal performance

125,000 m²
整体幕墙面积约
Overall curtain wall area

9 台
共设擦窗机数量
Window Scrubber

中信大厦幕墙系统采用四层双中空 Low-E 夹胶玻璃，大幅度提高了幕墙的保温节能的特性，同时幕墙隔音性能全面提升，窗墙比为 0.44，较国家标准节能 7.7%。其气密性 4 级，水密性能 3 级，抗风压性能 6 级，平面内变形性能 4 级，热工性能 6-8 级。最终实验结果全面超越要求的性能指标，部分性能远超目标。实现了"任风吹雨打，反正冰天雪地我也不怕"的状态。

施工中通过 BIM 技术三维建模进行参数化分析，将幕墙板块及面材进行近似整合，大大减少了板块种类，降低了材料生产厂商的难度，提高了生产及现场安装的效率。

幕墙单元体板块使用灯线分离可拆卸式装饰条系统，单元体板块全部预埋线管，工厂提前考虑做好预开孔，现场只装灯具，大大缩减了现场夜景照明施工工作。

由于中信大厦整体幕墙面积庞大（约 12.5 万 m²），需要多台擦窗机协同运行才能满足整栋楼的清洗和幕墙维护功能。中信大厦共设有 9 台擦窗机，分布在 F73 及屋顶两个部位。F73 的四台擦窗机用于 F4 至 F73 大厦圆角部位的幕墙清洁；设置

在屋顶的 5 台擦窗机则用于覆盖整栋建筑的主力擦窗及玻璃单元板更换。

The curtain wall system of CITIC Tower adopts four-layer double hollow Low-E glue sandwich glass, which greatly improves the thermal insulation and energy-saving characteristics of the curtain wall. At the same time, the sound insulation performance of the curtain wall is improved roundly. The ratio of window to wall is 0.44, which is 7.7% less than the national standard. Its air tightness is level 4, water tightness is level 3, wind load pressure resistance is level 6, in-plane deformation is level 4, and thermal performance is level 6 to 8. Finally, the experimental results completely exceed the required performance indicators, and some performance far exceeds the target. It realizes the state of "resisting all the rains and winds and standing firm in all conditions."

During the construction, parameterized analysis is carried out through BIM technology 3D modeling, and curtain wall plates and surface materials are approximately integrated, which greatly reduces the types of plates, reduces the difficulty of material manufacturers, and improves the efficiency of production and site installation.

The detachable decorative strip system with light and line separation is used in the curtain wall unit plate, and all the unit plates are pre-embedded with line pipes. The factory makes pre-opening with only lamps and lanterns are installed on the site, which greatly reduces the construction time of night scene lighting on the site.

Because of the huge curtain wall area of CITIC Tower (about 125,000m²), it needs several Window Scrubbers to work together to meet the cleaning and curtain wall maintenance functions of the entire building. CITIC Tower is equipped with 9 sets Window Scrubbers, which are located on the 73rd floor and on the roof of the building. Four Window Scrubbers on the 73rd floor at the corner of the building are used to clean the curtain wall from the 4th to the 73rd floor. The 5 sets Window Scrubbers installed on the roof are used to cover the main window-cleaning and glass unit plate replacement of the whole building.

中信大厦擦窗机系统
CITIC Tower Window Cleaning System

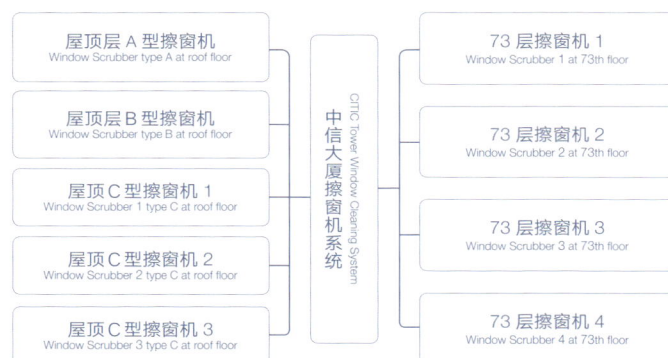

屋顶层 A 型擦窗机 Window Scrubber type A at roof floor		73 层擦窗机 1 Window Scrubber 1 at 73th floor
屋顶层 B 型擦窗机 Window Scrubber type B at roof floor	中信大厦擦窗机系统 CITIC Tower Window Cleaning System	73 层擦窗机 2 Window Scrubber 2 at 73th floor
屋顶 C 型擦窗机 1 Window Scrubber 1 type C at roof floor		73 层擦窗机 3 Window Scrubber 3 at 73th floor
屋顶 C 型擦窗机 2 Window Scrubber 2 type C at roof floor		73 层擦窗机 4 Window Scrubber 4 at 73th floor
屋顶 C 型擦窗机 3 Window Scrubber 3 type C at roof floor		

图片说明 Comments

1．擦窗机
Window Scrubber

2．工作中的擦窗机
Window Scrubber in operation

3．玻璃幕墙
Glass curtain wall

延伸阅读 Links

P184. P244

大数据 *Data*

4-104 层
单元式幕墙楼层位置
Unitized curtain wall floor location

14,400 块
幕墙板块数量
Number of curtain wall plates

幕墙系统 CURTAIN WALL SYSTEM

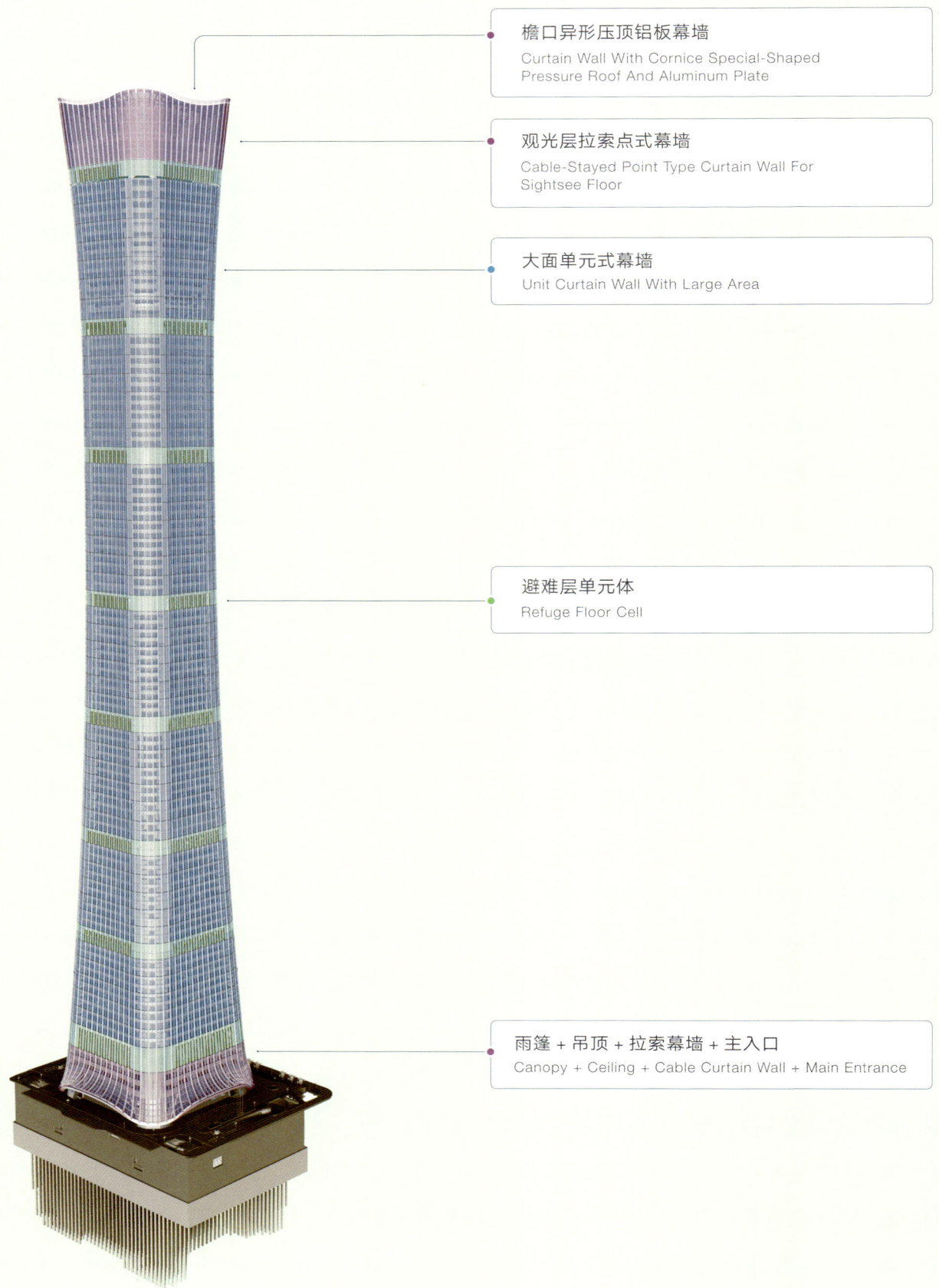

檐口异形压顶铝板幕墙
Curtain Wall With Cornice Special-Shaped
Pressure Roof And Aluminum Plate

观光层拉索点式幕墙
Cable-Stayed Point Type Curtain Wall For
Sightsee Floor

大面单元式幕墙
Unit Curtain Wall With Large Area

避难层单元体
Refuge Floor Cell

雨篷 + 吊顶 + 拉索幕墙 + 主入口
Canopy + Ceiling + Cable Curtain Wall + Main Entrance

双中空幕墙的热工性能表		双中空幕墙剖面图
（6 超白 +1.52PVB+6 超白 Low-E 特调（4 #）+12A+8+12A+8）mm 钢化夹层中空玻璃		
可见光透射比	0.46	
可见光反射比（室内测）	0.21	
可见光反射比（室内测）	0.30	
传热系数（W/m²·K）	1.33	
遮掩系数	0.36	
露点（℃）	- 60	

LED 超薄集成灯盘
Led Ultra-Thin Integrated Light Panel

标准办公区采用 41,500 个 LED 超薄集成灯盘。将照明、空调风口、烟感、喷淋、扬声器、监控摄像头和各类传感器等集成化创新设计，共 11 种模块组合，厚度 47mm，外观品质提升，安装检修便捷。拥有 17 项专利技术，并荣获 2018 第六届阿拉丁神灯奖：优秀产品奖。

LED 超薄集成灯盘光效达到 110lm/W，比普通荧光灯节能 50% 多。

The standard office area uses 41,500 LED ultra-thin integrated light panels. The lighting, air conditioning tuyere, smoke, spray, speaker, surveillance camera and various sensors are integrated and innovative design. A total of 11 modules are combined, the thickness is 47mm, the appearance quality is improved, and the installation and maintenance are convenient. It has 17 patented technologies and won the 2018 6th Aladdin Lamp Award - Outstanding Product Award.

LED ultra-thin integrated lamp panel achieves 110lm/W light efficiency, which is more than 50% energy saving than ordinary fluorescent lamps.

1

2

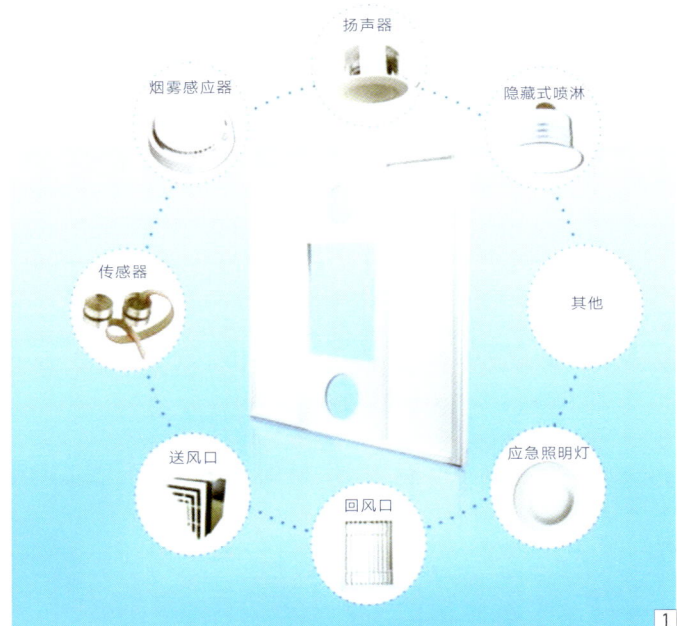

3

大数据 *Data*

41,500 个
超薄集成灯盘
Ultra-thin integrated light panel

11 种
模块功能集成
Module function integration

110 lm/W
LED 超薄集成灯盘光效
LED ultra-thin integrated lamp panel light effect

>50 %
比普通荧光灯节能
Energy saving than ordinary fluorescent lamps

图片说明 *Comments*

1. 11 种模块组合
LED 超薄集成灯盘
11 kinds of module combination
LED ultra-thin integrated lamp panel

2. LED 超薄集成灯盘
LED ultra-thin integrated light panel

3. 施工完成后的集成灯具吊顶
Integrated luminaire ceiling after construction

延伸阅读 *Links*

P188

环保与节能
Environment Protection & Energy Saving

大数据 *Data*

30%
变风量空调系统整体节能
VAV air-conditioning system overall energy saving by 30%

100Pa
风管连接处无需变径，可减少压力损失
The connection of the air duct does not need to be changed in diameter, which makes the pressure loss reduced by 100Pa

80%
贴附射流低温风口压损较通常减少
Less pressure loss at that cryogenic outlet of the attachment jets than usual by 80%

图片说明 *Comments*

1. 变风量系统送风温度控制
Air supply temperature control of VAV system

2. 区域新风和排风
Regional fresh air and exhaust air

3. AHU-Z0-4F-03 冬季运行数据
AHU-Z0-4F-03 winter operation data

4. VAV-FASU 系统
VAV-FASU system

延伸阅读 *Links*

P172. P179

6.5 变风量空调系统
Variable Air Volume (Vav) Air-Conditioning System

变风量空调系统整体节能 30%。

采用了第二代 VAV-BOX，叶轮式风速传感器避免了堵塞测速孔，测量数据稳定且精度高（1~12m/s），可感知微弱风速；可变式多孔叶片使气流稳定，整流效果好，具有线性控制性能；风管无需变径，压力损失小（可减少约 100Pa），运行更节能。

空调风系统末端采用风量平衡一体化送风系统（Flexible Air Supply Unit 简称 FASU），是一种辅助流量分配末端装置，FASU 单元工厂化生产，各支路风量平衡性好，安装方便，节约施工成本，节省工期，也有利于办公区二次装修的空调末端配合调整。

采用防结露、吊顶贴附射流型空调风口，可保证最佳的气流扩散，避免冷风直吹，舒适性更高；且风口压损低（比常规射流型低温风口减少 80%），更加节能。

应用了区域能量平衡控制、空调末端智慧控制、变静压和变送风温度整合控制、动态平衡变流量控制、环形风管系统联合控制和内外区相对独立控制等多项创新技术。

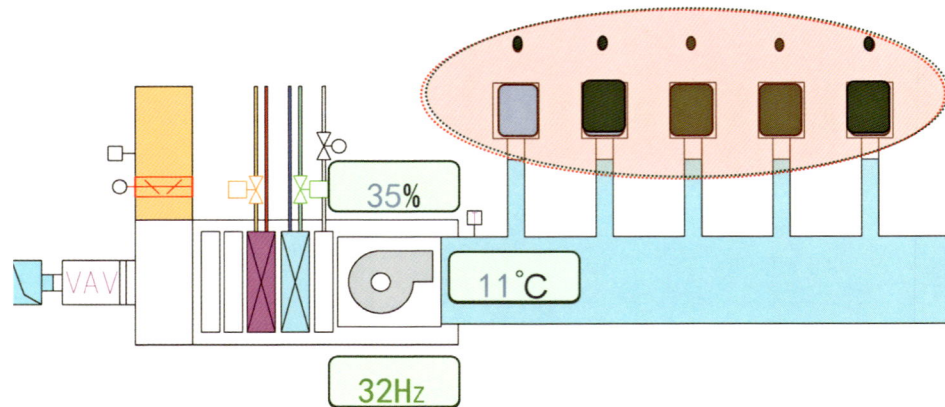

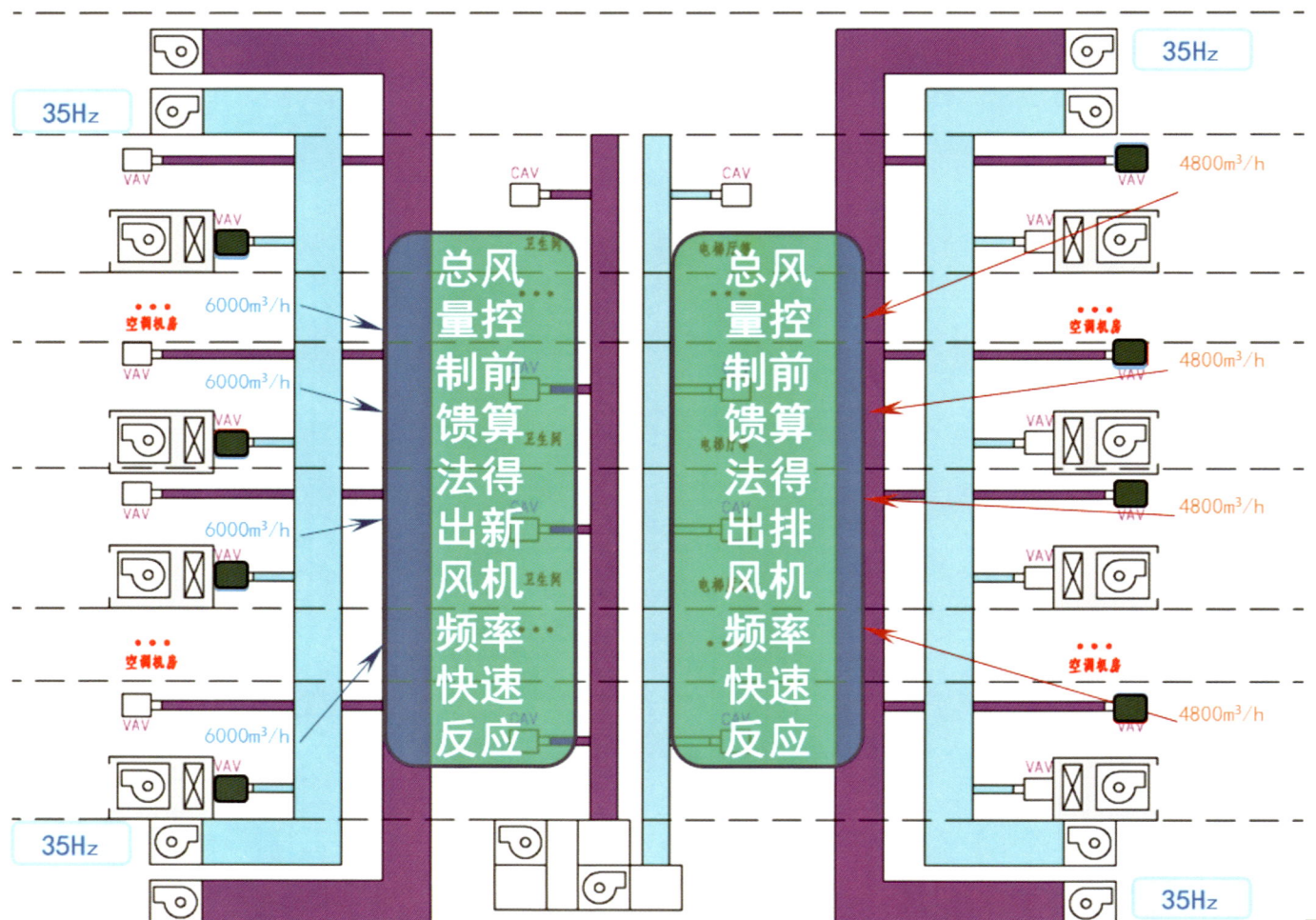

206   *BUILT BY CHINA*   CONSTRUCTION RECORD OF CITIC TOWER

VAV air-conditioning system overall energy saving by 30%

The second generation VAV-BOX is adopted. Its impeller type wind speed sensor avoids plugging the velocity measuring hole. And the measured data are stable and accurate (1-12m/s) to sense the weak wind speed. Variable multi-hole blade makes the air flow stable, well-rectified, and has the linear control performance; The air duct does not need to change in diameter, leading to small the pressure loss (reduced by 100Pa), and more energy-saving in operation.

Air-conditioning wind system terminal adopts Flexible Air Supply Unit (FASU), which is a kind of auxiliary flow distribution terminal device. FASU unit is manufactured. The air flow of each branch is well balanced and easy to install, which saves construction cost and construction period, and is also beneficial to the coordination and adjustment of the air conditioning terminal of the secondary decoration in the office area. Adopting anti-condensation and adhering jet type air outlet on the ceiling can ensure the best air diffusion, avoid the direct blowing of cold air, and improve the comfort of the air conditioner. And the pressure loss of the outlet is low (80% less than that of the conventional jet type outlet), so that the outlet is more energy-saving.

Many innovative technologies are applied, such as regional energy balance control, intelligent control at the air conditioner terminal, integrated control of variable static pressure and variable supply air temperature, dynamic balance and variable flow control, combined control of the annular air duct system and relatively independent control of inner and outer zones.

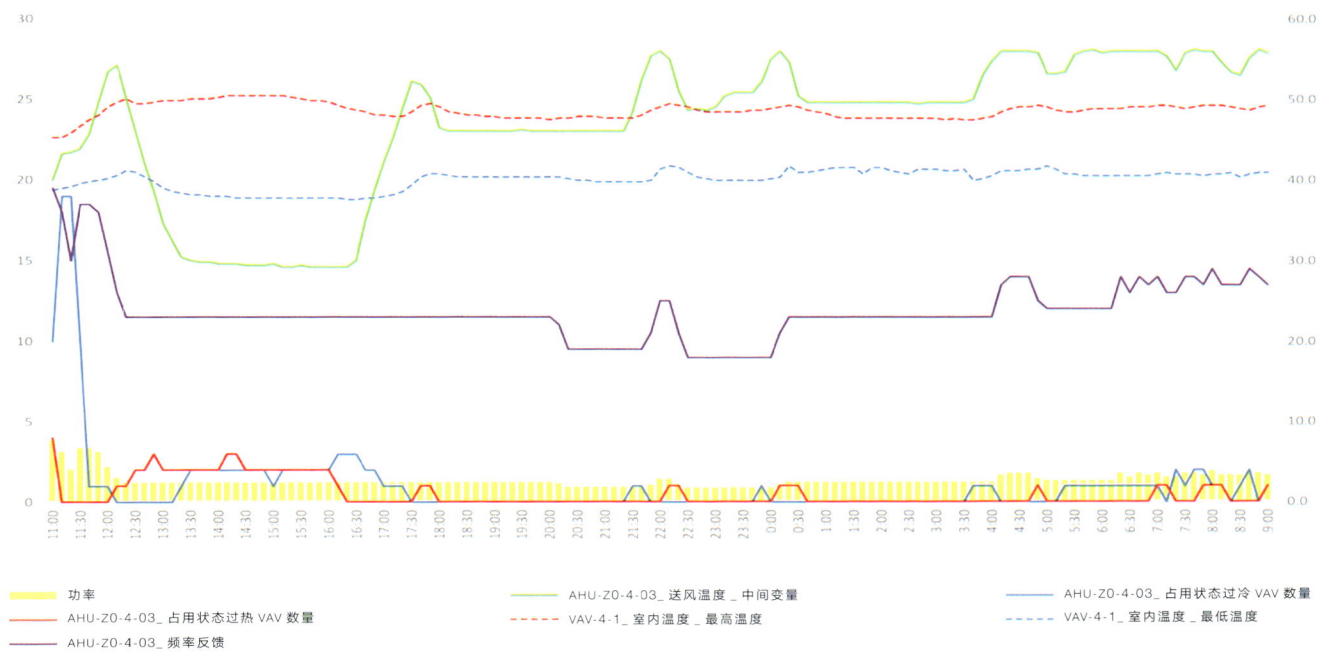

6.6 双轿厢高速电梯
Double-Car High-Speed Elevator

大数据 Data

21 台
双轿厢电梯
double-car elevators

35 %
能量反馈技术的
电能回收再利用率
Reutilization rate of energy recovery
by energy feedback technology

图片说明 Comments

1. 电梯碳纤维带示意图
Schematic diagram of
elevator carbon fiber belt

2. 电梯碳纤维绳
Carbon fiber traction
rope for elevator

延伸阅读 Links

P176. P218. P248

中信大厦配置了 21 台双轿厢电梯，共分为四个组，低层、中层、高层，观光层。这些电梯各自为高、中、低区的办公楼层提供一个便捷的转乘路径，在到达各高度的空中大堂并配合两组区间电梯，轻松把每天数以万计的乘客安全快捷地送到目的地。

其中观光层穿梭电梯是目前世界上提升高度最大的双轿厢电梯，也是利用碳纤维带曳引技术的最高行程电梯。结合了碳纤维带降低电梯总质量，实现了低功耗、高行程、高速度的新型电梯；由于电梯总质量的减少，实现了大楼承重荷载和支撑主体结构体量降低；还降低了能耗，实现了绿色环保。

双轿厢电梯对照单轿厢电梯的优势 ADVANTAGES OF DOUBLE-CAR ELEVATOR OVER SINGLE CAR ELEVATOR		
单轿厢电梯 Single-car elevator		双轿厢电梯 Double-car elevator

优化大楼的垂直交通

增加井道可用空间及改进运输效率

减少电梯井道数量，释放更多的可利用空间

紧凑的通力 EcoDisc 马达使机房布置更具灵活性

优化的派梯系统使客流体验更加完美

与传统单层轿厢相比，通力双层轿厢系统轴距更长，重量匹配更为均衡，乘坐舒适感更佳

双轿厢系统通过扶梯可以分流 50% 左右乘客，避免大厅在上下班高峰时的拥挤

提高大楼整体形象及品质

在同样井道数量下，运载能力更高；相同运载能力要求下，电梯数量更少

基于双轿厢系统运载率高且节省井道的特性，更多的面积可用于商用，带来更大商业价值

能量再生科技可以使电梯更加节能环保

采用能量反馈技术，35% 的电能可以回收再利用，加上双轿厢的高运输效率，减少电梯运行次数，电费更加节省

低流量时，电梯可以变更为单轿厢模式

Optimizing vertical traffic in buildings

Increase the available space of the shaft and improve the transportation efficiency

Reduce the number of elevator shafts and create more usable space.

Compact Kone EcoDisc motors make the layout in machine room more flexible

The optimized elevator system makes passenger flow experience more perfect.

Compared with the traditional single car, longer wheelbase and more balanced weight matching of the Kone double car v-type system make the riding comfort better.

The double-car system can divert about 50% of passengers through the escalator to avoid crowding in the lobby during rush hours.

Improve the overall image and quality of the building.

The carrying capacity is higher at the same number of shafts. Fewer elevators are required with the same capacity requirements.

Because of the high carrying capacity and shaft saving character of the double-car system, more area can be used for commercial use, which brings more commercial value.

Energy recycling technology can make elevators more energy-efficient and environmentally friendly.

Using energy feedback technology, 35% of the electricity can be recycled, coupled with the high transport efficiency of the double-car, reducing the number of elevator operation and saving more electricity costs.

The elevator can be switched into single car mode at low flow.

CITIC Tower is equipped with 21 double-car elevators, which are divided into four groups: low, middle, high rises and sightseeing floors. Each of these elevators provides a convenient interchange route for the office floors in the high, middle and low rise zones, easily delivering thousands of passengers a day to their destinations in the sky lobby at all heights and in conjunction with two sets of section elevators.

The sightseeing shuttle elevator is currently the world's largest double-car elevator, and it is also the highest-travel elevator using carbon fiber belt traction technology. Combined with the carbon fiber belt to reduce the total mass of the elevator, a new elevator with low power consumption, high stroke and high speed is realized; due to the reduction of the total mass of the elevator, the load-bearing load of the building and the volume of the supporting main structure are reduced; and the energy consumption is also reduced. Achieved green environmental protection.

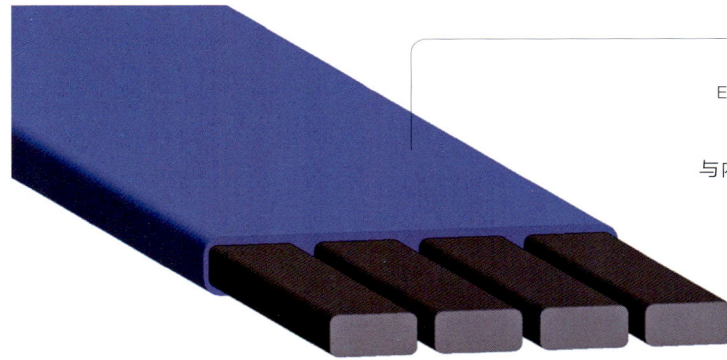

SPECIAL COATING 特殊涂层

保证高摩擦属性
Ensure a high friction resistance property
耐磨损 •
Stand wear and tear
与内置碳纤维结构与材料完美结合 •
Perfectly combine with built-in carbon fiber structure and materials
高摩擦涂层 •
High friction resistance coating

碳纤维曳引绳 CARBON FIBER TRACTION ROPE

STRUCTURE OF PATENT DESIGN 专利设计的结构

• 高强度碳纤维材料
High-strength carbon fiber material
• 高抗拉强度
High tensile strength
• 高强度
High strength
• 优秀的弯曲疲劳属性
Superior bending fatigue property
• 耐高温
High temperature resistanc

碳纤维为矩形结构
Carbon fiber is of rectangular structure
有效保证碳纤维结构的张力传递 •
Effectively ensure the tension transfer of carbon fiber structure

1

双轿厢电梯电梯分区性能表 SUBAREA PERFORMANCE TABLE OF DOUBLE-CAR ELEVATOR				
双轿厢电梯 Double-Car Elevator	低层穿梭梯 Shuttle Elevator In Low Rise	中层穿梭梯 Shuttle Elevator In Middle Rise	高层穿梭梯 Shuttle Elevator In High Rise	观光区穿梭梯 Shuttle Elevator In Sightseeing Area
服务层站 Service Floor	B1M,1 – 31,32	B1M, 1 – 59, 60	B1M, 1 – 90, 91	B1,B1M – 105, 105M
载重 (kg) Load (kg)	1600 x 2	1600 x 2	1600 x 2	1600 x 2
速度 (m/s) Speed (m/s)	7	9	10	10
提升高度 (m) Lifting Height (m)	159	286	429	503
曳引技术 Traction Technology	钢丝绳 Steel Rope	钢丝绳 Steel Rope	碳纤维绳 Carbon Fiber Traction Rope	碳纤维绳 Carbon Fiber Traction Rope
数量（台） Quantity (Set)	6	6	6	3

2

6.7 ALC 条板墙
Alc Slab Wall

大数据 Data

170,000 m²
地上 ALC 条板墙隔墙面积
Above-ground ALC board
wall partition area

图片说明 Comments

1. 施工安装中的 ALC 条板墙
ALC board wall in construction
and installation

中信大厦地上部分隔墙采用 ALC 条板墙，共 170,000m²。其尺寸准确、重量轻，在条板生产过程中，没有污染和危险废物产生。使用时，即使在高温下和火灾中仍无放射性物质和有害气体产生。各个独立的微气泡，使加气混凝土产品具有一定的抗渗性，可防止水和气体的渗透。施工时可大大地减少人力物力投入，能有效增加建筑的使用面积，降低地基造价，减少暖气、空调成本，达到节能效果。板材在安装时多采用干式施工法，大大减少现场湿作业，工艺简便、效率高，可有效地缩短建设工期。

The above-ground partition wall of CITIC Tower adopts ALC board wall with a total area of 170,000m². It is light and accurate in size, and has no pollution and hazardous waste during the production of the board. When used, no radioactive substances or harmful gases are produced, even at high temperatures and in fires. Each independent micro-bubble makes the aerated concrete products have certain impermeability and can prevent the permeation of water and gas. The construction can greatly reduce manpower and material resources input, can effectively increase the usable area of the building, lower the foundation cost, cut down the cost of heating and air conditioning, and achieve energy-saving effect. The dry construction method is often used in the installation of sheet metal, which greatly reduces the wet operation on site, is simple and efficient, and can effectively shorten the construction period.

ALC 条板墙技术参数 ALC BOARD WALL TECHNICAL PARAMETERS				
应用部位 Application site	地上核心筒内（包括边界）及设备避难层防火墙 Above-ground core tube (including boundary) and equipment refuge floor firewall	地上设备机房隔墙 Above-ground equipment machine room partition wall	地上电梯机房隔墙 Above-ground elevator machine room partition wall	地上电梯井道墙 Above-ground elevator shaft wall
燃烧性能 Combustion performance	≥ 4h	≥ 4h	≥ 4h	≥ 4h
抗压强度 Compressive strength	≥ 3.5MPa	≥ 3.5MPa	≥ 3.5MPa	≥ 7.5MPa
容重 Volume weight	≤ 550kg/m³	≤ 550kg/m³	≤ 550kg/m³	≤ 700kg/m³
收缩率 Shrinkage rate	≤ 0.3%	≤ 0.3%	≤ 0.3%	≤ 0.3%
抗冻性 Freezing resistance	质量损失% ≤ 5 Quality loss% ≤ 5 强度损失% ≤ 15 Strength loss% ≤ 15	质量损失% ≤ 5 Quality loss% ≤ 5 强度损失% ≤ 15 Strength loss% ≤ 15	质量损失% ≤ 5 Quality loss% ≤ 5 强度损失% ≤ 15 Strength loss% ≤ 15	质量损失% ≤ 5 Quality loss% ≤ 5 强度损失% ≤ 15 Strength loss% ≤ 15
导热系数 Heat conductivity coefficient	≤ 0.15w/m²	≤ 0.15w/m²	≤ 0.15w/m²	≤ 0.15w/m²
隔声量 Sound reduction index	≥ 50dB	≥ 50dB 需结合隔声做法 ≥ 50dB shall be combined with sound insulation	≥ 50dB 需结合隔声做法 ≥ 50dB shall be combined with sound insulation	≥ 50dB
抗冲击次数 Anti-impact times	≥ 7 次 ≥ 7 Times	≥ 7 次 ≥ 7 Times	≥ 7 次 ≥ 7 Times	≥ 12 次 ≥ 12 Times
抗风压 Wind load resistance				≥ 850Pa
层间位移 Story drift	≥ 1/500	≥ 1/500	≥ 1/500	≥ 1/500

永磁同步变频离心机
Permanent Magnet Synchronous Frequency Conversion Centrifuge

冰蓄冷系统的永磁同步变频离心机经第三方权威机构实测，高区基载冷机的制冷 COP 由设计要求的 5.6 提升为 6.46，提升了 15%，IPLV 由设计要求的 6.2 提升为 9.57，提升了 54%。比普通制冷机组节能 30%，达到国际领先水平。

通过永磁同步变频电机，使电机效率保持在 96% 以上，与定频机组相比节能 25%；

通过"宽频"的压缩机气动设计，实现压缩机在全工况下绝热效率达 86% 以上；

通过高速电机直驱叶轮，结构上取消了一对径向轴承和增速齿轮装置，机械效率达 99% 以上；

通过双级压缩补气增焓循环的设计使制冷循环效率比单级压缩提高 5%；

通过变频器四象限可控整流技术，功率因数高达 0.998，变频器效率达 97%。

The permanent magnet synchronous frequency conversion centrifuge in the ice storage system is measured by the third party authority, and it comes out that the refrigeration COP of the high zone base load cooler are increased from 5.6 to 6.46 by 15% according to the design requirements, and IPLV of that from 6.2 to 9.57 by 54%. Compared with ordinary refrigeration units, the overall energy saving rate is improved by 30%, reaching the international leading level.

Through permanent magnet synchronous frequency conversion electric motor, the electric motor efficiency is kept above 96%, the energy saving is improved by 25% compared with fixed frequency unit;

Through the pneumatic design of "broadband" compressor, the adiabatic efficiency of the compressor can reach more than 86% under all working conditions;

Through the direct drive impeller of high-speed electric motor, a pair of radial bearings and speed-increasing gear devices are eliminated in structure, which makes the mechanical efficiency over 99%;

The efficiency of the refrigeration cycle is increased by 5% compared with that of the single stage compression through the design of the two-stage compression air supply enthalpy increasing cycle;

Through four-quadrant controllable rectifier technology, the power factor is as high as 0.998, and the efficiency of frequency converter is up to 97%;

大数据 Data

30%
比普通定频冰蓄冷双工况机组节能
Ice storage duplex status unit consumes less energy than that of ordinary ones with constant frequency by 30%

25%
比定频机组相比节能
Saving 25% energy compared with constant frequency unit

99%
高速电机直驱叶轮机械效率
Mechanical efficiency of direct drive impeller of high speed electric motor

97%
四象限可控整流技术变频器效率
Four-quadrant controllable rectifier technology makes frequency converter efficiency up to 97%

图片说明 Comments

1．永磁同步变频离心机剖切面
Sectional plane of permanent magnet synchronous frequency conversion centrifuge

2．永磁同步变频离心机房
Permanent magnet synchronous frequency conversion centrifuge machine room

延伸阅读 Links

P172. P179

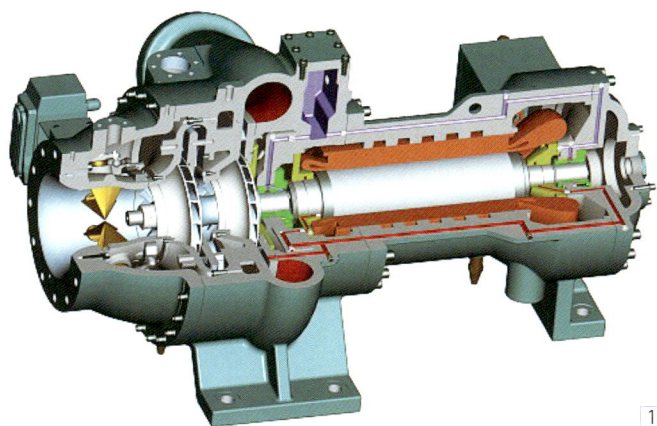

1

2

6.9 光伏发电系统
Photovoltaic Power Generation System

大数据 Data

640 块
145Wp CIGS 薄膜太阳能电池
640 pieces of 145Wp CIGS
thin film solar cell

92.8 kWp
太阳能总装机容量
Total installed capacity of
solar energy is 92.8Wp

15.43 %
光伏组件的光电转换效率
Photovoltaic conversion efficiency
of photovoltaic modules

15.5 %
太阳能光伏板面积占建筑基底
面积比例
Proportion of solar photovoltaic
panel area to building base area

图片说明 Comments

1. 塔冠太阳能光伏板 BIM 图
BIM diagram of solar photovoltaic
plate of tower crown

2. 夕阳里 528m 高空
的太阳能光伏组件
Solar photovoltaic module at
528m high in the sunset

3. 太阳能光伏组件
Solar photovoltaic modules

屋面共铺设 640 块 145Wp 的汉能集团旗下德国 Solibro 原装进口的 CIGS 薄膜太阳能电池，总装机容量为 92.8kWp，占用屋顶投影面积为 900 m²。光伏组件的光电转换效率为 15.43%，在同类产品中转换效率最高。等效太阳能光伏板面积占建筑基底面积比例为 15.5%。

光伏组件贴合中信大厦顶外形环绕排布，与大厦顶部融为一体。组件采用滑轨固定，在 528m 高空可承受更大的风荷载。比传统太阳能组件发电量提高 10% 以上。光伏系统引入大楼物联网系统，实现 24h 监控。

On the roof is in total placed 640 CIGS thin film solar cells of 145 Wp for each with the brand of German Solibro under Hanergy Group imported with the original packing. The total installed capacity is 92.8 kWp, occupying a projected area of 900m². The photovoltaic conversion efficiency of the photovoltaic module is 15.43%, which is the highest among the similar products. Proportion of equivalent solar photovoltaic panel area to building base area is 15.5%.

The photovoltaic modules are arranged around the top of CITIC Tower, and are integrated with the top of it. The assembly is fixed by slide rails and can withstand greater wind loads at an altitude of 528m.

The photovoltaic modules can generate electricity both in the hazy morning and the sunset of the evening, more than 10% increase in power generation capacity compared with the traditional solar modules. The photovoltaic system is introduced into the Internet of Things system of the building to realize 24-hour monitoring.

光伏系统原理图
Grid connection principle diagram of photovoltaic plate

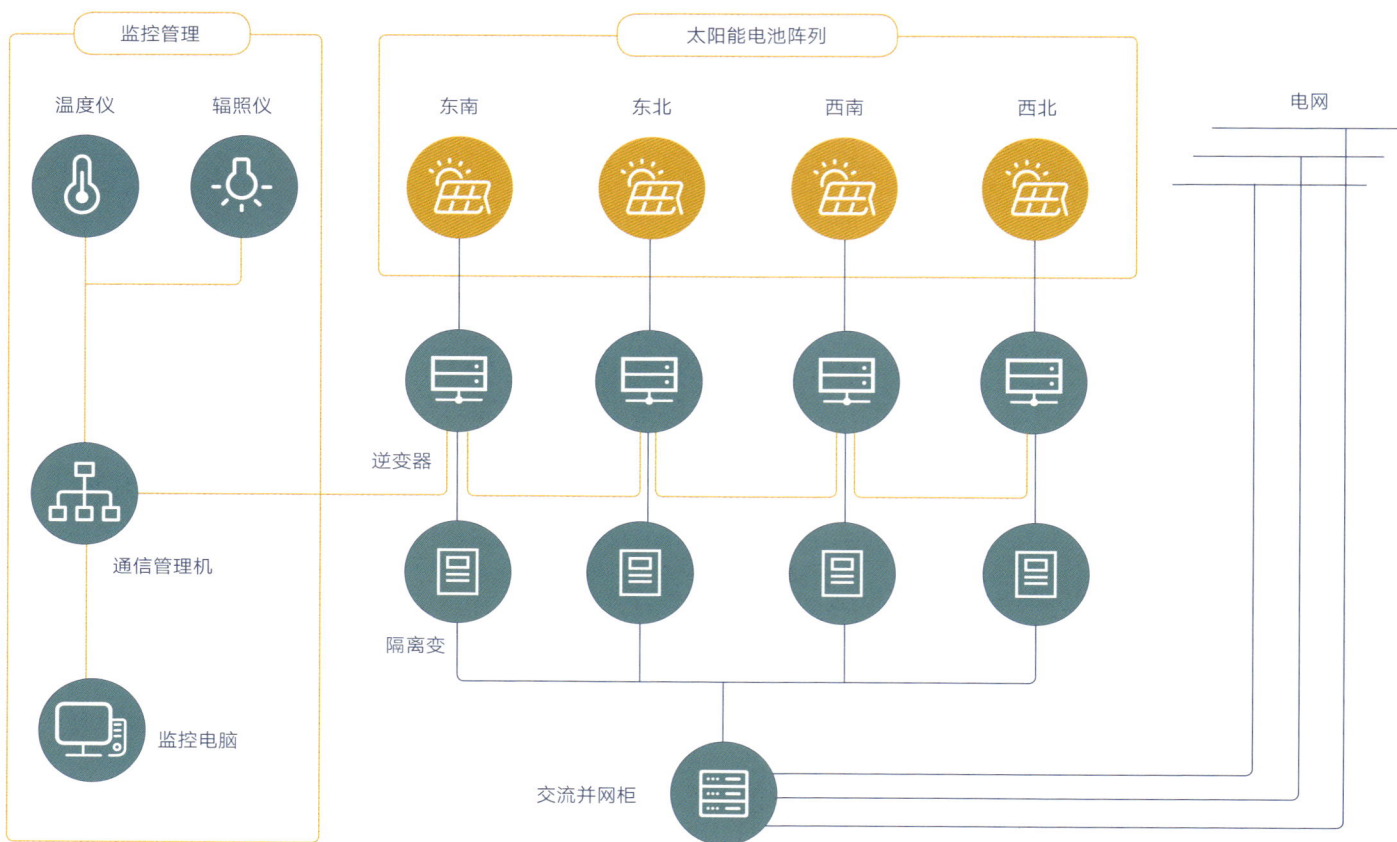

监控管理　　温度仪　辐照仪　　太阳能电池阵列　东南　东北　西南　西北　电网　逆变器　通信管理机　隔离变　监控电脑　交流并网柜

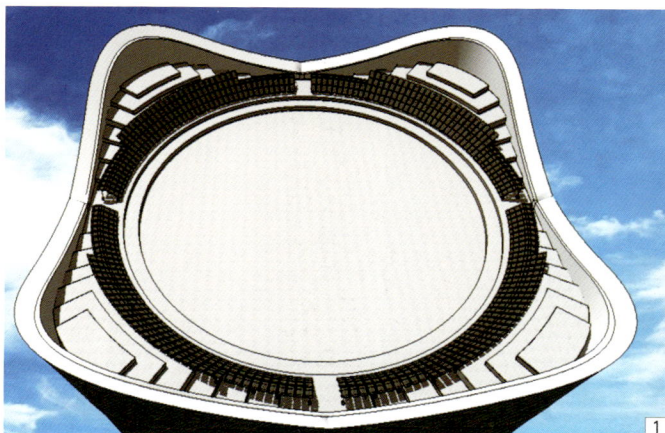

6.10 综合建筑能源管理系统
Integrated Building Energy Management System

中信大厦采用 Azbil 开发的综合建筑能源管理系统（BEMS），对庞大的机电系统进行动态的能耗分析与能效评估，对控制策略提供经济运行优化，提高能源使用效率。该系统共管理 60,000 个数据点；6,000 张分析图表。

内置专家经验，9 大评估系列；流程图导航，逐级分析问题，提供专家经验自动诊断运行设备，评分列表提示，便于快速处理运行数据积累，提供能耗预测空调机组设置智能阀与 BEMS 关联，实现流量控制、传感器和冷热量监测。

CITIC Tower adopts Azbil's Integrated Building Energy Management System (BEMS) to dynamically analysis energy consumption and evaluate the energy efficiency of the huge electromechanical system, and to optimize the economic operation of the control strategy and improve the energy efficiency. The system manages 60,000 data points and 6,000 analysis charts.

Built-in expert experience, 9 evaluation series; Flow diagram navigation, step-by-step analysis of problems, providing expert experience to automatically diagnose operation equipment, scoring list prompts, facilitating rapid processing of operation data accumulation, providing energy consumption prediction, air-conditioning unit setting intelligent valves associated with BEMS, and realizing flow control, sensor and cold and heat monitoring.

BEMS 的系统构成
BEMS compositions

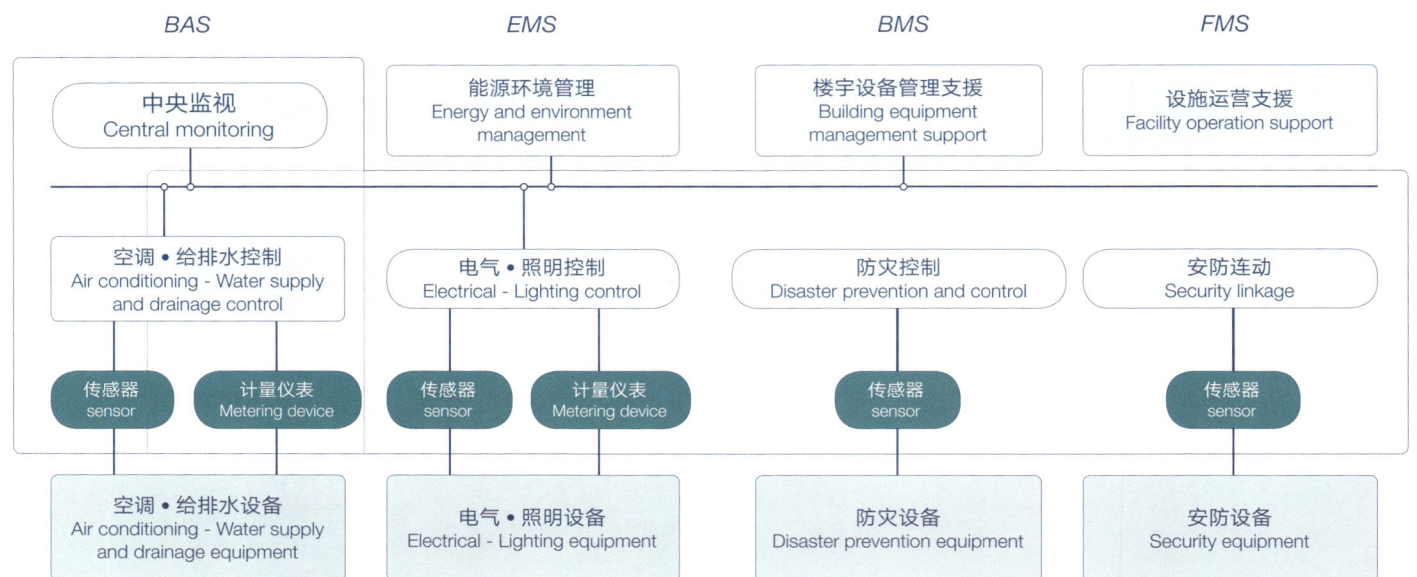

BAS	EMS	BMS	FMS
中央监视 Central monitoring	能源环境管理 Energy and environment management	楼宇设备管理支援 Building equipment management support	设施运营支援 Facility operation support
空调・给排水控制 Air conditioning - Water supply and drainage control	电气・照明控制 Electrical - Lighting control	防灾控制 Disaster prevention and control	安防连动 Security linkage
传感器 sensor / 计量仪表 Metering device	传感器 sensor / 计量仪表 Metering device	传感器 sensor	传感器 sensor
空调・给排水设备 Air conditioning - Water supply and drainage equipment	电气・照明设备 Electrical - Lighting equipment	防灾设备 Disaster prevention equipment	安防设备 Security equipment

建筑能源管理系统 9 大评估系列
Nine evaluation series of building energy management system

能源消费 Energy consumption 评估管理 Evaluation management	室内环境 indoor environment 评估管理 Evaluation management	设备运用 Equipment application 评估管理 Evaluation management
空调控制 Air conditioning control 评估管理 Evaluation management	热源控制 Heat source control 评估管理 Evaluation management	对策与验证 Countermeasures and verification 评估管理 Evaluation management
系统故障 System failure 评估管理 Evaluation management	蓄冰专项 Ice storage project 评估管理 Evaluation management	自然冷却 Natural cooling 评估管理 Evaluation management

能耗管理系统
Energy consumption management system

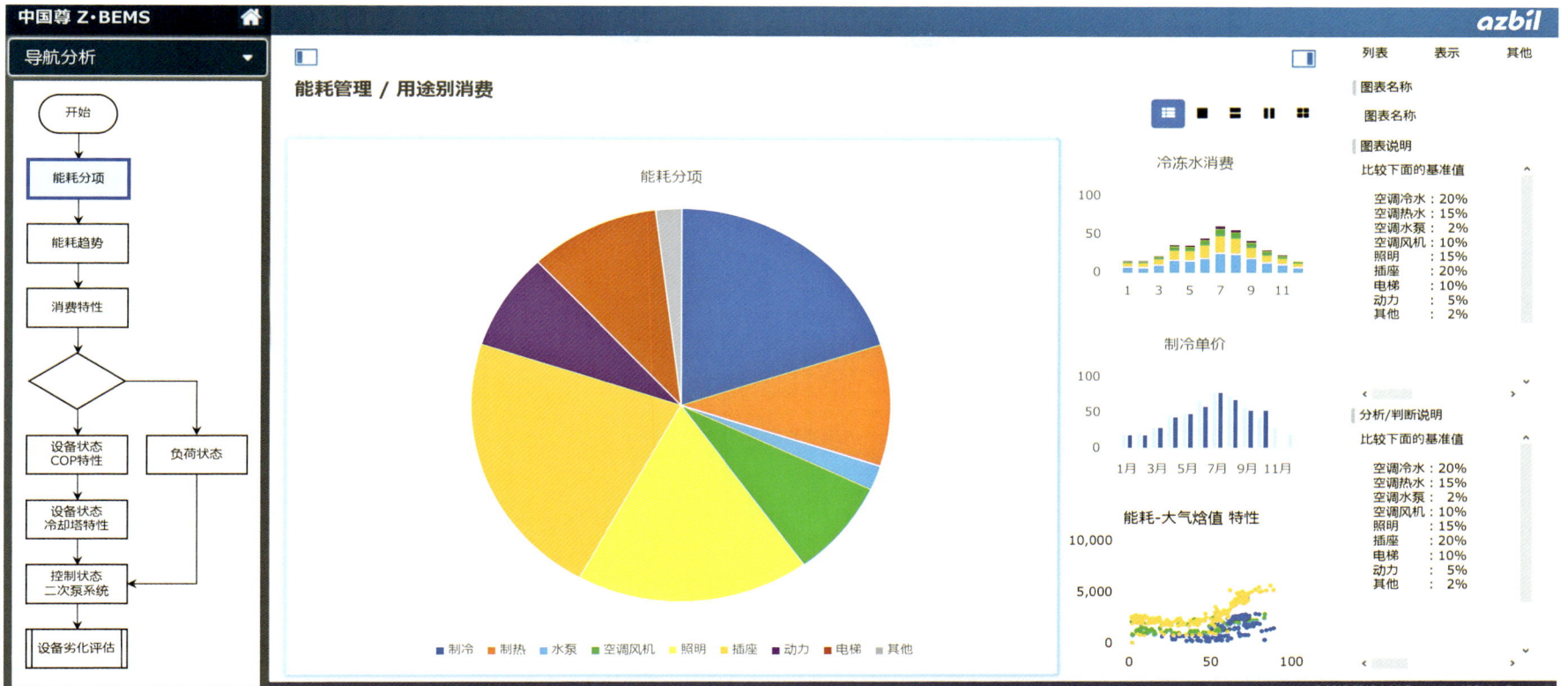

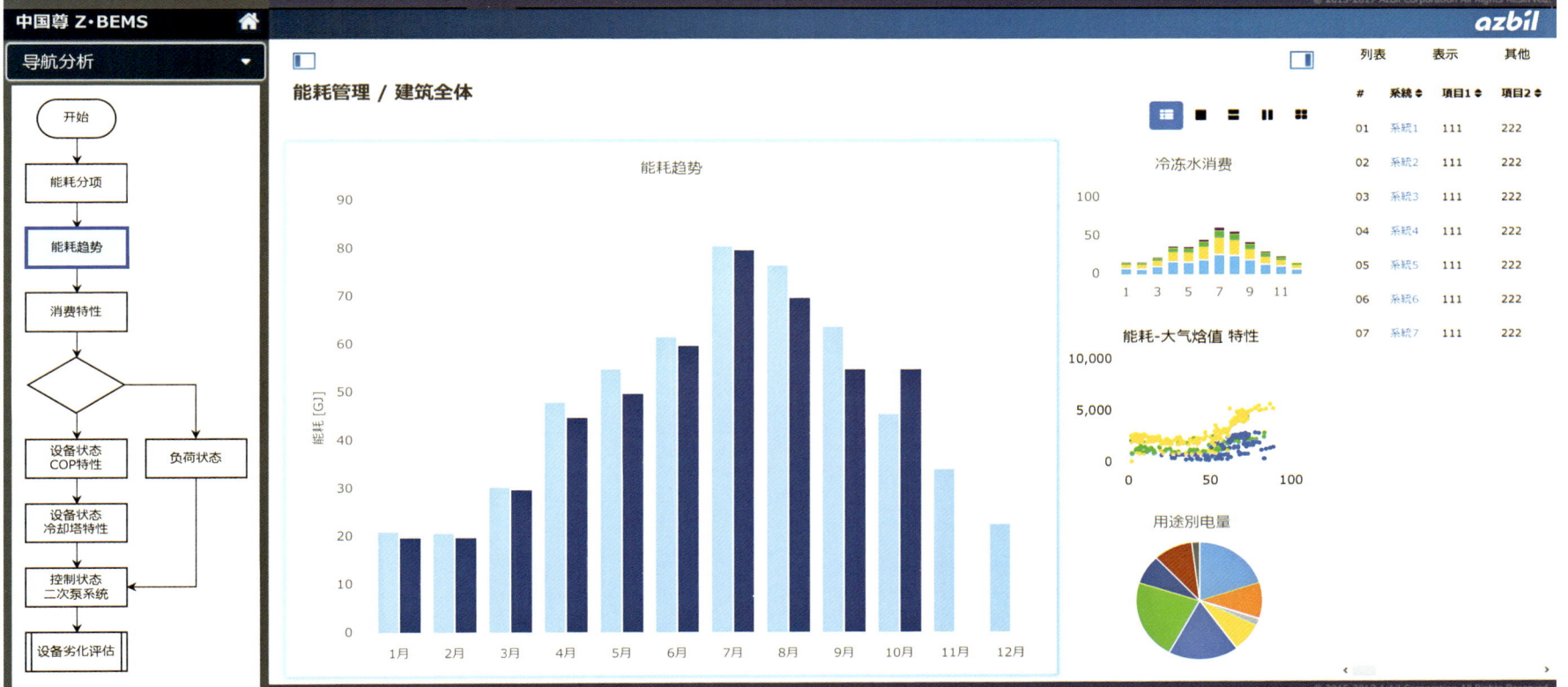

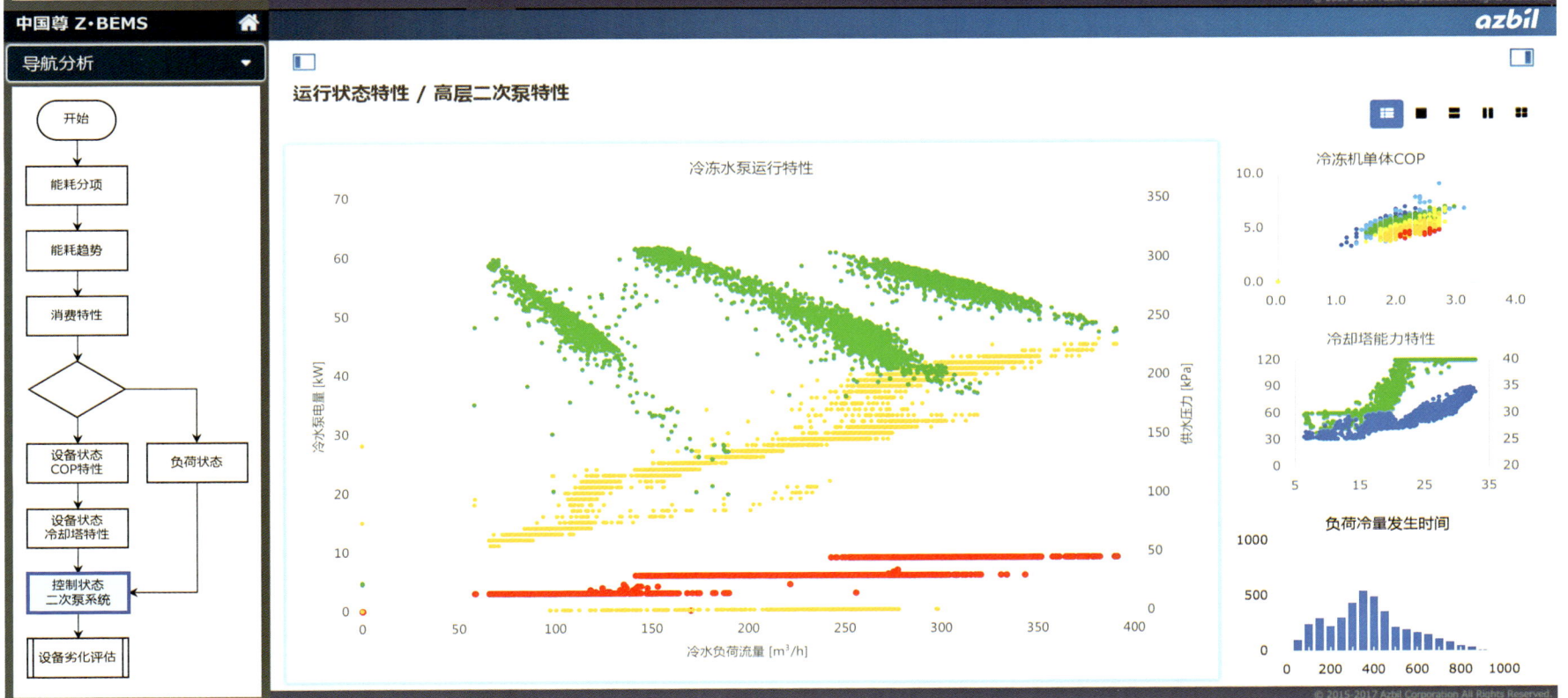

BUILT BY CHINA — CONSTRUCTION RECORD OF CITIC TOWER

7.1 抗震抗风措施
Anti-Seismic And Anti-Wind Measures

双重抗侧力体系和连梁阻尼器

为满足结构抗震与抗风的技术要求，中信大厦在结构上采用含有巨型柱、巨型斜撑及转换桁架的外框筒，以及含有组合钢板剪力墙的核心筒，形成内筒和外筒的双重抗侧力体系，可做到小震不坏、中震可修、大震不倒。同时，考虑到北京春、秋、冬季刚劲的风，设计中充分使用了有利的空气动力形状，可有效地减少风荷载。

中信大厦在 F59 至 F72，F74 至 F86，F89 至 F102 设置共 160 组软钢连梁阻尼器，以增加结构的耗能性。

电梯自动晃动控制器

轿厢上的 Base Isolation(智能基础隔震系统)技术采用了 4 个可调节的带有阻尼吸能装置的摆动系统，位于在轿厢每个角的底部。该摆动系统的调节长度可根据需要的频率来调节以达到最优化的阻尼。该系统也保证了轿厢地面的水平度达到一个完美的程度，由此可大大减少保持轿厢平衡的稳定作用力。

轿厢顶部稳定作用力的减少以及摆动系统上极其有限的作用力将持续、大幅降低传递到轿厢中的作用力，从而大大降低轿厢的水平振动（与地震类似）。这个系统在大幅度振动及小幅度振动中均能稳定工作。当电梯上下客时，该系统将保持锁定以保证轿厢内的稳定。系统另有一个额外的可调重量阻尼吸能装置来减小由于摆动频率共振造成的移动，其稳定的系统结构无需后续定期调整及维护，并为中信大厦超高速电梯的水平减震带来非常理想的效果。当摆幅过大时，电梯

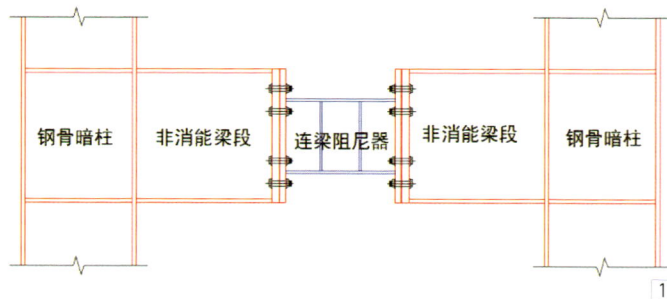

1

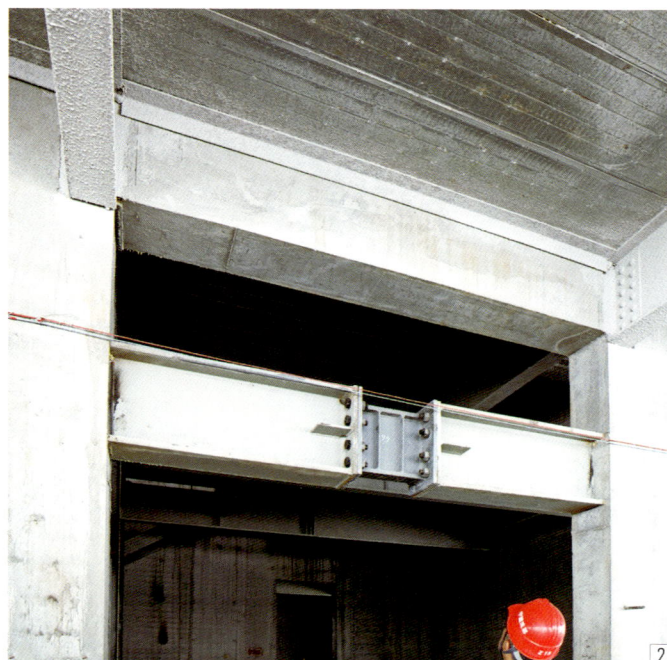

2

控制器将自动降速，如摆幅达到设计最大值时，电梯将在最近楼层停靠。

防摆钢梁

钢丝绳和电缆保护装置可防止它们在某些区域被钩住的危险。它还可以防止钢丝绳的过度磨损。实际需要安装的部件将依电梯布置图而定，包括在井道中的结构和位置。

Double lateral force system and connecting beam damper

In order to meet the technical requirements of structural anti-seismic and anti-wind, CITIC Tower adopts the outer frame tube with colossal columns, colossal diagonal braces and transfer trusses, and the core tube with composite steel plate shear wall to form the dual lateral force resisting system of inner tube and outer tube, to ensure that they will not break in small earthquakes, can be repaired in medium earthquakes, and will not be collapsed in large earthquakes. At the same time, considering the strong winds in spring, autumn and winter in Beijing, the favorable aerodynamic shape is fully used in the design, which can effectively reduce the wind load.

In order to increase the energy dissipation of the structure, 160 sets of mild steel coupling beam damper are installed on the 59 to 72 floors, 74 to 86 floors and 89 to 102 floors of CITIC Tower.

Elevator automatic sloshing controller

The Base Isolation technology in the car uses four adjustable swing systems with damping and energy absorption devices located at the bottom of each corner of the car. The adjustable length of the swing system can be adjusted according to the desired frequency to achieve the optimal damping. The system also ensures that the level of the car floor reaches a perfect degree, thereby greatly reducing the stabilizing force required to maintain the car balance. The reduction of the stabilizing force on the top of the car and the extremely limited force on the swing system will continue substantially reducing the force transmitted into the car, thereby greatly reducing the horizontal vibration of the car (similar to the earthquake). The system can work stably in both large and small amplitude vibration. When the elevator is loaded and unloaded, the system will remain locked to ensure stability in the car. The system also has an additional adjustable weight damper energy absorption device to reduce the movement caused by the oscillation frequency resonance, and its stable system structure does not need to be adjusted and maintained regularly, which brings a very ideal effect of the horizontal vibration reduction of the super-high-speed elevator of CITIC Tower. When the swing is too large, the elevator controller will automatically slow down, and if the swing reaches the design maximum, the elevator will stop at the nearest floor.

Anti-swing steel beam

Steel ropes and cable protectors prevent the danger of them being hooked in certain areas. It also prevents excessive wear of the steel rope. The actual components to be installed will depend on the elevator layout, including the structure and location of the shaft.

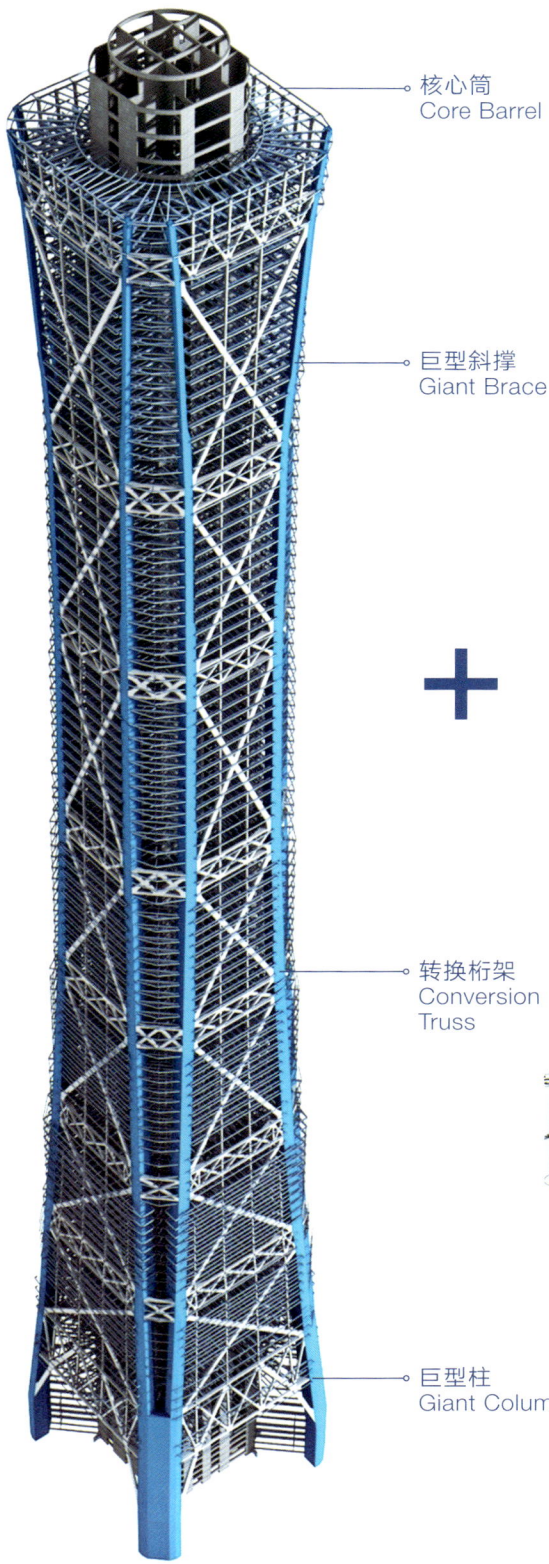

核心筒
Core Barrel

巨型斜撑
Giant Brace

转换桁架
Conversion Truss

巨型柱
Giant Colum

主体结构
Main Structure

+

塔冠 Tower Crown

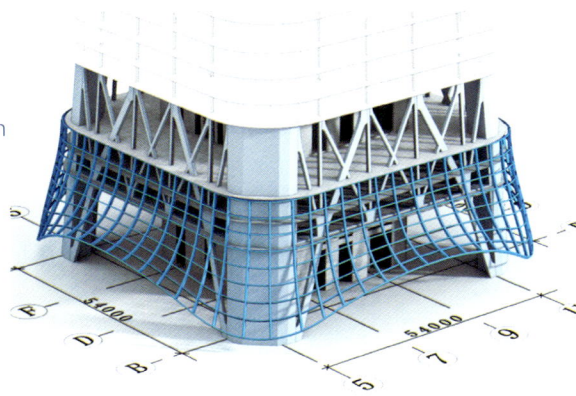

雨篷 Awning

=

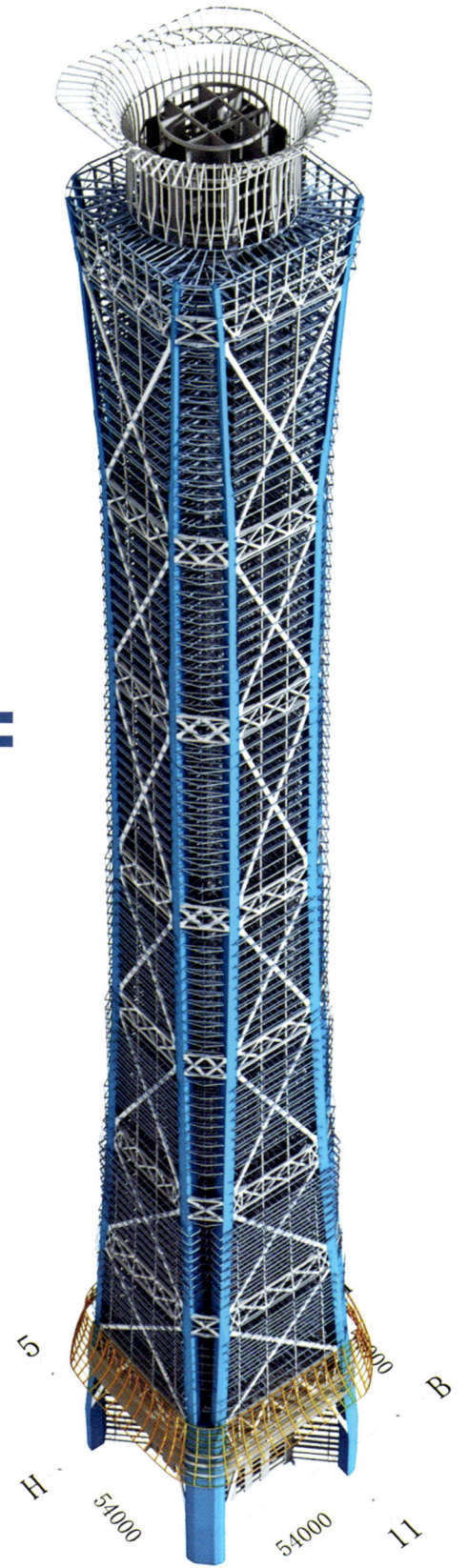

整体效果
Overall Effect

3

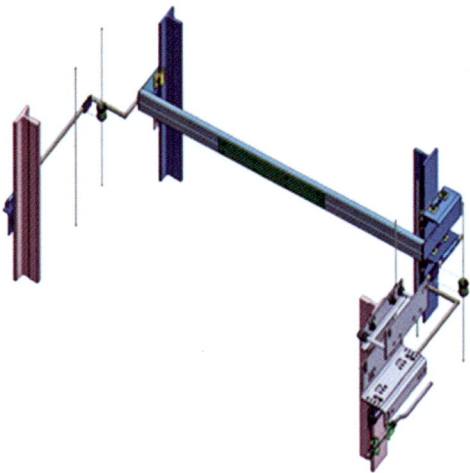

4

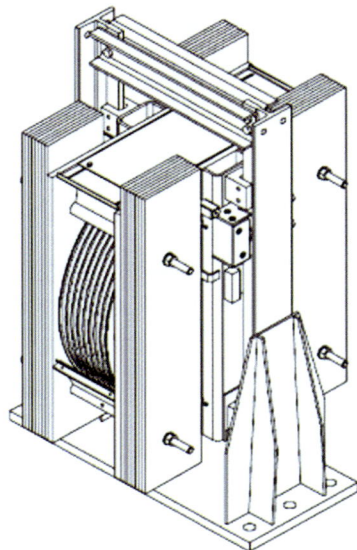

5

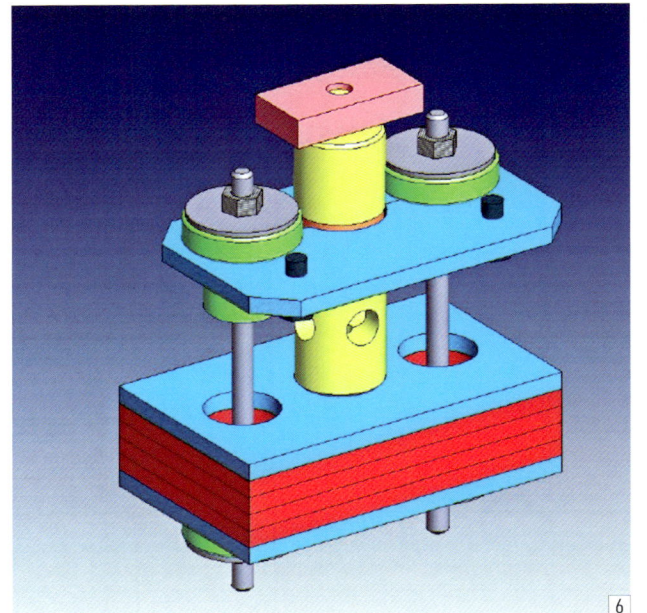

6

大数据 Data

4 类
消防设施体系
4 categories of fire-
fighting facility system

>20 个
消防设施体系子系统
Fire-fighting facility subsystems

45,000 个
火灾报警系统巡检点
fire alarm system patrol points

22,588 m²
避难层建筑总面积
Total building area of refuge floor

图片说明 Comments

1. 避难层分区楼层及面积
Subarea floor and area
of refuge floor

延伸阅读 Links

P090. P130. P135. P236

中信大厦属一类高层建筑，耐火等级一级。

建筑消防系统

1. 首层平面：大厦东侧、西侧、北侧均设置有消防车可以进出的出入口，且建筑北侧、东侧、西侧均布置可供消防车通行的 U 形环路，消防车道的宽度和净空高度满足规范要求，消防车道可以允许博浪涛 101m 登高平台消防车通行。

2. 避难层：根据建筑的功能分区，结合机电系统及消防要求，中信大厦在每区段底部结合结构腰桁架位置设置了为区段服务的设备区及避难区。设备、避难区共 8 个，避难区层高 3.5 m。总建筑面积为 22,588m²。在第一个避难层设置可供人员出入的救援通道，可以由室外进入避难层。避难区四周采用防火墙与设备区分隔开，设置正压送风。避难区设置采光窗，节约能源并减少疏散人员的心理压力。设置消防安防控制室或预留消防安防控制室及消防设施存放空间，为消防分区控制及救援扑救服务。各区段避难区均被核心筒分为两个区域，每个区域有 2 部疏散楼梯，2 个避难区可通过走廊连接，增加疏散的灵活性。所有装修材料均采用 A 级不燃材料。

3. 屋顶：屋顶的停机坪可以满足消防局的直升机运行要求，可以进行空中救援。停机坪直径大于 25m，设两个净宽度不

小于 0.9m 的疏散楼梯与建筑主体连接，并有两个设置了防冻措施的消火栓和主楼消火栓系统相连接。

4. 消防电梯：地上部分消防电梯设置严格按照每层建筑面积大于 1,500m² 但不大于 4,500m² 设置 2 部消防电梯，超过 4,500m² 的设置不少于 3 部消防电梯的原则设计。地下室共有 4 部消防电梯到达。B002、B003 至 B006 设置 1 部消防电梯。同时，场地北侧两部后勤服务电梯按照消防电梯级别设计，为消防救援提供可选择的竖向通道。

本项目由于消防电梯运行距离超长，电梯对于冲顶空间及机房空间的需求也不同于普通电梯。为防止电梯机房高出屋顶停机坪，本项目消防电梯在第 8 个避难层 F104 转换。转换后消防人员从首层到达 F108 的时间为 119.5s。配合服务电梯载重量需求，消防电梯的载重量都不低于 1,600kg。

5. 消防水系统：消防供水设施设置在 F103，水箱有效容积 690m³，B7 至 F96 采用重力常高压供水，F97 至 F108 采用临时高压供水。消防水箱补水采用生活补水和消防转输补水相结合的方式，在 B1/F18/F44/F74 设置消防转输供水设施。消防水系统涵盖了湿式水喷淋系统、预作用水喷淋系统、消火栓系统、IG541 气体灭火系统、大空间水炮灭火系统、屋

避难层 REFUGE FLOORS

R8	F104	3,186m²
R7	F088	2,428m²
R6	F074	2,235m²
R5	F058	2,485m²
R4	F044	2,861m²
R3	F030	3,264m²
R2	F018	3,732m²
R1	F006	2,397m²

		建筑构件材料防火措施 FIRE PROTECTION MEASURES FOR BUILDING COMPONENT MATERIALS		
	建筑构件	现行规范要求	性能化设计加强措施	本项目采用材料及其燃烧性能
1	防火墙	不燃烧体 3h		150 厚加气混凝土砌块墙，4h 150 厚加气混凝土板，4h 150 厚轻钢龙骨防火石膏板，3h
2	楼梯间墙	不燃烧体 2h	不燃烧体 2.5h	150 厚加气混凝土板，4h
3	电梯井道墙	不燃烧体 2h		200 厚加气混凝土板，4h
4	疏散走道两侧墙	不燃烧体 1h	地上防火分区大于 2,000m² 时， 采用耐火极限不小于 2h 的隔墙及乙级防火门设计	150 厚轻钢龙骨石膏板，1.5h 150 厚轻钢龙骨防火石膏板，2h
5	房间隔墙	不燃烧体 0.75h	地上防火分区大于 2,000m² 时， 采用耐火极限不小于 2h 的隔墙及乙级防火门设计	150 厚轻钢龙骨石膏板，1.5h 150 厚轻钢龙骨防火石膏板，2h
6	柱	不燃烧体 3h		
7	巨型外框筒		不燃烧体 3h	
8	梁	不燃烧体 2h	不燃烧体 2.5h	
9	楼板	不燃烧体 1.5h	不燃烧体 2h	
10	疏散楼梯	不燃烧体 1.5h		
11	屋顶承重构件	不燃烧体 1.5h		
12	吊顶	不燃烧体 0.25h		

顶停机坪低倍泡沫系统、柴油发电机房水喷雾系统。

6. 消防电系统：消防电系统设置 BIM 防控制中心、F30 和 F104 分控中心，采用双链路环形网络架构。消防电系统涵盖火灾自动报警系统、消防应急广播系统、消防直通电话系统、电气火灾监控系统、消防设备电源监控系统、防火门监控系统、智能疏散指示系统。

结构防火

钢结构防火涂料：钢结构构件耐火极限按照规范设计，楼板不小于 2h，柱不小于 3h，外框筒体系不小于 3h，楼梯梯板及休息平台板不小于 1.5h。主体钢结构体系防火涂料均采用厚涂型防火涂料。

CITIC Tower is a class I high-rise building, refractory grade 1.

Architectural fire protection systems:

1. First floor plane: The east, west and north sides of the building are provided with entrances and exits for fire engines, and the north, east and west sides of the building are provided with U-shaped loops for fire engines to pass through. The width and clearance height of the fire engines' lanes meet the requirements of the code. The fire engines' lanes can allow the fire engines on the 101m high platform of Bronto Airlift to pass through.

2. Refuge floor: According to the functional subarea of the building, combined with the electromechanical system and fire control requirements, CITIC Tower has set up the equipment area and refuge area serving for the section at the bottom of each section combined with the waist truss position of the structure. There are a total of 8 equipment and refuge areas, and the latter is 3.5m high in storey. Total building area of refuge floor is 22,588m². The first refuge floor is provided with a rescue passage for access, which can be accessed from outside. Around the refuge area, the firewall is used to separate the refuge area from the equipment area, and the positive pressure air supply is set. The refuge area is equipped with lighting windows to save energy and reduce the psychological pressure of evacuees. Set up the fire safety control room or reserve the storage space of the fire safety control room and fire protection facilities for fire subarea control and rescue services.
Each section of the refuge area is divided into two areas by the core tube with each has two evacuation stairs, and the two refuge areas can be connected through the corridor to increase the flexibility of evacuation. All decoration materials are made of Grade A non-combustible materials.

3. Rooftop: The apron on the roof can meet the operational requirements of helicopters put into use by Beijing Fire Bureau and can be used for air rescue. The diameter of the apron is greater than 25m. Two evacuation staircases with a net width of not less than 0.9m shall be connected with the main body of the building. Two fire hydrants with anti-freezing measures shall be connected with the fire hydrant system of the main building.

4. Fire elevator: Fire elevator setting: The above-ground part of the fire-fighting elevators shall be designed strictly in accordance with the principle that 2 fire-fighting elevators shall be set up on each floor with a floor area greater than 1,500m² but not greater than 4,500m², and no less than three fire elevators shall be set up on each floor with a floor area greater than 4,500m². There are 4 fire elevators in the basement reaching B002, and 1 fire elevator is set in B003-B006. At the same time, two logistic service elevators on the north side of the site are designed according to the level of the fire elevator to provide alternative vertical access for fire rescue.

5. Fire water system：The fire water supply facility is located on the F103 floor, the effective volume of the water tank is 690m³, the B7-F96 layer is gravity-based high-pressure water supply, and the F97-F108 layer is temporarily high-pressure water supply. The fire water tank hydration is combined with the life water supply and the fire water transfer and hydration, and the fire transfer water supply facilities are installed on the B1/F18/F44/74 floor. The fire water system covers the wet water sprinkler system, the pre-action water sprinkler system, the fire hydrant system, the IG541 gas fire extinguishing system, the large space water cannon fire extinguishing system, the roof apron low-expansion foam system, and the diesel generator house water spray system.

6. Fire-fighting system：The fire-fighting power system is equipped with a BIM control center, F30 and F104 sub-control centers, and adopts a dual-link ring network architecture. The fire-fighting system covers fire automatic alarm system, fire emergency broadcast system, fire-fighting direct telephone system, electrical fire monitoring system, fire equipment power monitoring system, fire door monitoring system, and intelligent evacuation indicator system.

Because of the extra-long running distance of the fire elevator in this project, the requirements of the elevator for over-running and machine room space are different from those of the ordinary elevator. In order to prevent the elevator machine room from rising above the roof apron, the fire elevator of this project is converted at the eighth refuge F104. The time from the first floor to the 108th floor was 119.5 seconds after conversion. In accordance with the load requirements of service elevators, the load capacity of fire elevators shall not be less than 1,600kg.

Structure fire protection:

Fire-proof coating on steel structures: The fire resistance limit of the steel structure components shall be designed according to the code. The floor slab shall not be less than 2 hours, the column shall not be less than 3 hours, the outer frame-tube system shall not be less than 3 hours, and the staircase ladder plate and the rest platform plate shall not be less than 1.5 hours. Thick fire retardant coatings are used in the main steel structure system.

7.3 数字安防
Digital Security

大数据 *Data*

1,689 个
出入口管理系统
Access control systems

1,776 个
视频监控系统
Video surveillance systems

840 个
红外探测器
Infrared detectors

延伸阅读 *Links*

P088. P089. P194. P232

中信大厦通过三个保障措施构建数字安防体系。

出入口管理系统（1,689 个门禁、速通门、车辆道闸、阻车器、安检设备等）和人脸识别系统实现对大楼的物理分区和隔离。

视频监控系统的 1,776 个各类摄像机和入侵报警系统的 840 个红外探测器对大楼出入口、周界、大堂、停车场、公共区域、重要设备机房等部位进行全覆盖实时监控和报警。

在线及离线式巡更设备相结合，安保人员定期按照巡更路线巡逻，机动灵活处理突发事件。

CITIC Tower builds a digital security system through three safeguards.

The access management system (1,689 access control machines, turnstiles, vehicle brakes, car arresters, security equipment, etc.) and the face recognition system provide physical partitioning and isolation of the building.

In the video surveillance system, the 1,776 video surveillance cameras and 840 infrared detectors of the intrusion alarm system provide real-time monitoring and alarm coverage of the entrance and exit of the building, perimeter, lobby, parking lot, public area, and critical equipment machine rooms.

Combining on-line and off-line patrol equipment, security personnel regularly patrol along routes and handle emergencies flexibly.

安防系统工作原理图
Security system working principle diagram

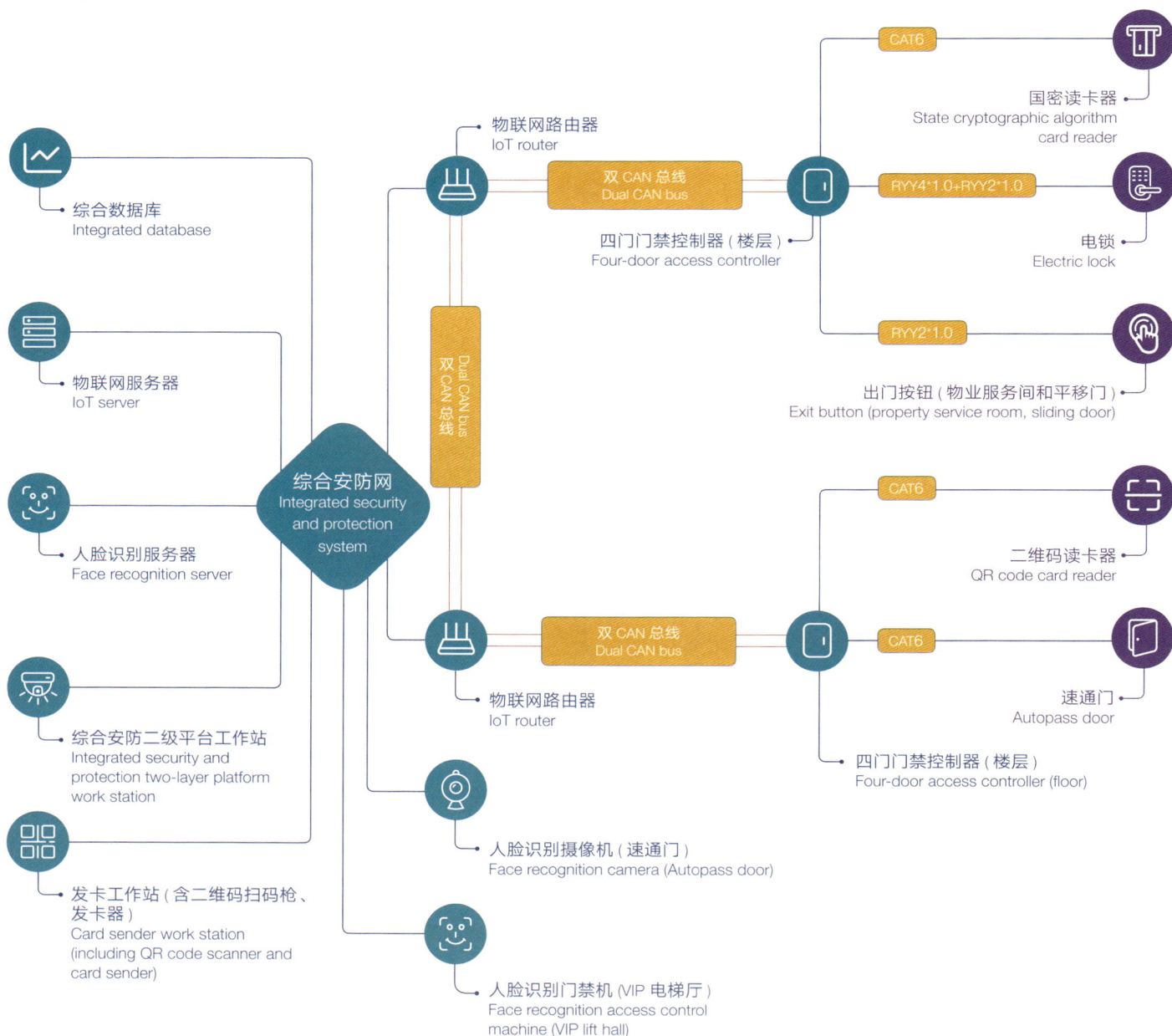

综合数据库
Integrated database

物联网服务器
IoT server

人脸识别服务器
Face recognition server

综合安防二级平台工作站
Integrated security and protection two-layer platform work station

发卡工作站（含二维码扫码枪、发卡器）
Card sender work station (including QR code scanner and card sender)

综合安防网
Integrated security and protection system

物联网路由器
IoT router

物联网路由器
IoT router

双 CAN 总线
Dual CAN bus

双 CAN 总线
Dual CAN bus

人脸识别摄像机（速通门）
Face recognition camera (Autopass door)

人脸识别门禁机（VIP 电梯厅）
Face recognition access control machine (VIP lift hall)

四门门禁控制器（楼层）
Four-door access controller

四门门禁控制器（楼层）
Four-door access controller (floor)

CAT6

国密读卡器
State cryptographic algorithm card reader

RYY4*1.0+RYY2*1.0

电锁
Electric lock

RYY2*1.0

出门按钮（物业服务间和平移门）
Exit button (property service room, sliding door)

CAT6

二维码读卡器
QR code card reader

CAT6

速通门
Autopass door

气密性加强措施
Airtightness Strengthening Measure

中信大厦是全球北纬 39° 以北地区的最高建筑。北方冬季室内外温差较大，对于超高层建筑而言，因建筑室内外温差较大，易产生烟囱和活塞效应，出现啸叫、电梯故障、建筑内部的穿堂风、门猛烈开合、紧急逃生通道阻塞等问题。

为减弱中信大厦的烟囱和活塞效应，在满足消防和功能要求的前提下，设计上针对重点楼层，建立风环境分区，定义分区属性，尽量减少分区边界上开门的数量，减少不必要的连通。根据位置、功能、气密性重要程度、产品性能等，选用适当的门型，例如普通气密门、防火气密门、旋转门、气密平滑自动门、应急平滑自动门等，少数薄弱的部位提出双门互锁、常闭等特殊要求；通过严控幕墙的气密性、优化节点构造做法，提高大厦外部围护系统气密性能；对电梯提出开关性能的要求以及防止啸叫现象的产生；通过加强通风系统的密封、控制阀门质量、对电梯井道进行温控、对给排水系统的预留管道进行封闭、加设地漏盖板等措施，提高机电系统的气密性；对幕墙与楼板结合部位、穿越各类保护区的管线洞口进行封堵。

CITIC Tower is the tallest building in the world of area beyond 39 degrees north latitude. The north is hot in summer and cold in winter, and the temperature difference between indoor and outdoor is large in winter. For super high-rise buildings, because of the large temperature difference between indoor and outdoor, stack effect is easy to occur, such as whistling, elevator failure, cross ventilation in the building, fierce opening and closing of doors, and blocking of emergency escape channels.

In order to weaken the stack effect of CITIC Tower, on the premise of meeting the requirements of fire control and function, the wind environment subarea is established for the key floors on the design, the zoning attribute is defined, the number of doors on the zoning boundary is reduced as far as possible, and the unnecessary connection is reduced. According to the position, function, the importance of air tightness and product performance, appropriate door types are selected, such as general air tight doors, fire-proof air tight doors, rotary doors, air tight smooth automatic doors, and emergency smooth automatic doors. A few weak parts have special requirements such as double door interlocking and normally closed; the airtightness of the exterior envelope system of the building can be improved by controlling the airtightness of the curtain wall and optimizing the joint construction; the requirements regarding switch performance of elevator are put forward to prevent the occurrence of whistling phenomenon; The air tightness of electromechanical system can be improved by strengthening the seal of ventilation system, controlling the quality of valves, controlling the temperature of elevator shafts, sealing the reserved pipe of water supply and drainage system, and installing the floor leakage cover; And the joints of curtain wall and floor and the convenient pipeline openings through all kinds of protected areas are blocked.

大数据 Data

10 倍

热膨胀率不燃性型封堵材料
Thermal expansion rate of non-combustible plugging material is improved by 10 times

图片说明 Comments

1. 风环境分区
Wind environment subarea

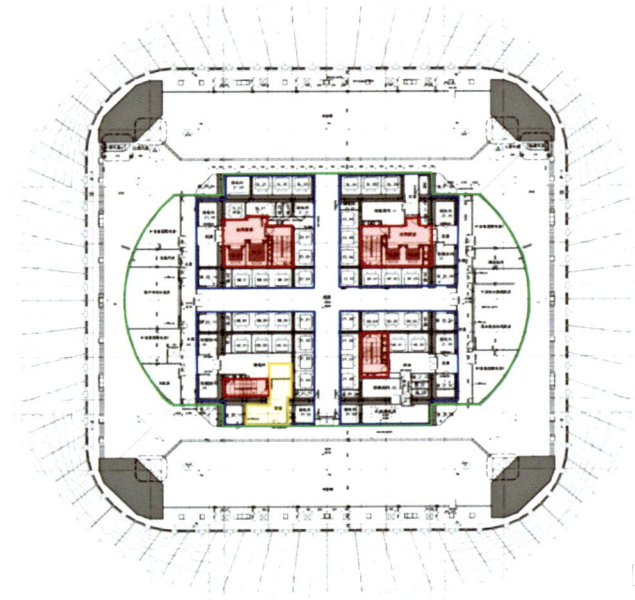

1. 保护对象：电梯（含机坑、机房）、楼梯、管井
2. 红区：保护区
3. 蓝区：控制区，用于设置阻碍气流的屏障
4. 绿区：过渡区，风环境为半室外空间

1. Protected objects: Elevator (including machine pit and machine room), staircase and tube well
2. Red zone: Protected areas
3. Blue zone: Control area for providing a barrier to air flow
4. Green zone: Transition area in which wind environment is semi-outdoor space

1

门分区索引表
GATE PARTITION INDEX TABLE

序号	标段设置位置	设置楼层	类别／数量（樘）			备注 Remarks
			自动平滑门	自动平滑互锁门	旋转门	
1	地下通道	B002	3	4		
		B001	6	2	4	
		B001M	7		2	
2	首层大堂	F001、F002	6			
3	Z3 空中大堂	F031	2			
		F032	2			
		F033	4			
4	Z5 空中大堂	F059	5			
		F060	5			
		F061	4	4		
5	Z7 空中大堂	F089	1	4		
		F090	3	4		
		F091	4			
6	Z8 空中大堂	F105	1	2		
		F105M	1	2		
		F106				
7	合计		54	24	6	

7.5

紧急及备用电源系统
Emergency And Standby Power Supply System

位于大厦 B1 的柴油发电机房设置 2 台 2,500kVA/10kV 高压和 3 台 1,500kVA/0.4kV 低压柴油发电机组，总发电量 9,500kVA。机组通过全数字化 ADEC 发动机电子管理系统、共轨燃油喷射系统及涡轮增压器和两段式冷却水循环等先进技术，保障了柴发机组的安全稳定性。市电网停电，柴油机组单机接到启动信号，8s 内达到额定值，并投入供电，10s 内达 100% 额定负载。多台并机完成输出电力时间 25s。室外埋地储油罐共 15m³，5 台机柴发机组在满负荷运行下，可持续供电 8.5h。确保大厦在断电或火灾等突发状态下重要负荷的供电安全。

2 sets of 2,500 kVA/10kV high-voltage and 3 sets of 1,500 kVA/0.4kV low-voltage diesel generating sets are installed in the diesel generator room located on the B1 of the building, with a total power generation capacity of 9,500kVA. The safety and stability of the diesel generator unit are guaranteed by advanced technologies such as fully digital ADEC engine electronic management system, common rail fuel injection system, turbocharger and two-stage cooling water cycle. Once there is a power failure in the mains supply, diesel engine unit will receive the start signal to reach the rating within 8 seconds, put into power supply, and achieving 100% rated load within 10 seconds. Multiple parallel machines complete power output time of 25s. 15m³ outdoor buried oil storage tanks in total can provide electricity for a sustainable period of eight and a half hours in five diesel generator units operating at full capacity. The safety of power supply for important loads are assured in case of power failure or fire.

中信大厦柴油发电机组供电示意图
Schematic diagram of the power supply of CITIC Tower diesel generator set

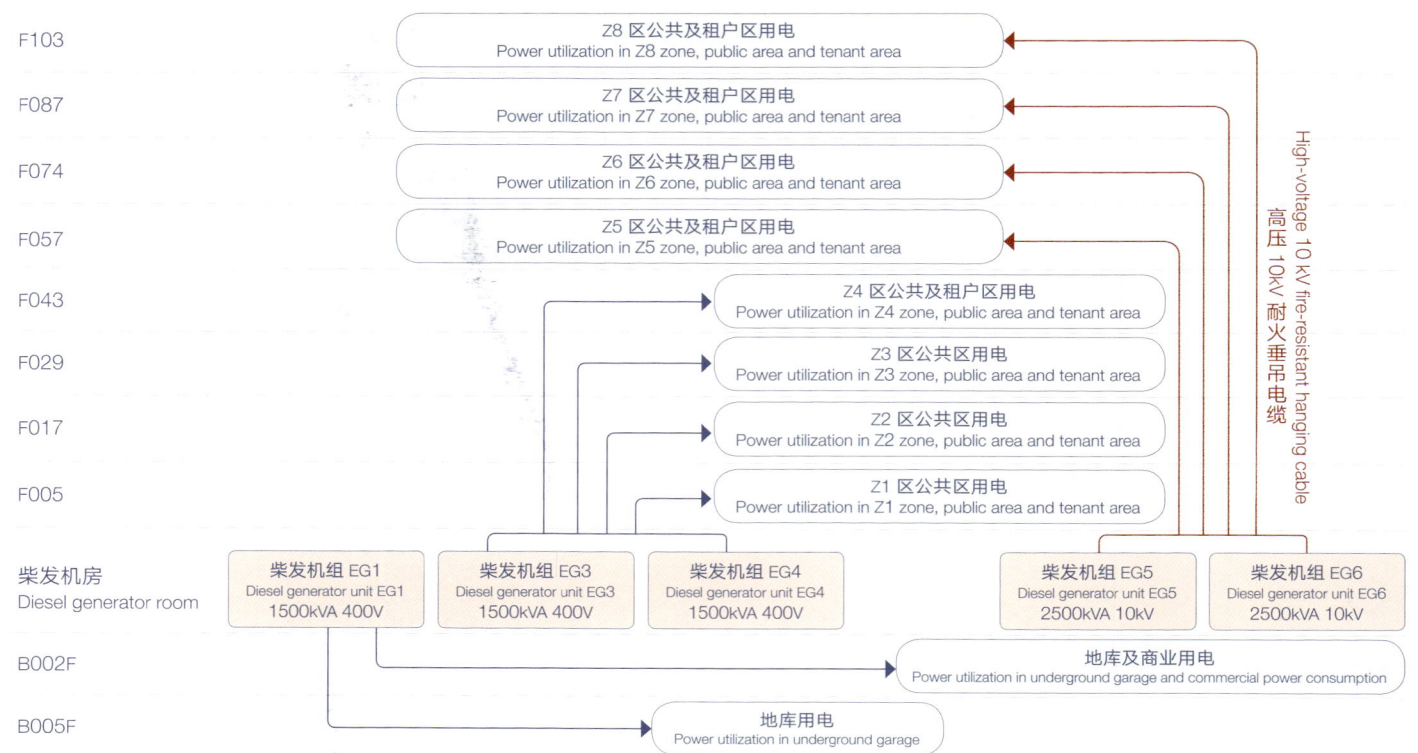

F103 — Z8 区公共及租户区用电 / Power utilization in Z8 zone, public area and tenant area

F087 — Z7 区公共及租户区用电 / Power utilization in Z7 zone, public area and tenant area

F074 — Z6 区公共及租户区用电 / Power utilization in Z6 zone, public area and tenant area

F057 — Z5 区公共及租户区用电 / Power utilization in Z5 zone, public area and tenant area

F043 — Z4 区公共及租户区用电 / Power utilization in Z4 zone, public area and tenant area

F029 — Z3 区公共区用电 / Power utilization in Z3 zone, public area and tenant area

F017 — Z2 区公共区用电 / Power utilization in Z2 zone, public area and tenant area

F005 — Z1 区公共区用电 / Power utilization in Z1 zone, public area and tenant area

High-voltage 10 kV fire-resistant hanging cable
高压 10kV 耐火垂吊电缆

柴发机房 / Diesel generator room

柴发机组 EG1 Diesel generator unit EG1 1500kVA 400V	柴发机组 EG3 Diesel generator unit EG3 1500kVA 400V	柴发机组 EG4 Diesel generator unit EG4 1500kVA 400V	柴发机组 EG5 Diesel generator unit EG5 2500kVA 10kV	柴发机组 EG6 Diesel generator unit EG6 2500kVA 10kV

B002F — 地库及商业用电 / Power utilization in underground garage and commercial power consumption

B005F — 地库用电 / Power utilization in underground garage

建成 CITIC TOWER ‖ A NEW ICON FOR BEIJING

COMPLETION

8.1

高低压变、配电站
High-Voltage And Low-Voltage Transformers And Distribution Stations

8.2 中央控制室及副监测中心
Central Control Room And Monitoring Sub-Center

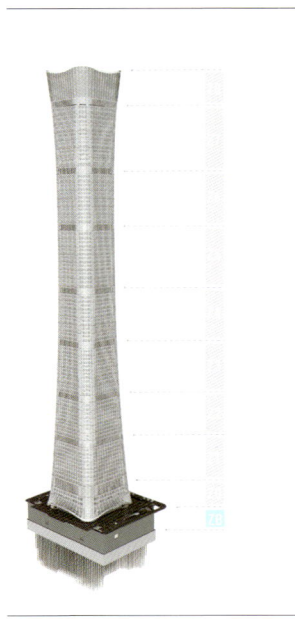

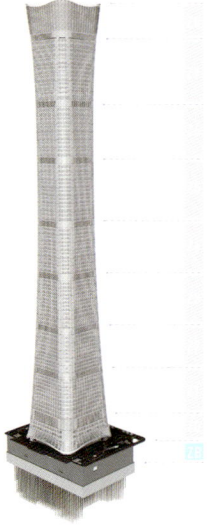

8.3　冷水机组及冰蓄冷装置
Chiller And Ice Storage Device

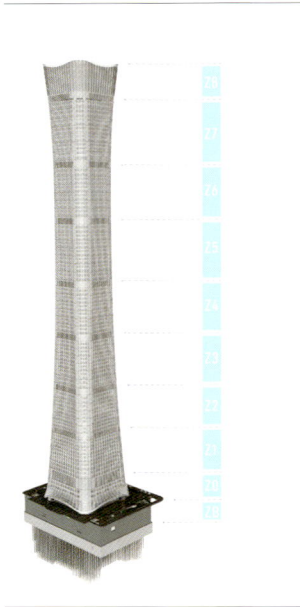

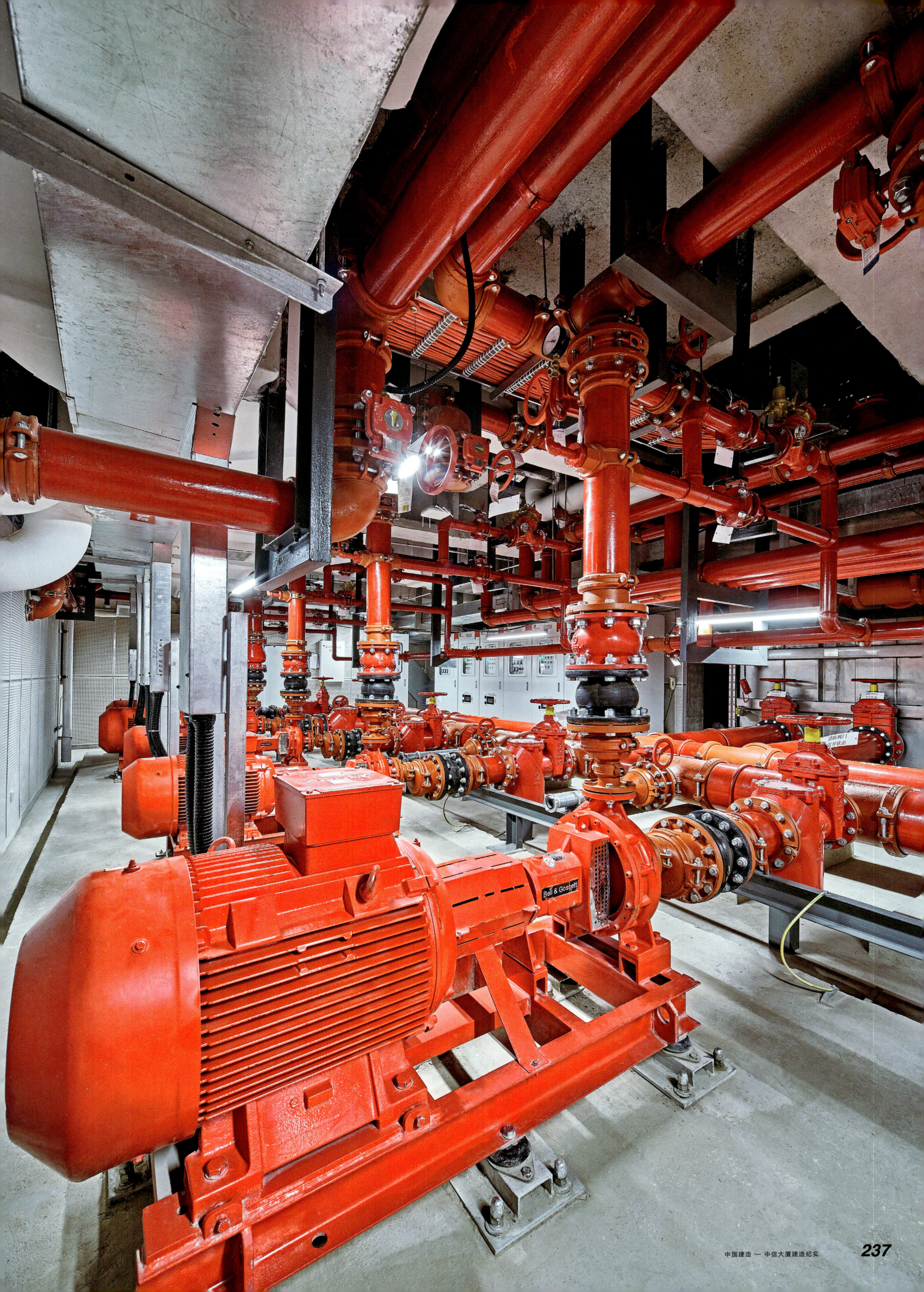

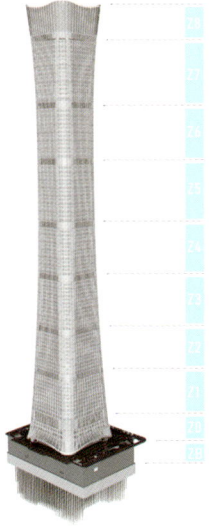

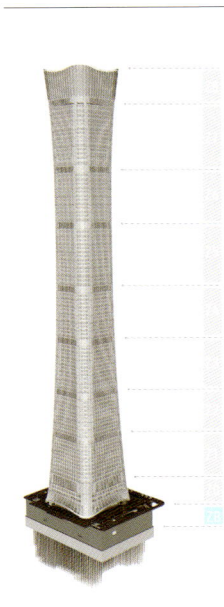

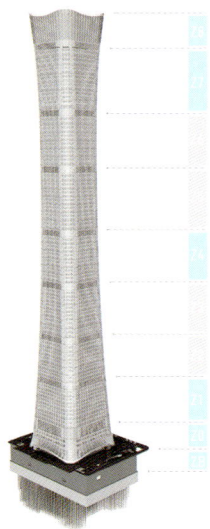

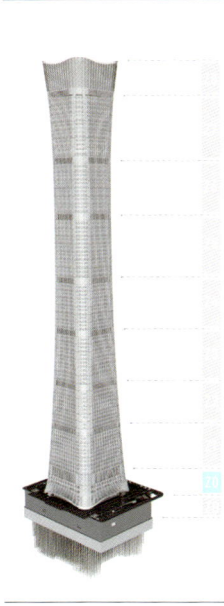

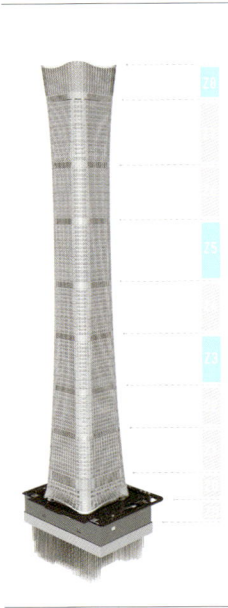

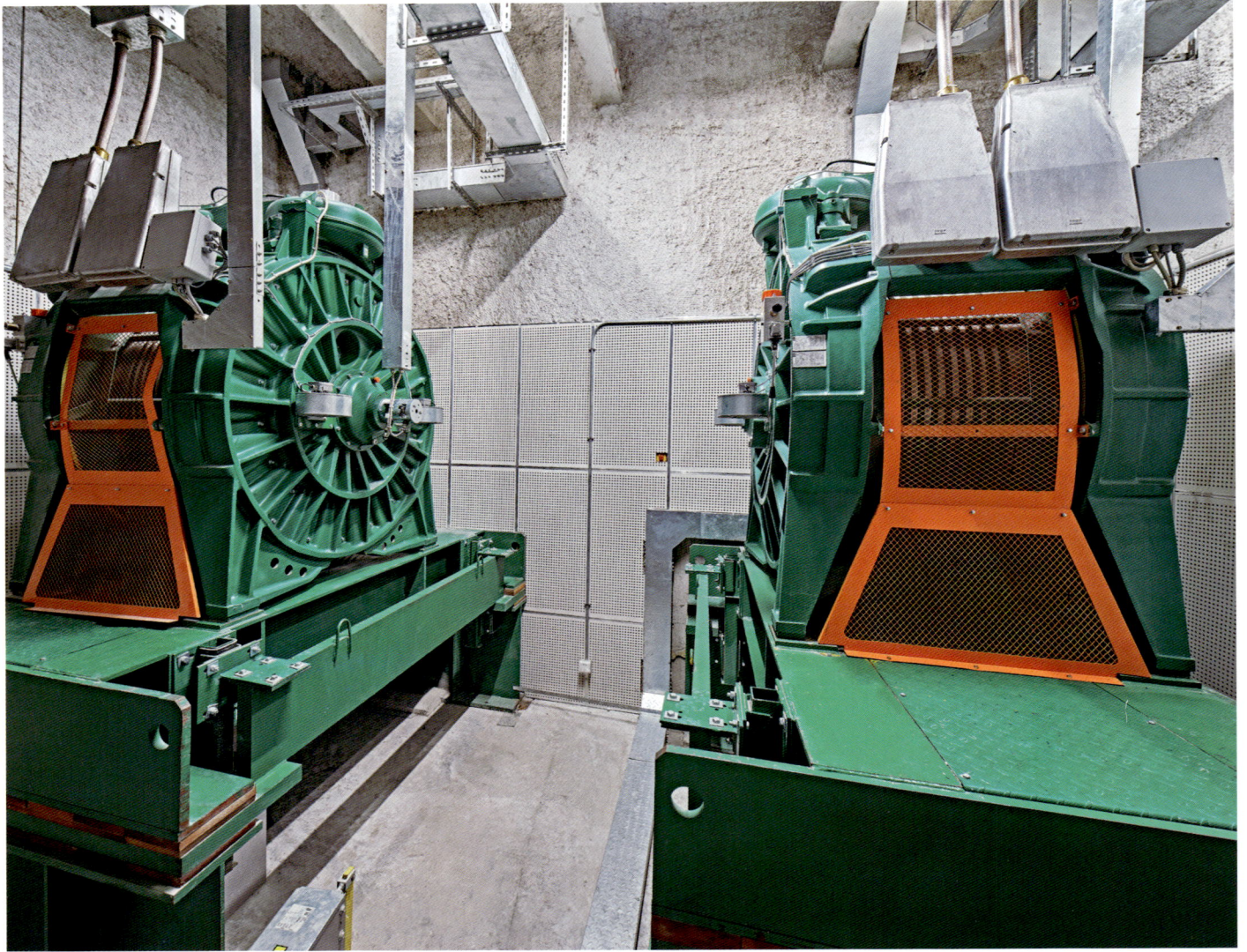

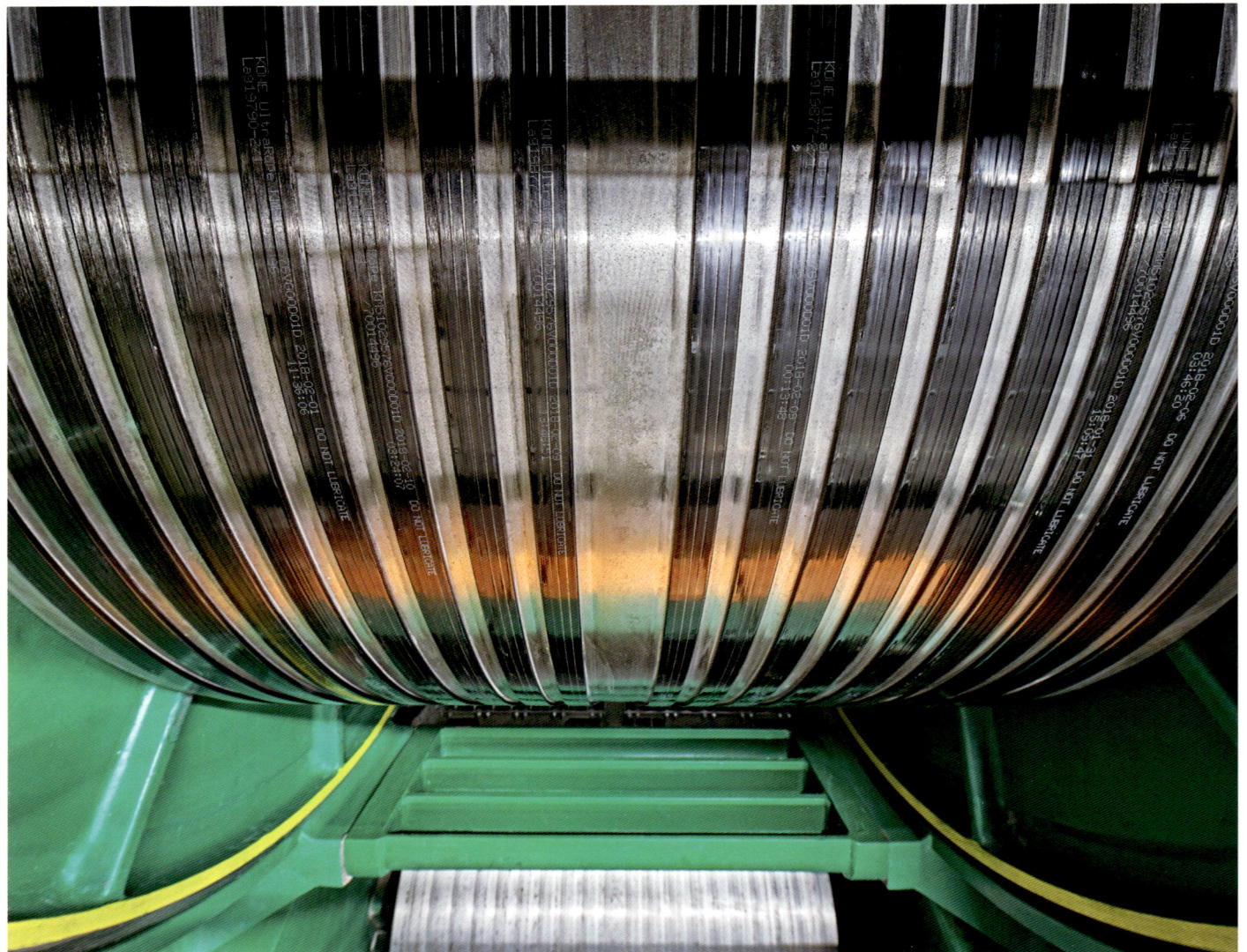

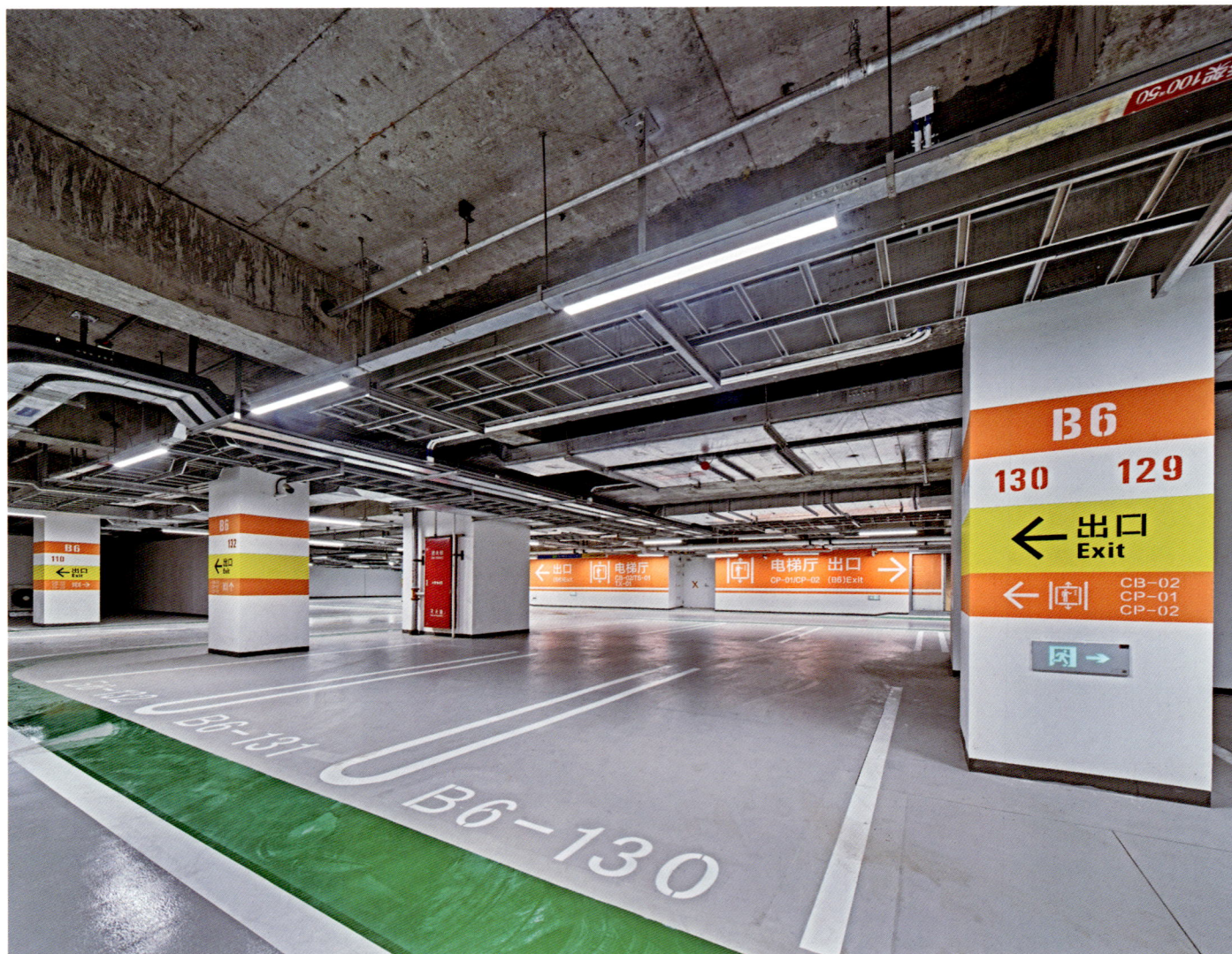

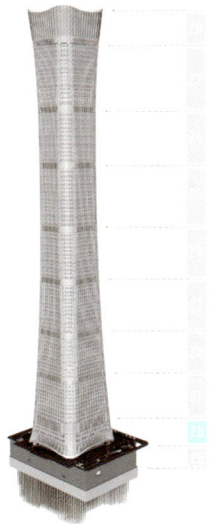

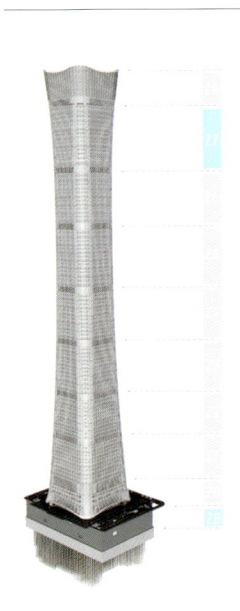

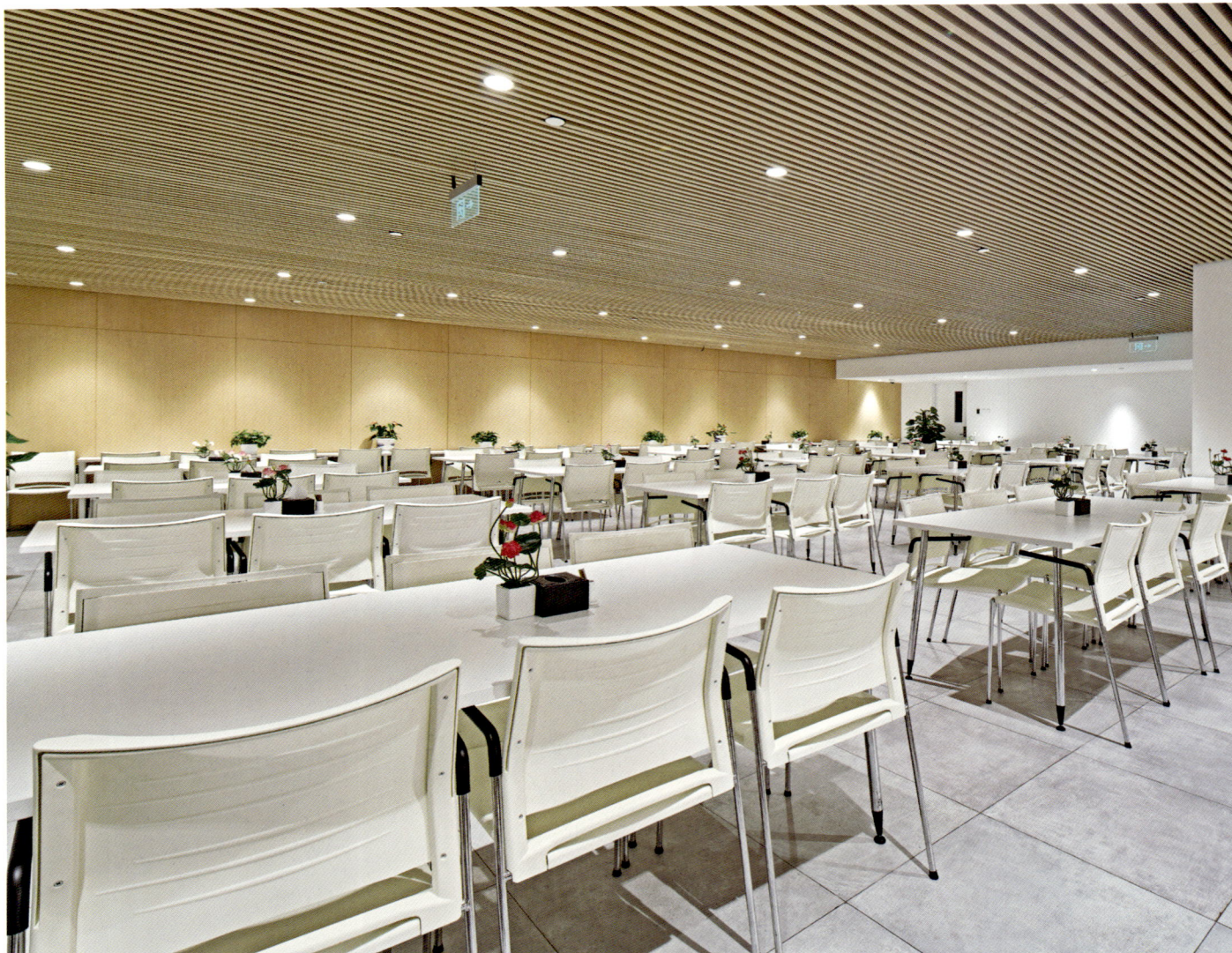

公共空间
Public Space

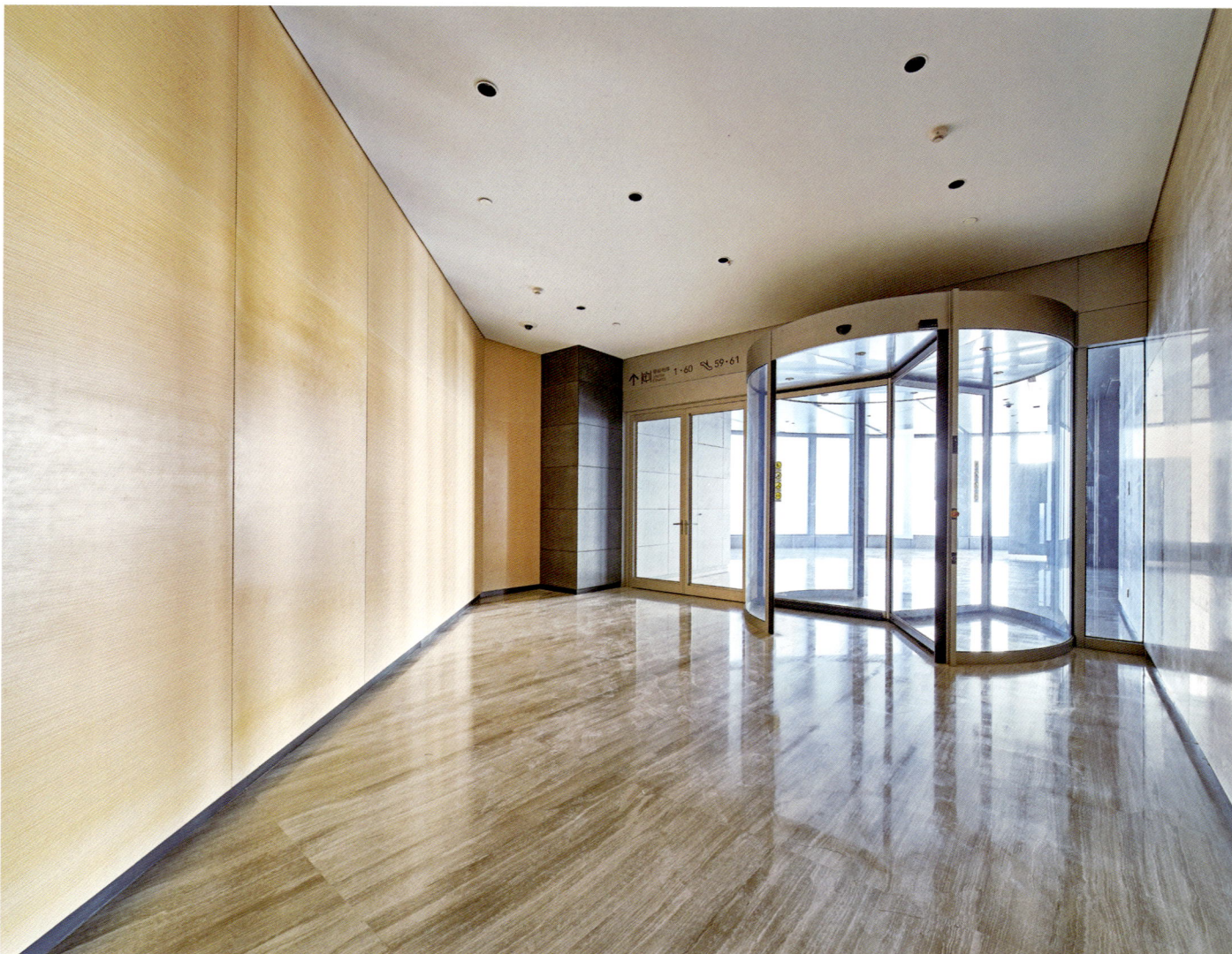

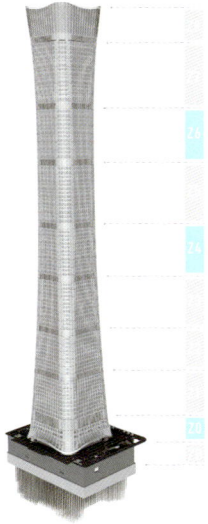

中信银行
CHINA CITIC BANK

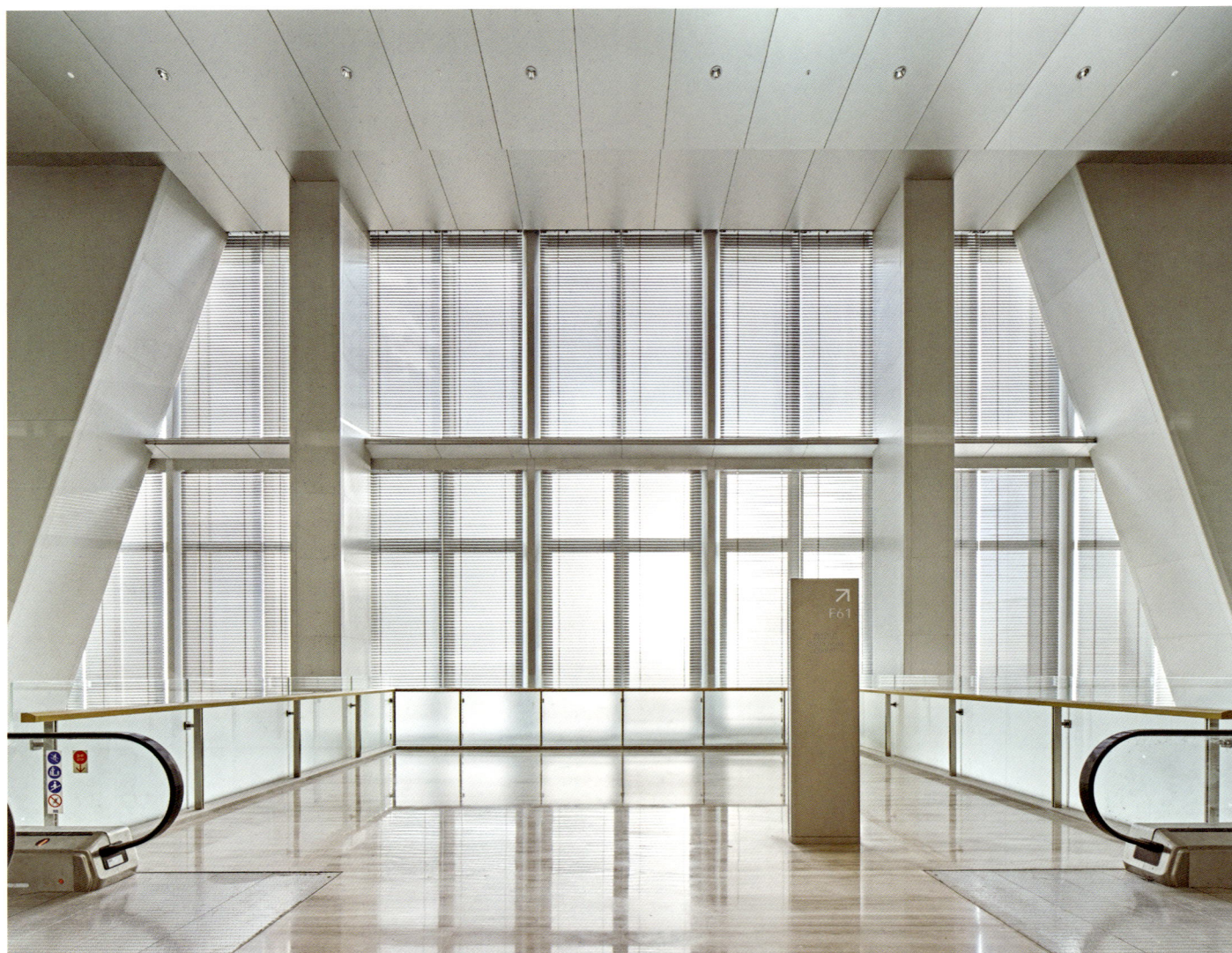

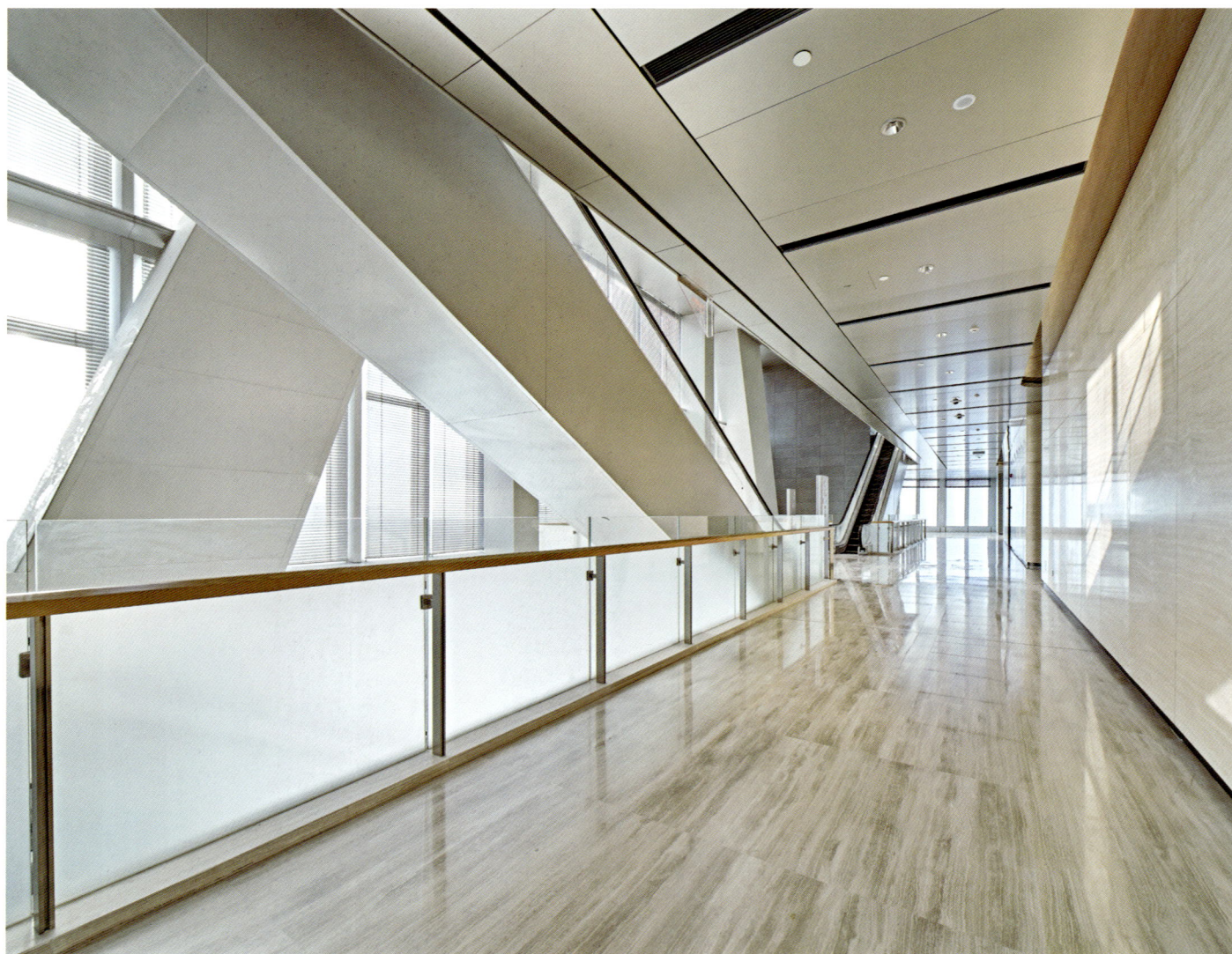

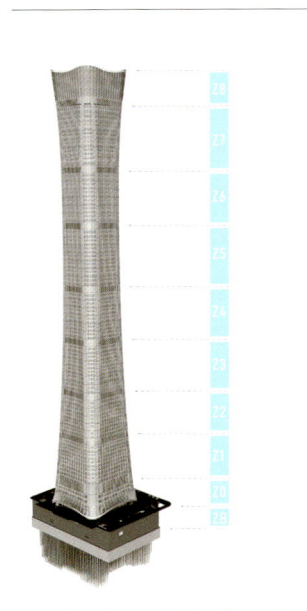

9.6 **楼梯间**
Staircase

电梯厅与 VIP 通道
Elevator Hall And VIP Passage

穿梭电梯
Shuttle Elevator
91·1

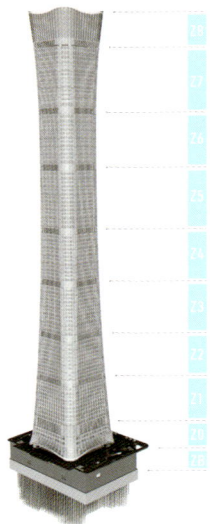

9.9 会议中心
Conference Center

9.10　办公区
Office Area

9.11　商务中心
Business Center

回顾 CITIC TOWER ⫾ A NEW ICON FOR BEIJING
REVIEW

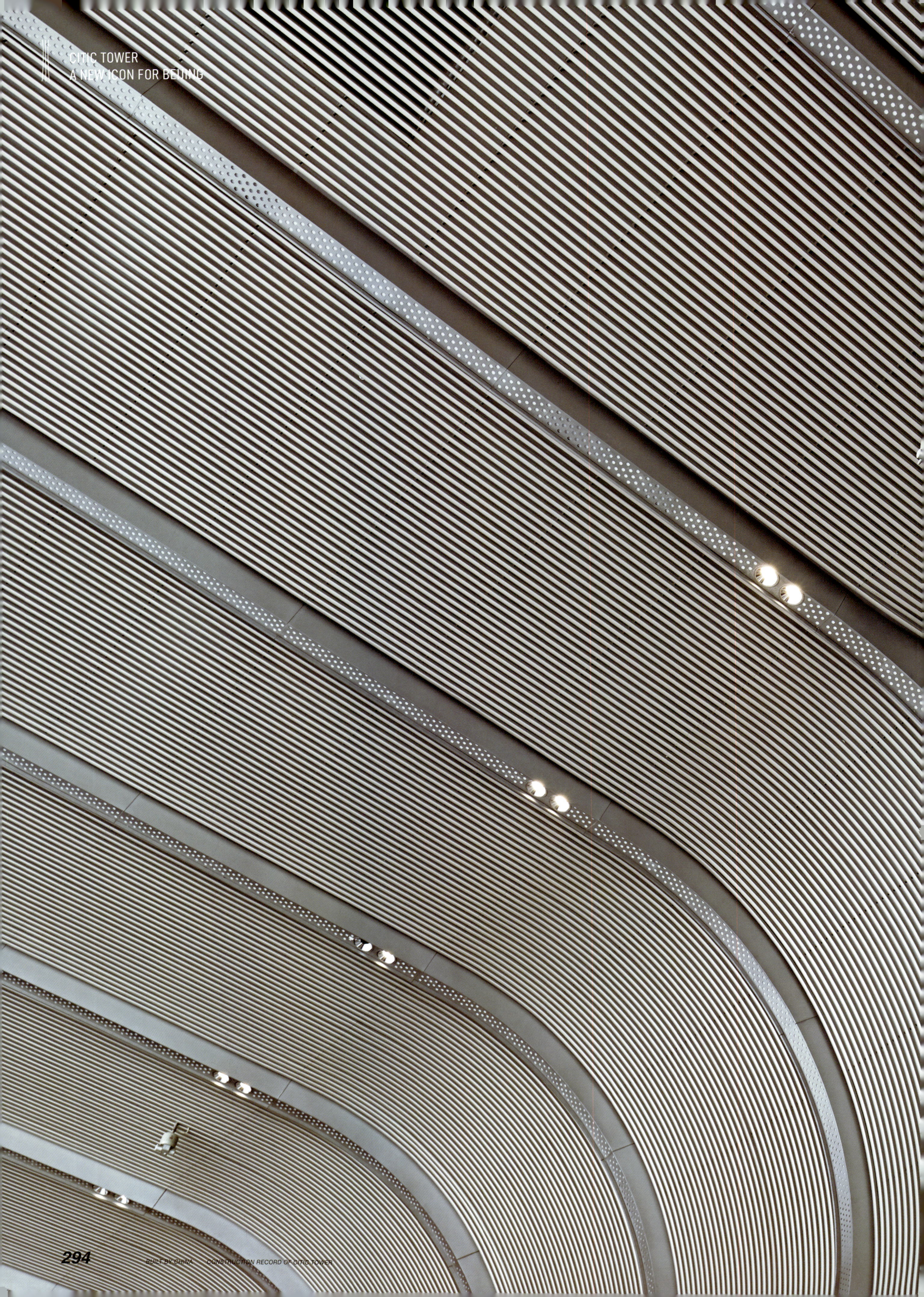

北纬 **39** 度以北全球最高建筑
The tallest building in the world north of 39 degrees north latitude

管道总长度（包括空调水、消防水、给排水）**793,500**m，可绕北京五环 **8** 圈
Total length of pipelines (including air-conditioning water, fire-fighting water, water supply and drainage) is 793,500m, It can go around Beijing fifth ring-road eight times

电缆、电线、控制线总长度 **9,594,800**m，是北京到上海直线距离的 **9** 倍，可绕北京五环 **97** 圈
The total length of cables, wires and control lines is 9,594,800 m, which is 9 times the distance from Beijing to Shanghai as the crow flies, or can go around Beijing fifth ring-road 97 times

办公区 LED 超薄集成灯盘 **41,500** 套
Office area LED ultra-thin integrated light disc 41,500 sets

风管面积为 **334,900**m², 相当于 **47** 个标准足球场
The air duct area is 334,900m². Equivalent to 47 standard football fields

大楼总重量约 **723,000** t（不含基础），幕墙面积约 **125,000** m²，幕墙板块数为 **28,376** 块
The total weight of the building is about 723,000t (excluding the basic), The curtain wall area is about 125,000m², and the number of curtain wall plates is 28,376

世界首个在抗震设防烈度 **8** 度区建造的 **500**m 以上超高层建筑大楼
The world's first super high-rise building in more than 500m with an 8 degree earthquake-resistant fortification intensity

夜景照明 LED 灯具 **72,208** 套，LED 线性灯总长度超过 **100,000**m，总像素点 **380,000** 个
Night scene lighting LED lamps 72,208 sets, LED linear lamp total length exceeds 100,000m, total pixel points 380,000

单筒楼梯踏步 **3,510** 阶，**100** 部直梯，电梯井道总长度约 **16,000** m
There are 3,510 steps of single cylinder stairs and 100 elevators. The total length of elevator shaft is about 16000m

混凝土总用量约 **390,000** m³，约 **975,000**t
The total amount of concrete is about 390,000m³, about 975,000t

底板采用直径为 **40**mm 的 HRB**500** 级钢筋，总用量约 **17,000**t，总长度超过京广线长度，可绕北京五环 **25** 圈
The bottom plate adopts HRB500 rebar with a diameter of 40mm, and the total consumption is about 17,000t.Its total length exceeds that of the Beijing-Guangzhou Line, which equals to 25 times of the length of Fifth Ring Road of Beijing

深化设计图纸 **97,805** 张，**41** 家单位协同搭建 BIM 模型平台，综合模型 **2,773** 个，数据量 **218.13**GB
There are 97,805 drawings of deepening design, 41 units cooperate to build BIM model platform, 2,773 comprehensive models, and the data amount is 218.13GB

93h 连续浇筑 **56,000**m³ 底板混凝土，创北京市单体民用建筑大体积底板混凝土施工新纪录
Continuously pouring 56,000m³ floor concrete for 93 hours, setting a new construction record for mass bottom plate concrete of monomer civil buildings in Beijing

项目管理过程中安全交底 **53,000** 份
先后对施工人员进行 **2,672,000** 人次的安全教育培训
53,000 safety disclosures during project management
2,672,000 safety education and training sessions were conducted for construction personnel

本工程编审主要施工方案 **751** 项，专家论证会 **386** 次
There are 751 main construction schemes and 386 expert demonstration meetings for the project

11 国参与建造：中国、美国、法国、德国、西班牙、意大利、日本、瑞典、丹麦、比利时、芬兰
11 countries participated in the construction: China, the United States, France, Germany, Spain, Italy, Japan, Sweden, Denmark, Belgium and Finland

参建单位 **274** 家，共 **8,160,000** 个工日，高峰期现场施工人员超 **4,000** 人
There were 274 contractors, 8,160,600 working days in total,
Over 4,000 construction personnel on site during peak period

智能化集成点数 **650,505** 个点，数字化安防摄像头个数 **1,776** 个
650,505 intelligent integration points，There are 1,776 digital security cameras

应用三维激光扫描技术，辅助施工质量控制，地上所有楼层的结构三维扫描数据容量达 **2.3**TB
The 3D laser scanning technology is applied to assist the construction quality control, and the 3D
scanning data capacity of all floors above ground reaches 2.3TB

基坑深 **38**m，为北京第一深坑，土方开挖总量约 **440,000** m³，
The foundation pit is 38m deep, which is the deepest pit in Beijing. The total amount of earth excavation is
about 440,000m³. The total number of engineering piles is 896, a total of 37,463.6m 工程桩总数量为 **896** 根，总长 **37,463.6** m

世界首次将 **2** 台 M900D 大型动臂塔机与施工平台相结合，同步顶升，节约工期 **56**D
For the first time in the world, two M900D large boom tower cranes are combined with
the construction platform to synchronize jacking, saving 56 days of construction period

10.2 中信大厦的创新统计
Innovative Statistics Of CITIC Tower

世界之最

1　北纬 39° 以北全球最高建筑（528m）。

2　全球容积率最高（35.0）的超高层建筑。

3　全球超高层建筑中观光层室内空间面积最大（2,800㎡）、净高最高的室内观光平台（净高约 18m 的无柱观光空间，可 360° 俯瞰北京城）。

4　全球地下室层数最多的超高层建筑（地下 8 层）。

5　全球使用面积（得房率）最高的超高层建筑（标准层最大达 70%）。

6　全球首座施工中采用跃层电梯技术（提升高度达 514m；运行速度达 4m/s）的超高层建筑。

7　全球首个建立消防设施临 / 永结合自救体系的超高层建筑。

8　电梯运行总长度与大厦高度比（30:1）、垂直运力全球最大的超高层建筑。

9　全球配备双轿厢高速电梯（10m/s），穿梭高度最高（508m）的超高层建筑。

10　全球首座采用"双中空"超白玻璃单元幕墙 500m 以上的超高层建筑。

11　全球 500m 以上超高层建筑中玻璃幕墙"窗墙比 0.44"最小（最节能）的大厦。

12　全球人均新风量最大（50 m^3/h.p）的超高层建筑。

13　全球超高层施工中配备的承载力最高（4,800t）、面积最大（1,849m^2）、智能化程度最高的顶升作业钢平台。

14　全球配备厚度最薄、综合指标最优的"超静音一体化窗边空调机组"的超高层建筑。

15　全球大厦首层面积最大（6,084m^2）的超高层建筑。

16　全球观光运力最高的电梯配置（大于 1,400 人 / 小时）。

17　全球竖向行程最长（504m）的幕墙擦窗机及配套系统。

18　全球超高层建筑中最节能的夜景照明系统（220V 交流电替代 24V 直流供电，线损最低）。

　　……

中国之最

1　中国第一个由业主主导 EPCO 管理模式建造的超高层建筑。

2　中国第一个采用"设计联合体"模式的特大型房建工程设计管理体系。

3　中国第一个采用"双总包"施工管理体系的超高层建筑。

4　中国类似规模超高层建造中开发周期最短（93 个月）的大厦。

5　中国第一个由业主主导的"全生命周期"系统运用 BIM 技术的超高层建筑。

6　中国室内空气净化系统性能及综合指标最优的超高层建筑。

7　中国风振舒适度最高的办公类超高层建筑。

8　中国第一个采用碳纤维电梯曳引绳（相比钢丝绳更耐火、无谐振、节能、长寿）的超高层建筑。

9　中国设计平均能耗最低的超高层建筑（128VA/m^2）。

10　中国第一个采用 PLC 控制系统，实现大厦空调舒适度高、最节能精密控制的超高层建筑。

11　中国第一个取消大厦核心筒与裙房"后浇带"的超高层建筑。

12　中国单元玻璃幕墙模数最少（702 个）的超高层建筑。

13　中国室内办公区域空间净高最高（3m/3.5m）的超高层商业办公建筑。

　　……

北京之最

1　北京市最高建筑（528m）。

2　北京市最高的观景餐厅（513m）。

3　北京市最高的观光平台（503m）。

4　北京市第一个"分段（分 4 段）"获得《建筑工程施工许可证》的城市综合体开发项目。

5　北京市大厦基坑开挖深度最深（40m）的纪录创造者。

6　北京市房建工程中最深的地质勘察孔（钻探深度地下 180m）纪录的创造者。

7　北京市房建工程中工程桩单桩静载最大加载（40,000kN）纪录的创造者。

8　北京市最厚的建筑混凝土基础底板（6.5m）的纪录创造者。

9　北京市第一个采用重力消防水供水（最可靠）的超高层建筑。

WORLD RECORDS

1   The tallest building in the world (528 m) in the area beyond 39 degrees north latitude.

2   The world super high-rise buildings with the highest plot ratio (35.0)

3   Of the super high-rise buildings in the world, the largest interior space of the sightseeing floor (2,800 m²), and the interior sightseeing platform with the highest net height (the column less sightseeing space with a net height of about 18 m, overlooking Beijing at 360°).

4   Having the world's largest number of basements (8 floors underground) for super high-rise buildings.

5   The world's super high-rise building with the highest usable area (The max space availability rate is 70% at standard floors).

6   The world's first super high-rise building to be constructed with jump lift technology (lifting height up to 514 m; running speed up to 4 m/s).

7   The first super high-rise building in the world to establish a combination of temporary/formal self-rescue system for fire-fighting facilities.

8   The highest ratio of elevator length to building height (30:1), maximum vertical capacity super high-rise building in the world.

9   Equipped with the highest shuttle (508 m) elevator with double-car high-speed (10 m/s) of super high-rise buildings in the world.

10 The world's first super high-rise building with "double hollow" ultra-white glass unit curtain wall of more than 500 m.

11 The smallest (the most energy-efficient) "window-wall ratio is 0.44" of glass curtain wall buildings among global super high-rise building at 500 m above.

12 The world's largest fresh air volume per capita (50 m³/p.h) of super high-rise buildings.

13 The world's super high-rise construction equipped with the highest bearing capacity (4,800 t), the largest area (1,849 m²), the most intelligent jacking operation steel platform.

14 Super high-rise buildings with the smallest thickness and the best comprehensive index, and equipped with "Super-mute Integrated Window-side air-conditioning unit" in the world.

15 The world's largest first floor (6,084 m²) in super high-rise buildings.

16 The world's largest sightseeing elevator configurations (greater than 1,400 person per hour)

17 The world's longest vertical length (504 m) of curtain wall Window Scrubbers and supporting systems.

18 The world's most energy-efficient nightscape lighting system (220 V alternating current instead of 24 V direct current with minimum line loss) for super high-rise buildings.

...

THE BEST IN CHINA

1   China's first super high-rise building built by owner-led EPCO management model.

2   The first large-scale building engineering design management system adopting the "design consortium" model in China.

3   The first super high-rise building to adopt the "double general contractor" construction management system in China.

4   The shortest develop cycle (93 months) of its super high-rise buildings on a similar scale in China.

5   China's first owner-led "Full Life Cycle" system using BIM technology for the construction of the super high-rise building.

6   Super high-rise buildings with the best performance and comprehensive index of indoor air purification system in China.

7   Office super high-rise buildings with the highest wind-induced vibration comfort in China.

8   The first super high-rise building using carbon fiber elevator tractor ropes (which are more fire-resistant, non-resonant, energy-efficient and long-lived than steel ropes) in China.

9   Super high-rise building with the lowest average energy consumption (128 VA/m²) in China.

10 The first super high-rise building that adopts PLC control system to achieve high comfort of air conditioning and the most energy-saving precision control in China.

11 The first super high-rise building in China to eliminate the "post-cast strip" of the building's core tube and podium.

12 The super high-rise building with a minimum of modulus (702) of unit glass curtain wall in China.

13 Super high-rise commercial office building with the largest net height (3 m/3.5 m) of indoor office area in China.

...

THE BEST IN BEIJING

1   The tallest building in Beijing (528 m).

2   Beijing's tallest viewing restaurant (513 m).

3   Beijing's tallest viewing platform (503 m).

4   The first urban complex development project in Beijing that has been granted the Construction Permit for Construction Projects in "Sections (4 Sections)".

5   Record creator of the deepest foundation pit (40 m) of buildings in Beijing.

6   Creator of the record of the deepest geological survey borehole (underground drilling depth 180 m) in Beijing's housing construction project.

7   The creator of the record of maximum static loading (40,000 kN) of single pile in Beijing engineering pile.

8   Record creator of Beijing's thickest concrete foundation bottom plate (6.5 m).

9   The first super high-rise building to use gravity fire water supply (the most reliable).

序号	专利、工法及著作权名称	类型	编号
1	一种集垂直运输设备及模板为一体的自顶升施工平台	发明专利	ZL201110032815.4
2	大钢模板的退模和合模装置及其安装方法	发明专利	ZL201110032819.2
3	具有模板功能的凸起式可周转混凝土承力件及其施工方法	发明专利	ZL201210047627.3
4	一种用于超高层建筑施工的智能型施工平台	发明专利	ZL201210338024.9
5	异型多腔体巨型钢结构的焊接方法	发明专利	ZL201310500262.X
6	适用于巨型柱的自爬式操作平台	发明专利	ZL201310526959.1
7	通过2维坐标选取3维场景中坐标点的方法和系统	发明专利	ZL201410084905.1
8	锚栓套架的安装方法	发明专利	ZL201410090628.5
9	锚栓套架的设计方法	发明专利	ZL201410093817.8
10	钢梁吊装紧固式专用夹具	发明专利	ZL201410138738.4
11	梁吊装紧固式专用夹具	发明专利	ZL201410138740.1
12	一种基于网络的设备协同处理系统及方法	发明专利	ZL201510631462.8
13	一种实现不同厂商设备对接的方法和系统	发明专利	ZL201510655994.5
14	异形多腔体巨型柱焊接变形控制方法	发明专利	ZL201610044795.5
15	方便扩展以及拆装的LED灯盘面框	发明专利	ZL201610299478.8
16	一种可移动式动臂塔吊大梁端头约束结构及其拆装方法	发明专利	ZL201610622890.9
17	一种点式激光抄平笔	发明专利	ZL201720090614.2
18	一种空调机房管段预制化安装方法	发明专利	ZL201811110535.9
19	一种可拼接式焊接机器人轨道	实用新型专利	ZL201320499209.8
20	锚栓套架	实用新型专利	ZL201520290835.5
21	拆卸式防风棚	实用新型专利	ZL201520290965.9
22	收缩式钢梁焊接火盆	实用新型专利	ZL201520291511.3
23	移动式防护结构	实用新型专利	ZL201520291513.2
24	用于深基坑混凝土浇筑的串管、溜槽组合体系	实用新型专利	ZL201521056339.X
25	一种型钢混凝土剪力墙结构拉结模板的工具	实用新型专利	ZL201521056346.X
26	零件板坡口加工设备	实用新型专利	ZL201620064928.0
27	一种中厚板坡口加工设备	实用新型专利	ZL201620065259.9
28	超高层建筑核心筒构件新型智能吊装系统	实用新型专利	ZL201620067525.1
29	智能顶升钢平台桁架影响下钢板墙吊装滑梁	实用新型专利	ZL201620069297.1
30	智能顶升钢平台桁架影响下钢板墙分段分节结构	实用新型专利	ZL201620100408.0
31	LED面板灯的面框结构	实用新型专利	ZL201620630240.4
32	一种LED灯具高空固定装置	实用新型专利	ZL201621166984.1
33	一种超高层建筑用楼层内倒运小车	实用新型专利	ZL201621284601.0
34	钢筋接地跨接线机械加工装置	实用新型专利	ZL201621288843.7
35	机电模块化拼装式卫生间体系	实用新型专利	ZL201721470936.6
36	一种超高层建筑核心筒墙体施工人员逃生爬笼	实用新型专利	ZL201721485277.3
37	多功能测量仪器基座	实用新型专利	ZL201721485279.2
38	一种用于超高层建筑施工时的楼层截水排水装置	实用新型专利	ZL201721485857.2
39	一种运行工况自动切换的变风量末端系统	实用新型专利	ZL2018202125236

40	一种管道接头	实用新型专利	ZL201821110235.9
41	一种双控式防晃卡具	实用新型专利	ZL201821338121.7
42	一种双吊型辅助吊具	实用新型专利	ZL201821338511.4
43	一种可调式重型电缆放线架	实用新型专利	ZL201821338565.0
44	一种电缆防摆动装置	实用新型专利	ZL201821338836.2
45	一种单吊型辅助吊具	实用新型专利	ZL201821338898.3
46	一种多用途中心轴	实用新型专利	ZL201821338932.7
47	一种可柔性调节的电缆导向装置	实用新型专利	ZL201821339033.9
48	用于解决超高层建筑跃层电梯防水、防砸物的工具	实用新型专利	ZL201821892413.5
49	用于超高层建筑施工阶段的消防水系统	实用新型专利	ZL201821955875.7
50	一种建筑外立面施工时与吊篮配套使用的无配重悬挂机构	实用新型专利	ZL201822196681.X
51	一种用于超高层建筑跃层电梯垂直运输的工具	实用新型专利	ZL201822892412.0
52	一种用于狭小空间的灯具安装固定装置	实用新型专利	ZL201920331791.4
53	一种 3D 可视化交底仪	实用新型专利	ZL206226606U
54	SAFTOP ACS4000-8AL40/E 八防区报警控制器软件 V2.8	著作权	2014SR008227
55	SAFTOP ACS4000-RD4/E 四门门禁读卡控制器软件 V4.3	著作权	2014SR008232
56	SAFTOP S6000-LC8R 八回路智能电气开关控制器软件 V8.6	著作权	2014SR008235
57	SAFTOP ACS4000-8I8O 通用输入输出控制器软件 V3.2	著作权	2014SR009687
58	SAFTOP S6000-CLUSTER 服务器软件 V1.1	著作权	2014SR053962
59	SAFTOP S6800-HT 通讯服务控制器软件 V1.1	著作权	2014SR053969
60	SAFTOP 门禁管理模块软件 V1.0	著作权	2016SR191861
61	SAFTOP S6800-RT 物联网路由控制器软件 V1.0.20.36	著作权	2016SR193804
62	SAFTOP ACS-RD-GM 国密读卡器软件 V1.1	著作权	2018SR260637
63	SAFTOP S6800-RD2/E 两门门禁控制器软件 V1.0	著作权	2018SR262313
64	SAFTOP S6000-IOT 物联网配置系统软件 V1.0	著作权	2018SR573060
65	SAFTOP S6000-BIM-PM BIM 物业管理系统软件 V1.0	著作权	2018SR573401
66	SAFTOP SBC-CLOUD 智慧建筑云平台软件 V1.0	著作权	2018SR573800
67	SAFTOP S6000-SECURE 综合安防系统软件 V1.0	著作权	2018SR573885
68	SAFTOP WEB-PORTAL 门户网站软件 V1.0	著作权	2018SR574493
69	SAFTOP BIGDATA 大数据支撑平台软件 V1.0	著作权	2018SR574580
70	SAFTOP S6000-EMERGENCY 智能应急系统软件 V1.0	著作权	2018SR574587
71	SAFTOP S6000-SCADA BIM 组态模块软件 V1.0	著作权	2018SR574594
72	SAFTOP S6000-RT-OPC OPC 接口软件 V1.0	著作权	2018SR652671
73	超高层可拆装式锚栓套架施工工法	工法	省部级工法
74	超高层结构复杂多腔体异型巨柱柱脚大体量灌浆施工工法	工法	省部级工法
75	管井立管模块化施工工法	工法	局级工法
76	卫生间一体化模块施工技术研究与应用工法	工法	局级工法
77	基于 BIM 技术应用的超高层建筑倾斜立管施工工法	工法	局级工法
78	"临时/永久"相结合消防系统施工工法	工法	局级工法
79	超高层跃层电梯的应用技术施工工法	工法	局级工法

NO.	PATENT, CONSTRUCTION AND COPYRIGHT NAMES	TYPES OF	NO.
1	Self-lifting construction platform integrating vertical transportation equipment and template	Patent	ZL201110032815.4
2	Demolition and clamping device for large steel formwork and installation method thereof	Patent	ZL201110032819.2
3	Convex type reversible concrete bearing member with template function and construction method thereof	Patent	ZL201210047627.3
4	Intelligent construction platform for super high-rise building construction	Patent	ZL201210338024.9
5	Welding method of shaped multi-cavity giant steel structure	Patent	ZL201310500262.X
6	Self-climbing operating platform for giant columns	Patent	ZL201310526959.1
7	Method and system for selecting coordinate points in 3D scene by 2D coordinates	Patent	ZL201410084905.1
8	Anchor bolt sleeve installation method	Patent	ZL201410090628.5
9	Design method of anchor bolt frame	Patent	ZL201410093817.8
10	Steel beam lifting and fastening special fixture	Patent	ZL201410138738.4
11	Beam hoisting fastening special fixture	Patent	ZL201410138740.1
12	Network-based device collaborative processing system and method	Patent	ZL201510631462.8
13	Method and system for realizing docking of devices of different manufacturers	Patent	ZL201510655994.5
14	Welding control method for deformed multi-cavity giant column	Patent	ZL201610044795.5
15	Easy to expand and disassemble LED light panel	Patent	ZL201610299478.8
16	Removable boom tower crane beam end restraint structure and disassembly and assembly method thereof	Patent	ZL201610622890.9
17	Point laser pen	Patent	ZL201720090614.2
18	Prefabrication installation method for pipe section of air conditioner room	Patent	ZL201811110535.9
19	A splicable welding robot track	Utility model patents	ZL201320499209.8
20	Anchor bolt frame	Utility model patents	ZL201520290835.5
21	Detachable wind shelter	Utility model patents	ZL201520290965.9
22	Shrinking steel beam welding brazier	Utility model patents	ZL201520291511.3
23	Mobile protective structure	Utility model patents	ZL201520291513.2
24	Series of tubes and chutes for concrete pouring in deep foundation pits	Utility model patents	ZL201521056339.X
25	Tool for drawing profile of steel reinforced concrete shear wall structure	Utility model patents	ZL201521056346.X
26	Part plate groove processing equipment	Utility model patents	ZL201620064928.0
27	Medium and thick plate groove processing equipment	Utility model patents	ZL201620065259.9
28	New intelligent lifting system for core tube components of super high-rise buildings	Utility model patents	ZL201620067525.1
29	Intelligent uplift steel platform truss influences steel plate wall hoisting sliding beam	Utility model patents	ZL201620069297.1
30	Sectional sectional structure of steel plate wall under the influence of intelligent jacking steel platform truss	Utility model patents	ZL201620100408.0
31	Face frame structure of LED panel light	Utility model patents	ZL201620630240.4
32	LED lamp high altitude fixing device	Utility model patents	ZL201621166984.1
33	Inverted transport car in the floor of super high-rise building	Utility model patents	ZL201621284601.0
34	Reinforced grounding jumper machining device	Utility model patents	ZL201621288843.7
35	Electromechanical modular assembly toilet system	Utility model patents	ZL201721470936.6
36	A super high-rise building core tube wall construction personnel escape climbing cage	Utility model patents	ZL201721485277.3
37	Multi-function measuring instrument base	Utility model patents	ZL201721485279.2
38	Floor water intercepting and drainage device for super high-rise building construction	Utility model patents	ZL201721485857.2
39	Variable air volume end system for automatic switching of operating conditions	Utility model patents	ZL2018202125236

40	Pipe joint	Utility model patents	ZL201821110235.9
41	Double control anti-sway fixture	Utility model patents	ZL201821338121.7
42	Double hanging type auxiliary spreader	Utility model patents	ZL201821338511.4
43	Adjustable heavy-duty cable pay-off rack	Utility model patents	ZL201821338565.0
44	Cable anti-swing device	Utility model patents	ZL201821338836.2
45	Single hanging type auxiliary spreader	Utility model patents	ZL201821338898.3
46	Multipurpose central axis	Utility model patents	ZL201821338932.7
47	Flexible adjustable cable guiding device	Utility model patents	ZL201821339033.9
48	A tool for waterproofing and anti-smashing of high-rise buildings	Utility model patents	ZL201821892413.5
49	Fire water system for construction stage of super high-rise buildings	Utility model patents	ZL201821955875.7
50	A counter weightless suspension mechanism used in conjunction with a hanging basket during construction of a building facade	Utility model patents	ZL201822196681.X
51	A tool for vertical transportation of high-rise buildings	Utility model patents	ZL201822892412.0
52	Lamp mounting fixture for narrow space	Utility model patents	ZL201920331791.4
53	A 3D visualization instrument	Utility model patents	ZL206226606U
54	SAFTOP ACS4000-8AL40/E eight zone alarm controller software V2.8	Copyright	2014SR008227
55	SAFTOP ACS4000-RD4/E four-door access control card controller software V4.3	Copyright	2014SR008232
56	SAFTOP S6000-LC8R eight-circuit intelligent electrical switch controller software V8.6	Copyright	2014SR008235
57	SAFTOP ACS4000-8I8O universal input and output controller software V3.2	Copyright	2014SR009687
58	SAFTOP S6000-CLUSTER Server Software V1.1	Copyright	2014SR053962
59	SAFTOP S6800-HT Communication Service Controller Software V1.1	Copyright	2014SR053969
60	SAFTOP access control management module software V1.0	Copyright	2016SR191861
61	SAFTOP S6800-RT IoT Routing Controller Software V1.0.20.36	Copyright	2016SR193804
62	SAFTOP ACS-RD-GM National Secret Card Reader Software V1.1	Copyright	2018SR260637
63	SAFTOP S6800-RD2/E two-door access controller software V1.0	Copyright	2018SR262313
64	SAFTOP S6000-IOT IoT Configuration System Software V1.0	Copyright	2018SR573060
65	SAFTOP S6000-BIM-PM BIM Property Management System Software V1.0	Copyright	2018SR573401
66	SAFTOP SBC-Cloud Smart Building Cloud Platform Software V1.0	Copyright	2018SR573800
67	SAFTOP S6000-SECURE Integrated Security System Software V1.0	Copyright	2018SR573885
68	SAFTOP WEB-Portal portal software V1.0	Copyright	2018SR574493
69	SAFTOP BIGDATA Big Data Support Platform Software V1.0	Copyright	2018SR574580
70	SAFTOP S6000-Emergency intelligent emergency system software V1.0	Copyright	2018SR574587
71	SAFTOP S6000-SCADA BIM configuration module software V1.0	Copyright	2018SR574594
72	SAFTOP S6000-RT-OPC OPC interface software V1.0	Copyright	2018SR652671
73	Construction method of super high-rise detachable anchor bolt sleeve	Construction method	Provincial and ministerial Construction method
74	Massive grouting construction method for super high-rise structures with complex multi-cavity and special-shaped giant columns at the foot of the column	Construction method	Provincial and ministerial Construction method
75	Modular construction method for pipe well riser	Construction method	Bureaus Construction method
76	Research and application of toilet integrated module construction technology	Construction method	Bureaus Construction method
77	Construction method of inclined riser for super high-rise building based on BIM technology application	Construction method	Bureaus Construction method
78	"Temporary/permanent" combined fire system construction method	Construction method	Bureaus Construction method
79	The application technology construction method of super high - rise jump - floor elevator	Construction method	Bureaus Construction method

序号	专业	环保节能技术、新产品应用	备注说明
1	暖通	永磁同步变频离心机	通过采用永磁同步变频电机、"宽频"的压缩机气动设计、高速电机直驱叶轮、双级压缩、变频器四象限可控整流等关键技术，机械效率达 99% 以上。经第三方权威机构实测高区基载冷机的制冷 COP 由设计要求的 5.6 提升为 6.46，提升了 15%，IPLV 由设计要求的 6.2 提升为 9.57，提升了 54%。整体比普通制冷机组节能 30%，达到国际领先水平
2	暖通	纳米复合冰盘管蓄冷系统	冰盘管采用聚合物基纳米高分子复合材料，结冰层厚度更薄，为 18mm~20mm，结冰、融冰性好，系统效率更高；换热面积大，表面不结垢；耐腐蚀，使用寿命达 30 年以上；复合材料重量轻，韧性好，减小蓄冰槽底部的承重，安装维护更加方便
3	暖通	双层聚脲防水和内外侧保温蓄冰槽	国内第一次在混凝土冰槽内部采用双层聚脲防水保温，在冰槽机房一侧的外部墙面增加了保温，减小冷量的损失
4	暖通	引风式冷却塔和阵列式消声器	区别于传统的鼓风式冷却塔，采用引风式冷却塔，结构紧凑，满足设备层安装高度；引风机横向直接排风，风阻小，更加节能；塔体内负压，有效降低飘水率，提高冷却塔的热力性能，有效防止细菌滋生，运行环境更加清洁
5	暖通	三级换热、大温差低温空调系统	空调水通过三级换热，由冰槽供冷 3.3℃，经换热板依次换热成 4.5℃、5.7℃ 和 6.9℃ 的空调冷水，且采用 10℃ 大温差供回水，能源利用率高，可减少空调机组送风量，冰蓄冷合理利用峰谷电价，降低运行费用
6	暖通	一体化集成空调机组、过渡季新风旁通技术	空调机组风机段与消声器集成设计，上下叠放，整体消声，20,000m³/h 的空调机组噪声为 52.1dB(A)，创国际先进。机组内设旁通阀，过渡季运行更节能。整体漏风率 0.32%，远低于国家标准 0.5%
7	暖通	第二代叶轮式 VAV-BOX	变风量系统采用了第二代 VAV-BOX，叶轮式风速传感器避免了堵塞测速孔，测量数据稳定且精度高（1 ~ 12m/s），可感知微弱风速；可变式多孔叶片使气流稳定，整流效果好，具有线性控制性能；风管无需变径，压力损失小（可减少约 100Pa），运行更节能
8	暖通	一体化 FASU 系统	空调风系统末端采用风量平衡一体化送风系统（Flexible Air Supply Unit 简称 FASU），是一种辅助流量分配末端装置，FASU 单元工厂化生产，各支路风量平衡性好，安装方便，节约施工成本，节省工期，也有利于办公区二次装修的空调末端配合调整
9	暖通	贴附射流型防结露低温风口	采用防结露、吊顶贴附射流型空调风口，可保证最佳的气流扩散，避免冷风直吹，舒适性更高；且风口压损低（比常规射流型低温风口减少 80%），更加节能
10	暖通	一体化窗边风机盘管系统	采用创新定制的具有超薄、低噪声、低能耗、低风速的风机盘管，将同层水管敷设优化为下层水管敷设，通过风机盘管和窗台面板工厂化预制装配，使窗边风机盘管系统由原设计 500mm 压缩到 288mm，为大厦增加净使用面积 4,399m²
11	暖通	变水量智慧阀空调水系统（VWV）	将原设计"动态平衡电动调节阀 + 冷热量表"的阀门组整合，采用一体化智慧阀，具备动态平衡、最大流量设定电动调节和冷热量计量等功能，减少安装空间，减少管道接口，便于能源计量及管理
12	暖通	耦合催化空气净化技术	空调机组中加设耦合催化板，利用室外新风和静电产生的臭氧作为能量与催化板进行催化耦合反应，降解臭氧，高效分解 TVOC（苯、甲醛、二甲苯等有机挥发物），提高室内空气品质
13	暖通	室内空气质量控制系统	针对日益严重的环境问题，空调机组设置 PM2.5 控制功能，本着过滤器高效低阻、运行能耗低、便于维护清洗、成本经理等原则，采用"新风端设 G4 板式初效过滤，空调机组送风端设 G4 板式初效过滤 + 双驱静电中效过滤 +F7 中效袋式过滤"的过滤方式，PM2.5 过滤效率可达 99.8%，使室内 PM2.5 可控制在 50μg/m³ 内
14	暖通	过渡季节变新风比运行	采取过渡季节加大新风比的措施，以达到节能目的。大堂区全空气系统在过渡季增大新风比运行，最大新风比可达 60%~70%。办公区、会议室等区域采用变风量空调系统，新风来源于设置在设备层的集中新风机组，新风加压风机在过渡季可以增大新风量运行，最大新风比可达 70%
15	暖通	空调热回收系统	核心筒十字区设置独立的热回收机组。排风以核心筒部分卫生间排风为主，新风主要供给核心筒内电梯厅和预留房间。热回收装置采用热管非接触式，送风侧压力高于排风侧，避免排风污染新风，且有效降低空调能耗
16	暖通	环形空调风管分区温度控制	办公区空调风系统主管道采用环形布置，提高空调系统容错率，降低风管静压，实现系统最小阻力运行，更加节能；AHU 与 VAV-BOX 对应控制逻辑清晰，实现分区域送风温度控制和分时段空调使用功能
17	暖通	大空间分层空调系统	首层大堂高度约 20m，在保证建筑风格的前提下，利用建筑的四周巨柱和核心筒外墙侧实现了分层空调送风的设计，保证大空间空调舒适性的同时有效节约能耗

18	暖通	冷却塔直接供冷系统	过渡季采用冷却塔作为冷源，冷却水与冷冻水通过板换直接换热，达到过渡季节能运行的目的
19	暖通	柴油发电机组兼用冷却系统	柴油发电机组采用水冷冷却，与空调冷却塔合用，节约冷却塔占地面积
20	暖通	根据 CO_2 浓度进行新风量调节	办公区变风量系统均在回风管道设置 CO_2 浓度传感器，可根据室内 CO_2 浓度监测值，实现可变新风量的控制方式
21	暖通	根据 CO 浓度进行车库通风量调节	地下车库排风系统风口部位设置 CO 浓度传感器，当 CO 达到设定浓度后启动相应送、排风机和诱导风机，降低风机能耗
22	暖通	冬季烟囱效应预防系统	为预防超高层建筑冬季电梯井道的烟囱效应，对电梯井道采取加压送风措施，抑制冬季烟囱效应；为了控制电梯厅正压，在卫生间排风增加了定风量阀门，精准控制排风量；空调回风延伸至末端，避免回风短路，保证了电梯厅的风压稳定，有效缓解烟囱效应
23	暖通	节能防结露电控一体阀	节能防结露电控一体阀通过控制空气调节水泵启停、消除停泵水锤、防止泵逆转，泵的启动轴功率可减少 50%，具有零流量启动水泵、零流量关闭水泵、意外停电保护、节能特性、驱动气缸防结露等功能
24	强电	10kV 高压垂吊电缆及敷设	高压垂吊电缆将电力从底层高压变所直供至高层的副变，更加靠近负荷中心，有效缩短低压电缆供电路径，缩短供电线路，降低电能损耗，减少电压损失，也改善了供电质量，确保供电安全可靠，同时还能够节约有色金属，让建筑更加环保、智能
25	强电	Smart Panel 电力监控系统	Smart Panel 低压配电智能系统方案是业界首个将框架式断路器、塑壳断路器、终端配电以及多功能表计整合在一起的针对楼宇能源管理的一体化系统解决方案；通过收集能源使用和设备信息，与能源管理软件相结合，储存并分析数据，进而采取有针对性的优化举措实现节能增效
26	强电	低谐波变频装置	动力变频设备采用低谐波变频方案，比原方案谐波畸变率更低，提高大厦的用电品质
27	强电	AC220V 夜景照明供电系统	通过精确设计灯具亮度；将原设计的 DC24V 供电改为 AC220V 供电方式，简化供电回路划分，降低线路的工作电流；取消格栅提高灯具的发光效率等措施，优化后的夜景照明系统总运行功率比设计降低了 350kW，线路损耗也大幅度降低，达到节能环保的要求，全年节能约为原设计的 33%
28	强电	LED 超薄集成灯盘	LED 超薄集成灯盘光效达到 110lm/W，比普通荧光灯节能 50% 多；将照明、空调风口、烟感、喷淋、扬声器、监控摄像头和各类传感器等集成化创新设计，厚度 47mm，外观品质提升，安装检修便捷
29	强电	智能照明监控系统	可实现多种照明控制方式，如现场面板控制、亮度感应控制、智能时钟控制、触摸屏控制、中控室集中控制，从而节约照明能耗
30	强电	DUPS 飞轮储能系统	采用 DUPS 飞轮储能系统替代传统 UPS 蓄电池，更加安全可靠、环保
31	强电	光伏发电系统	采用进口的 145Wp CIGS 薄膜太阳能电池，总装机容量为 92.8kWp，占用屋顶投影面积为 900 ㎡；光伏组件的光电转换效率为 15.43%，在同类产品中转换效率最高；等效太阳能光伏板面积占建筑基底面积比例为 15.5%
32	强电	施工期临永供电措施	施工阶段将临时电接入正式电系统，有效保证现场用电稳定，节约临时电系统投入成本
33	给排水	密闭式污水提升装置	将设计的地下室卫生间及垃圾间的集水坑污水泵优化为密闭式污水提升装置，增加了密闭性，减少异味扩散，提升大楼品质
34	给排水	施工期临永排水措施	施工阶段将临时排水系统与正式排水系统接通，有效保证现场排水通畅，节约临时排水系统投入成本
35	智能化	综合能源管理系统	综合能源管理系统旨在建立动态的能耗分析与能效评估系统，实时监控与分析各类能源的使用情况，对能耗数据进行管理、查询和分析，使运行管理者能直观、方便、快速地了解能源使用情况，优化经济运行控制策略，减少能源浪费，提高能源使用效率，实现低能耗绿色建筑的目标
36	智能化	数据机房冷通道	机柜采用"背靠背、面对面"摆放，设置空调送回风 + 精密空调 + 冷通道，使整个机房气流组织合理，提高了机房精密空调的利用率，进一步提高制冷效果
37	预制化	工厂化预制建造技术	本项目的楼梯、大型水管井、空调机房水管及阀部件、风管、桥架、集成灯盘、一体化窗边风机盘管系统、一体化 FASU 系统、一体化卫生间等大量应用工厂化预制建造，机械化程度高，施工质量和效率高，节约工期和成本，减少现场焊接作业，更加安全环保

NO.	PROFESSIONAL	ENVIRONMENTAL PROTECTION AND ENERGY SAVING TECHNOLOGY, NEW PRODUCT APPLICATION	REMARKS
1	HVAC	Permanent magnet synchronous frequency conversion centrifuge	Through the use of permanent magnet synchronous variable frequency motor, "wideband" compressor pneumatic design, high speed motor direct drive impeller, two-stage compression, inverter four-quadrant controllable rectification and other key technologies, the mechanical efficiency is over 99%. The refrigeration COP measured by the third-party authoritative organization in the high-area base-loaded chiller was upgraded from the design requirement of 5.6 to 6.46, an increase of 15%, and the IPLV was upgraded from the design requirement of 6.2 to 9.57, an increase of 54%. The overall energy saving is 30% than that of ordinary refrigeration units, reaching the international leading level
2	HVAC	Nano composite ice coil storage system	The ice coil adopts polymer-based nano-polymer composite material, the thickness of the ice layer is thinner 18mm-20mm, the icing and melting properties are good, the system efficiency is higher; the heat exchange area is large, the surface is not scaled; corrosion resistance, The service life is more than 30 years; the composite material is light in weight, good in toughness, and reduces the load-bearing capacity at the bottom of the ice storage tank, making installation and maintenance more convenient
3	HVAC	Double-layer polyurea waterproof and inner and outer thermal storage ice storage tank	For the first time in the country, double-layer polyurea waterproof insulation was used inside the concrete ice trough, and the external wall on one side of the ice tank machine room was added with heat preservation to reduce the loss of cooling capacity
4	HVAC	Induced draft cooling tower and array muffler	The difference is from the traditional blast-type cooling tower, which adopts the induced draft cooling tower, which is compact in structure and meets the installation height of the equipment layer. The induced draft fan directly discharges air in the horizontal direction, and the wind resistance is small, which is more energy-saving; the negative pressure in the tower body effectively reduces the floating water rate. Improve the thermal performance of the cooling tower to prevent bacteria from growing and the operating environment is cleaner
5	HVAC	Three-stage heat exchange, large temperature difference low temperature air conditioning system	The air-conditioning water passes through the three-stage heat exchange, and is cooled by the ice trough by 3.3°C. The heat exchange is exchanged into 4.5°C, 5.7°C and 6.9°C air-conditioning cold water through the plate exchange, and the large temperature difference of 10°C is used for the return water, and the energy utilization rate is high. It can reduce the air supply volume of the air conditioning unit, and use the peak and valley electricity price reasonably to reduce the operating cost
6	HVAC	Integrated integrated air conditioning unit, transitional season new bypass technology	The fan section of the air conditioning unit is integrated with the muffler, stacked on top of each other, and the overall noise is eliminated. The noise of the air conditioning unit of 20,000m³/h is 52.1dB(A), which is internationally advanced. The bypass valve is installed in the unit, which is more energy efficient during the transition season. The overall air leakage rate is 0.32%, which is much lower than the national standard of 0.5%
7	HVAC	Second generation impeller type VAV-BOX	The variable air volume system adopts the second generation VAV-BOX. The impeller type wind speed sensor avoids blocking the speed measuring hole, the measurement data is stable and high precision (1~12m/s), and the weak wind speed can be sensed; the variable porous blade makes the air flow stable. Good rectification effect, linear control performance; no need for variable diameter of the air duct, low pressure loss (can reduce about 100Pa), and more energy-saving operation
8	HVAC	Integrated FASU system	The end of the air-conditioning air system adopts the air-balanced air supply system (FASU), which is an auxiliary flow distribution terminal device. The FASU unit is factory-produced. The air flow balance of each branch is good, the installation is convenient, and the construction cost is saved. , saving the construction period, is also conducive to the adjustment of the air conditioning end of the second renovation of the office area
9	HVAC	Attached to the jet type anti-condensation low temperature air outlet	Anti-condensation and ceiling-mounted jet-type air-conditioning air vents ensure optimal airflow diffusion, avoid cold air blowing, and have higher comfort; and low air pressure loss (80% less than conventional jet-type low-temperature air vents), more energy-saving
10	HVAC	Integrated window side fan coil system	The innovative and customized fan coil with ultra-thin, low noise, low energy consumption and low wind speed optimizes the same layer of water pipe laying for the lower layer of water pipe, and prefabricates the wind turbine coil and window sill panel to make the window fan coil The system is compressed from the original design of 500mm to 288mm, adding a net use area of 4,399m² for the building
11	HVAC	Variable water quantity intelligent valve air conditioning water system (VWV)	The original design "dynamic balance electric control valve + cold heat meter" valve group is integrated, using integrated intelligent valve, with dynamic balance, maximum flow setting electric regulation and cold heat metering, reducing installation space and reducing pipeline Interface for easy energy metering and management
12	HVAC	Coupled catalytic air purification technology	A coupling catalytic plate is added to the air-conditioning unit to utilize the ozone generated by fresh air and static electricity as a catalytic coupling reaction between the energy and the catalytic plate, degrading ozone, and efficiently decomposing TVOC (organic volatiles such as benzene, formaldehyde, xylene, etc.) to improve indoor air quality
13	HVAC	Indoor air quality control system	In response to increasingly serious environmental problems, the air conditioning unit is equipped with PM2.5 control function. Based on the principle of high efficiency and low resistance of the filter, low energy consumption, easy maintenance and cleaning, and cost manager, the new wind end is equipped with G4 plate type primary filter, air conditioner. The filter type of G4 plate type primary effect filter + double drive static medium effect filter + F7 medium effect bag filter is provided at the air supply end of the unit. The PM2.5 filtration efficiency can reach 99.8%, so that the indoor PM2.5 can be controlled at 50μ g/m³ Inside
14	HVAC	Changing seasons into newer winds than running	Take measures to increase the proportion of fresh air in the transitional season to achieve energy-saving purposes. The full wind system in the lobby area increases the fresh air ratio during the transition season, and the maximum fresh air ratio can reach 60%-70%. The variable air volume air conditioning system is adopted in the office area and conference room. The fresh air comes from the centralized new air blower unit installed in the equipment layer. The fresh air pressure air blower can increase the fresh air volume operation during the transition season, and the maximum fresh air ratio can reach 70%
15	HVAC	Air conditioning heat recovery system	An independent heat recovery unit is provided in the core tube cross section. Exhaust air is mainly based on the core of the bathroom, and the fresh air is mainly supplied to the elevator hall and reserved room in the core tube. The heat recovery device adopts a non-contact type of heat pipe, and the pressure on the air supply side is higher than that on the exhaust side to avoid polluting the fresh air, and effectively reducing the energy consumption of the air conditioner
16	HVAC	Ring air conditioning duct zone temperature control	The main pipeline of the air-conditioning wind system in the office area adopts a circular arrangement to improve the fault tolerance of the air-conditioning system, reduce the static pressure of the air duct, achieve the minimum resistance operation of the system, and save energy. The control logic of AHU and VAV-BOX is clear, and the function of sub-zone air supply temperature control and time-division air-conditioning is realized

17	HVAC	Large space stratified air conditioning system	The height of the first floor lobby is about 20m. Under the premise of ensuring the architectural style, the design of the tiered air conditioning and air supply is realized by using the surrounding giant pillars and the outer wall side of the core tube to ensure the comfort of the large space air conditioner and effectively save energy
18	HVAC	Cooling tower direct cooling system	In the transition season, the cooling tower is used as a cold source, and the cooling water and the chilled water are exchanged directly through the plate to achieve the purpose of running in the transitional season
19	HVAC	Diesel generator set combined cooling system	The diesel generator set is cooled by water and used together with the air conditioning cooling tower to save the cooling tower floor space
20	HVAC	New air volume adjustment based on CO_2 concentration	The variable air volume system in the office area is equipped with a CO_2 concentration sensor in the return air duct, and the variable fresh air volume control mode can be realized according to the indoor CO_2 concentration monitoring value
21	HVAC	Garage ventilation adjustment based on CO concentration	The CO concentration sensor is arranged in the tuyere of the underground garage exhaust system. When the CO reaches the set concentration, the corresponding sending, exhausting and inducing fans are started to reduce the energy consumption of the fan
22	HVAC	Winter Chimney Effect Prevention System	In order to prevent the chimney effect of the elevator shaft in the winter of super high-rise buildings, pressurized air supply measures are taken for the elevator shaft to suppress the winter chimney effect. In order to control the positive pressure of the elevator hall, a fixed air volume valve is added to the exhaust air in the bathroom to accurately control the air exhaust volume. The air-conditioning return air extends to the end to avoid return air short circuit, which ensures the wind pressure of the elevator hall is stable and effectively mitigates the chimney effect
23	HVAC	Energy-saving anti-condensation electronic control integrated valve	The energy-saving anti-condensation electronic control integrated valve can control the air-regulating water pump to start and stop, eliminate the pump water hammer, prevent the pump from reversing, the pump's starting shaft power can be reduced by 50%, with zero flow start water pump, zero flow shut-off water pump, and accidental power failure protection. , energy-saving features, drive cylinder anti-condensation and other functions
24	Strong electricity	10kV high voltage hanging cable and laying	The high-voltage hanging cable directly supplies power from the bottom high-voltage substation to the high-level sub-change, closer to the load center, effectively shortens the low-voltage cable power supply path, shortens the power supply line, reduces power loss, reduces voltage loss, and improves power supply quality. Ensure that the power supply is safe and reliable, while also saving non-ferrous metals, making the building more environmentally friendly and intelligent
25	Strong electricity	Smart Panel Power Monitoring System	The Smart Panel low-voltage power distribution intelligent system solution is the industry's first integrated system solution for building energy management that integrates frame circuit breakers, molded case circuit breakers, terminal power distribution and multi-function meters. By collecting energy use and equipment information, combined with energy management software, storing and analyzing data, and then taking targeted optimization measures to achieve energy efficiency
26	Strong electricity	Low harmonic frequency conversion device	The power frequency conversion equipment adopts a low harmonic frequency conversion scheme, which has lower harmonic distortion rate than the original scheme and improves the power quality of the building
27	Strong electricity	AC220V night lighting power supply system	By precisely designing the brightness of the lamp; changing the DC24V power supply of the original design to the AC220V power supply mode, simplifying the division of the power supply circuit, reducing the working current of the line; eliminating the measures of improving the luminous efficiency of the lamp by the grille, and optimizing the total operating power ratio of the night lighting system after optimization The design has been reduced by 350kW, the line loss has also been greatly reduced, and the energy saving and environmental protection requirements have been met. The annual energy saving is about 33% of the original design
28	Strong electricity	LED ultra-thin integrated light panel	LED ultra-thin integrated lamp panel achieves 110lm/W light efficiency, which is more than 50% energy saving than ordinary fluorescent lamps. Integrating innovative design with lighting, air conditioning vents, smoke, spray, speakers, surveillance cameras and various sensors, thickness 47mm, improved appearance quality, easy installation and maintenance
29	Strong electricity	Intelligent lighting monitoring system	A variety of lighting control methods can be realized, such as on-site panel control, brightness sensing control, intelligent clock control, touch screen control, and central control room centralized control, thereby saving lighting energy consumption
30	Strong electricity	DUPS flywheel energy storage system	The DUPS flywheel energy storage system replaces the traditional UPS battery, which is safer, more reliable and environmentally friendly
31	Strong electricity	Photovoltaic power generation system	It adopts imported 145Wp CIGS thin film solar cell with a total installed capacity of 92.8kWp and an occupied roof projection area of 900m². The photoelectric conversion efficiency of photovoltaic modules is 15.43%, which has the highest conversion efficiency among similar products. The equivalent solar photovoltaic panel area accounts for 15.5% of the building base area
32	Strong electricity	Linyong power supply measures during construction period	Temporary electricity is connected to the official electrical system during the construction phase, which effectively ensures the stability of on-site electricity consumption and saves the input cost of the temporary electrical system
33	Drainage	Closed sewage lifting device	Optimize the sump sewage pump in the designed basement toilet and garbage room into a closed sewage lifting device, which increases the airtightness, reduces the spread of odor and improves the quality of the building
34	Drainage	Linyong drainage measures during construction period	During the construction phase, the temporary drainage system is connected to the official drainage system, which effectively ensures the smooth drainage of the site and saves the input cost of the temporary drainage system
35	Intelligent	Integrated energy management system	The integrated energy management system aims to establish a dynamic energy analysis and energy efficiency assessment system to monitor and analyze the use of various energy sources in a timely manner, manage, query and analyze energy consumption data, so that the operation manager can intuitively, conveniently and quickly Understand energy use, optimize economic operation control strategies, reduce energy waste, improve energy efficiency, and achieve the goal of low-energy green buildings
36	Intelligent	Data room cold aisle	The cabinet adopts "back to back, face to face" placement, and the air conditioner is sent back to the wind + precision air conditioner + cold passage, which makes the airflow organization of the whole machine room reasonable, improves the utilization rate of the precision air conditioner in the machine room, and further improves the cooling effect
37	Prefabricated	Factory prefabricated construction technology	The project's staircase, large water pipe well, air conditioning machine room water pipe and valve components, air duct, bridge frame, integrated lamp panel, integrated window side fan coil system, integrated FASU system, integrated toilet and other large-scale application factory prefabrication, High degree of mechanization, high construction quality and efficiency, saving time and cost, reducing on-site welding operations, and making it safer and more environmentally friendly

BUILT BY CHINA   CONSTRUCTION RECORD OF CITIC TOWER

11.1　中信和业简介
CITIC Heye Profile

中信和业投资有限公司
CITIC HEYE INVESTMENT CO., LTD.

中信和业投资有限公司成立于 2011 年 5 月 4 日，是中信集团全资一级子公司。公司主营业务包括房地产开发、EPCO 工程管理、项目投资、物业管理、专业承包、工程咨询等。目前主要负责中信大厦的开发建造与运营管理。

八年来，中信和业秉承"诚信、创新、凝聚、融合、奉献、卓越"的中信文化，形成了"责任心、执行力、创新力"的企业核心价值观，孜孜以求，大胆创新，以专业求品质，以创新铸精品，在中信大厦开发建设、规划设计、施工建造和运营管理等方面不断探索钻研新技术、践行新理念，刷新了国内外超高层建筑领域的多项纪录，先后获得了 2017~2018 年度北京市结构工程长城杯金质奖、2018 年度 MIPIM Asia 中国最佳未来大型项目金奖、CTBUH 2019 年度全球最佳高层建筑杰出奖、中国钢结构金奖杰出工程大奖等荣誉，得到了社会各界的高度关注和认可。

公司以超高层开发建设为发端，采用"开发建设运维一体化"模式，打造超高层建筑运维管理核心能力，带动资产管理及工程咨询业务，力求成长为超高层建筑管理、商业地产管理和资产管理领域，受人尊敬的、业界知名的、一流的综合服务商。

CITIC Heye Investment Co., Ltd. was registered and established on May 4, 2011, a wholly-owned subsidiary of CITIC Group. Main businesses of the company cover real estate development, EPCO project management, project investment, property management, professional contractor, engineering consulting and the like. At present, the company is mainly responsible for the development, construction and operation management of CITIC Tower.

During the past eight years, CITIC culture of "integrity, innovation, cohesion, integration, dedication and excellence" has been carried by CITIC Heye to form the core values of "responsibility, execution and innovation". With assiduous pursuit, bold innovation, guaranty of high quality based on professional competency, and construction of excellent products based on innovation, CITIC Heye has been continuously exploring new technology and practicing new ideas in terms of development, planning, design, construction and operation management of CITIC Tower, refreshing records in the field of super high-rise buildings at home and abroad. It has won the gold award of Beijing Structural Great Wall Cup of the year 2017-2018, MIPIM Asia Award "Best Chinese Futura Mega Project" of the year of 2018, CTBUH Awards of Excellence for Best Tall Building and Gold Award of China's Construction Engineering Steel Structure, obtaining great praise and recognition of all levels of the society.

With the initiative of super high-rise building development and construction and the mode of "integration of development, construction, operation and maintenance", the company builds core competence of super high-rise building operation and maintenance management, drives services of asset management and engineering consultation, and strives to grow into a respected, well-known and first-class integrated service provider in the fields of super high-rise building management, commercial real estate management and asset management.

施工双总包单位简介
Construction Double General Contractor Profile

中国建筑集团有限公司（简称中国建筑），正式组建于 1982 年，是我国专业化发展最久、市场化经营最早、一体化程度最高、全球规模最大的投资建设集团。集团主要以上市企业中国建筑股份有限公司为平台开展经营管理活动，拥有上市公司 7 家，二级控股子公司 100 余家。

中国建筑营业收入平均每 12 年增长 10 倍。2018 年，公司新签合同额 2.63 万亿元人民币，营业收入、利润总额在 97 家中央企业中分别名列第 4 位、第 8 位，第 13 次获得中央企业负责人经营业绩考核 A 级，位居 2019 年度《财富》世界 500 强第 21 位，《财富》中国 500 强第 3 位，全球品牌价值 500 强第 44 位，连续获得标普、穆迪、惠誉等国际三大评级机构信用评级 A 级，为全球建筑行业最高信用评级。

中国建筑的经营业绩遍布国内及海外一百多个国家和地区，业务布局涵盖投资开发（地产开发、建造融资、持有运营）、工程建设（房屋建筑、基础设施建设）、勘察设计、新业务（绿色建造、节能环保、电子商务）等板块。在我国，中国建筑投资建设了 90% 以上 300m 以上的摩天大楼、3/4 重点机场、3/4 卫星发射基地、1/3 城市综合管廊、1/2 核电站，每 25 个中国人中就有一人使用中国建筑建造的房子。

中国建筑将深入学习贯彻习近平新时代中国特色社会主义思想和党的十九大精神，以"拓展幸福空间"为使命，秉承"品质保障、价值创造"的核心价值观和"诚信、创新、超越、共赢"的企业精神，贯彻"五位一体"总体布局和"四个全面"战略布局，落实新发展理念，致力于打造具有全球竞争力的世界一流企业，力争成为世界投资建设领域的第一品牌和中国建筑业改革发展与推动我国城镇化建设的一面旗帜，为实现中华民族伟大复兴的中国梦不断奋进。

China State Construction Engineering Corporation (hereinafter referred to as China State Construction), formally established in 1982, is now an investment and construction group featuring the longest professional development, the earliest market-oriented operation, the highest degree of integration, and the largest scope in the world. China State Construction mainly carries out business management activities through public enterprise- China State Construction Engineering Corporation Ltd., with 7 listed companies and more than 100 secondary holding subsidiaries.

Operating revenue of China State Construction has increased tenfold every twelve years on average. The Corporation saw new contract value hit 2.63 trillion yuan, and ranked fourth and eighth among the 97 central enterprises in terms of operating income and total profits respectively in 2018. It has won the A-level appraisal of the operating performance of the head of the central enterprise for 13 times, ranking 21st in Fortune Global 500, 3rd in Fortune China 500 and 44th in the Brand Finance Global 2019. It was rated A by S&P, Moody's and Fitch in 2018, the highest credit rating in the global construction industry.

China State Construction has been doing business in more than 100 countries and regions in the world, covering investment and development (real estate, construction financing, holding and operation), construction engineering (housing and infrastructure), as well as survey and design, new business (green construction, energy conservation and environmental protection, and e-commerce) and other sectors. In China, China State Construction has built more than 90% of the skyscrapers above 300m, three-quarters of the key airports, three-quarters of the satellite launching bases, one-third of the urban utility tunnels and half of the nuclear power plants, and one out of every 25 Chinese lives in the house built by China State Construction.

China State Construction will further study and implement Xi Jinping Thought on Socialism with Chinese Characteristics for a New Era and the spirit of the 19th National Congress of the CPC, with the mission of "expanding the happiness space", adhering to the core values of "quality assurance and value creation" and the corporate spirit of "honesty, innovation, transcendence and win-win". The Corporation will implement the overall layout of five-sphere integrated plan and the four-pronged comprehensive strategy, implement the new development philosophy, strive to build a world-class enterprise with global competitiveness, strive to become the first brand in the field of investment and construction in the world and a banner for the reform and development of China's construction industry and the promotion of China's urbanization, and constantly strive to realize the Chinese dream of the great rejuvenation of the Chinese nation.

中建三局集团有限公司
CHINA CONSTRUCTION THIRD ENGINEERING BUREAU GROUP CO.,LTD.

中建三局集团有限公司（简称中建三局），全国首家行业全覆盖盖房建施工总承包新特级资质企业，同时拥有市政公用工程施工总承包特级资质和公路工程施工总承包特级资质，排名中国建筑业竞争力两百强企业榜首。

近年来，中建三局坚持建造、投资"两轮"驱动发展战略，市场遍布全国31个省、区、市，海外在建工程遍及巴基斯坦、阿尔及利亚、印度尼西亚、越南、柬埔寨等12个国家。

成立54年来，中建三局秉持"敢为天下先，永远争第一"的"争先"精神，创造了三天一层楼的"深圳速度"，书写了中国改革开放的代名词。先后承建、参建包括上海环球金融中心（492m）、天津117大厦（597m）、中信大厦（528m）在内的全国20个省、区、市第一高楼，累计获得鲁班金像奖（国家优质工程奖）218项、专利1,400多项。中建三局发挥规划设计、投资开发、基础设施、房建总包"四位一体"优势，参与城市建设，开发品质楼盘，不断拓展建筑工业化、地下空间、水利水务、节能环保等新兴业务，企业实现高质量发展。

如今，中建三局发展成为年合同额超4,000亿元、营业收入超2,000亿元的现代企业集团，主要经济指标多年排名中建集团工程局第一名、湖北百强企业第二名，先后蝉联三届全国文明单位，四次捧回全国五一劳动奖状，企业正朝着"成为最具价值创造力的世界一流投资建设集团"的目标不懈奋斗。

China Construction Third Engineering Bureau Co., Ltd. is the first new special-grade qualification enterprise in the field of general construction contractor of housing construction, has the special-grade qualification in general construction contractor of municipal public works and highway engineering, ranking first among the top 200 competitiveness enterprises in the competitiveness of the construction industry in China.

In recent years, China Construction Third Engineering Bureau Co., Ltd. has adhered to the "two-wheel" drive development strategy of construction and investment, with markets in 31 provinces, autonomous regions and municipalities across the country. And overseas construction projects are under way in 12 countries, including Pakistan, Algeria, Indonesia, Viet Nam and Cambodia.

In the past 54 years since its establishment, the China Construction Third Engineering Bureau Co., Ltd. has been adhering to the spirit of "always striving for the first place in the world", and has created a "Shenzhen speed" of one story in three days, becoming a model of China's reform and opening up. It has successively undertaken and participated in the construction of the tallest building in 20 provinces, autonomous regions and municipalities, including the Shanghai World Financial Center (492m), Tianjin 117 Building (597m) and CITIC Tower (528m). It has won 218 Luban Construction Award (National High Quality Project Award) and more than 1,400 patents. China Construction Third Engineering Bureau Co., Ltd. has taken advantage of the "four-in-one" advantages of planning and design, investment and development, infrastructure, and general contractor of housing construction, participated in urban construction, developed quality buildings, continuously expanded new businesses such as building industrialization, underground space, water conservancy and water affairs, energy conservation and environmental protection, and realized high-quality development of the enterprise.

At present, China Construction Third Engineering Bureau Co., Ltd. has developed into a modern enterprise group with an annual contract value exceeding 400 billion yuan and operating income exceeding 200 billion yuan. It has been ranked first by the China Construction Engineering Bureau and the second by the top 100 enterprises in Hubei Province for many years. It has successively won three National Civilized Units and four National May 1st Labor Certificate. And it is striving unremittingly to become "the most valuable and creative world-class investment and construction group".

中建安装集团有限公司简称中建安装，前身是中国人民解放军基建工程兵，1983 年集体整编为中建八局工业设备安装公司，2012 年更名为中建安装工程有限公司，2018 年 12 月 18 日更名为中建安装集团有限公司，是中建集团旗下最具影响力和核心竞争力的专业公司，是中建集团主业发展的强力专业支撑。

公司以"做强主业、做优名片、行业领先、国际一流"为战略目标。做强石化工程、机电安装两大核心主业，对标世界一流水准、全球最高水平。做优"中建装备制造""中建电子""中建轨道交通电气化"三张名片，成为中建集团在装备制造、智慧领域、基础设施等产业链发展的重要专业支撑力量。

公司通过了 ISO9001、ISO14001、GB/T28001"三位一体"，美国工程师协会压力容器阿斯米认证、国家高新技术企业认证，截至目前获 48 项鲁班奖、38 项国家优质工程奖、5 项詹天佑奖，260 多项省部级以上优质奖，14 项中国安装之星，并获得国家科技进步奖一等奖 1 项、二等奖 1 项、全国钢结构金奖 3 项。公司先后被授予全国优秀施工企业、全国建筑业 AAA 级信用企业、全国守合同重信用企业、全国创先争优先进基层党组织、全国五一劳动奖状等 160 余项省级以上荣誉称号。

公司具有化工石油工程总承包特级，房屋建筑、机电安装、市政公用工程、机电安装等施工总承包一级资质，钢结构、消防、化工石油设备管道、机电设备安装等多项专业承包一级资质，铁路铺轨架梁工程专业承包贰级资质以及特种设备及压力容器制造许可、化工石化医药行业、建筑行业（建筑工程）甲级设计、咨询资质，先后承建了近 400 项国内外重点工程项目，在大型公建和高层（超高层）建筑机电安装、异形钢结构制作安装、化工、石油、燃气、水务、油气贮备、电子、轨道、造纸、医药、装备制造、精细化工等领域积聚了独特、领先的施工技术优势，综合实力稳居国内安装行业领先地位，盛有"铁军"的美誉。

公司驻地南京，聚焦京津冀一体化、长江经济带、珠三角城市群，核心区域市场，实现营销派出机构与所在区域公司的协同联动，不断拓展东北业务，联合西北、西南、新疆，引导全公司做好中西部市场，以中建安装独特的专业差异优势，为非洲、中西亚、东南亚、美洲等海外区域筑造幸福空间，促进当地就业，形成产业链带动效应。

China Construction Industrial & Energy Engineering Group Co., Ltd., or CCEEG, is formerly known as the Infrastructure Engineer Force of the Chinese People's Liberation Army. In 1983, it was collectively reorganized as the Industrial Equipment Installation Company of China Construction Eighth Bureau, then renamed as China Construction Installation Engineering Co., Ltd. in 2012, and renamed as China Construction Industrial & Energy Engineering Group on December 18, 2018. It is the most influential professional company with strongest core competitiveness of China Construction Group and the prominent professional support for the development of the main business of China Construction Group.

The Group takes "strengthen the main business and optimize business card to be industry-leading and first-tier in the world" as the strategic goal. It aims to strengthen the two main businesses of petrochemical engineering as well as electromechanics installation, and aims at the world-class standards and the highest level in the world. It strives to win three "business cards" of China Construction Equipment Manufacturing, CSCEC Electronic and China Construction Rail Transit Electrification", and becomes an important professional support force for the development of the industrial chain of China Construction Group in the fields of equipment manufacturing, intelligence and infrastructure.

The Group has passed the certification of three standards of ISO9001, ISO14001 and GB/T28001, ASME BPVC, and National High-tech Enterprise Certification. So far it has won 48 Luban Awards, 38 National High-quality Engineering Awards, 5 Tien-yow Jeme Civil Engineering Prizes, more than 260 provincial and ministerial-level high-quality awards, 14 China Installation Star Awards, and won the 1 first prize and 1 second prize of National Science and Technology Progress Award, and 3 times of Gold Award of China's Construction Engineering Steel Structure The Group has been awarded the National Excellent Construction Enterprises, the National Construction Industry AAA Credit Enterprise, the National Contract-abiding Enterprises, the National Advanced Grass-roots Party Organizations, the National May 1st Labor Certificate and other 160 provincial-level honorary titles.
The Group has the special grade qualification of general contractor of chemical and petroleum engineering, Grade I qualification for general construction contractor of building construction, electromechanics installation, and municipal public works, Grade I qualifications for professional contractor of steel structure, fire control, chemical and petroleum equipment pipelines, electromechanical equipment installation, Grade II for professional contractor qualification of railway track laying and girder erection, manufacturing license for special equipment and pressure vessel, Grade A qualification for design and consultation in chemical, petrochemical, pharmaceutical and construction (construction engineering) industries. We have undertaken nearly 400 key projects at home and abroad, and have accumulated a unique, leading construction technology advantages in large-scale public buildings and high-rise (super high-rise) building electromechanical installation, special-shaped steel structure production installation, chemical, petroleum, gas, water, oil and gas reserves, electronics, rail, paper, medicine, equipment manufacturing, fine chemicals and other fields, possessing leading position in the domestic installation industry on comprehensive strength, and wining "Iron Army" reputation.

The Group is based in Nanjing, focusing on Beijing-Tianjin-Hebei integration, the Yangtze River Economic Belt, the Pearl River Delta urban agglomeration, and other core regional markets. To realize the cooperative linkage between the marketing dispatched offices and the regional companies, we expand the northeast business, unite the northwest, southwest and Xinjiang Province, and guide the whole group to do a good job in the central and western markets. Build a happy space for Africa, Central and Western Asia, Southeast Asia, America and other overseas regions with the unique professional difference advantages of CCEEG, to promote local employment, and to form the driving effect of the industrial chain.

2013

2013 年 2 月 5 日中信大厦结构设计正式通过超限审查
Structural design of CITIC Tower officialiy passed the review for exceed-code high-rise building residence irreguiarity structure

2013 年 4 月 10 日中信大厦建筑工程设计合同签约
Construction engineering contract of CITIC Tower was signed

2013 年 7 月 9 日地下结构工程施工单位签约仪式
Signing ceremony for underground structure engineering construction unit

1 2 3 4 5 6 7 8 9 10 11 12

2013 年 7 月 29 日工程正式开工
The project commenced officially

2013 年 10 月 18 日消防性能化设计方案通过专家组评审
Firefighting performance design scheme passed expert group review

2013 年 12 月 31 日工程桩施工完成
Construction of engineering piles was completed

2014

2014 年 2 月 28 日土方完成
Earthwork excavation is completed

2014 年 3 月 13 日底板防水施工完成
Completion of bottom slab waterproof construction

2014 年 4 月 27 日基础底板混凝土浇筑完成
Completion of concrete pouring in foundation slab

1 2 3 4 5 6 7 8 9 10 11 12

2014 年 5 月 15 日钢结构首吊
First hoisting of steel structure

2014 年 12 月 10 日地下结构施工完成
Completion of underground structure construction

2014 年 12 月 19 日地上工程正式启动
Commencement of above-ground engineering

2015

2015 年 2 月 25 日 M900D 塔吊安装
Installation of M900D tower crane

2015 年 3 月 27 日北侧钢平台拆除
Removal of steel platform at north side

2015 年 4 月 22 日机电总承包签约
Electromechanical general contract was signed

1 2 3 4 5 6 7 8 9 10 11 12

2015 年 4 月 12 日智能化施工装备集成平台投入使用
Intelligent construction equipment integrated platfor was put into use

2015 年 9 月 28 日主体混凝土结构高度超 100m
The height of main concrete structure exceeded 100 m

2015 年 12 月 25 日玻璃幕墙开始安装
Commencement of glass curtain wall installation

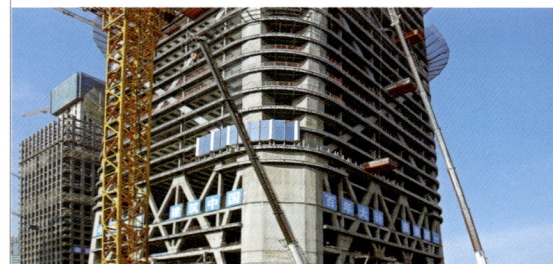

2016

2016 年 3 月 31 日主体混凝土结构高度超 200m
The height of main concrete structure exceeded 200 m

2016 年 7 月 17 日第一批变压器 M2 吊装完成
Completion of the first batch of transformers M2 lifting and installation

2016 年 8 月 17 日跃层电梯投入使用
Jump lift was put into use

1 2 3 4 5 6 7 8 9 10 11 12

2016 年 8 月 18 日主体混凝土结构高度成为北京第一高
The main concrete structure became the highest building in Beijing

2016 年 9 月 1 日临永结合消防系统启用
The temporary and permanent combined firefighting system was enabled

2016 年 11 月 30 日主体混凝土结构高度超 400m
The height of main concrete structure exceeded 400m

2017

2017 年 5 月 25 日机电双工况冷水机组首吊
First hoisting of electromechanical dual working condition water cooler

2017 年 6 月 3 日主体混凝土结构高度超 500m
The height of main concrete structure exceeded 500 m

2017 年 6 月 25 日钢结构最后一根巨柱安装完成
Completion of installation of the last large column in steel structure

1 2 3 4 5 6 7 8 9 10 11 12

2017 年 8 月 18 日主体结构全面封顶
Completion of main steel structure

2017 年 11 月 29 日柴油发电机组吊装仪式
Ceremony of diesel generator lifting and installation

2017 年 12 月 20 日幕墙标准层单元板块安装完成
Completion of installation of standard floor unit plates

2018

2018 年 8 月 22 日全球最长的垂吊电缆吊装完成
Completion of lifting and installation of the longest hanging cable in the world

2018 年 9 月 19 日幕墙外立面全部封闭
Completion of all façades of curtain walls

2018 年 9 月 20 日签署补充协议确明竣工日期
The supplementary agreement was signed to confirm the date of completion

1 2 3 4 5 6 7 8 9 10 11 12

2018 年 11 月 19 日 中信大厦正式供暖
Commencement of heating supply in CITIC Tower

2018 年 12 月 19 日 消防验收通过
Fire protection was accepted

2018 年 12 月 28 日 举行大厦初步接收证书颁发仪式
Ceremony for issuing the preliminary acceptance certificate was held

期 待 刷 新 世 界 建 筑 高 度 的 超 越 者 们
TRANSCENDENT PIONEERS LOOKING FORWARD TO REFRESHING THE WORLD'S BUILDING HEIGHTS

中信大厦建造大事记（摘要）
CITIC Tower Construction Memorabilia (summary)

2010 年

• 12 月 27 日，中信集团在公开竞标中获胜，就 Z15 地块与北京市国土资源局签署了《国有建设用地出让合同》、与北京市土地整理储备中心和商务分中心签署了《朝阳区东三环北京商务中心区（CBD）核心区 Z15 土地开发建设补偿协议》。

2011 年

• 3 月 7 日，中信集团与九龙仓集团（香港）签署了《关于合作开发北京 CBD 核心区 Z15 地块的框架协议》。

• 4 月 22 日，中信集团注资 1 亿人民币注册北京中信和业投资有限公司（以下简称中信和业）。

• 9 月 19 日（朝阳区商务节），中信集团与中国建设银行举办了"CBD 核心区 Z15 地块项目启动仪式"。

• 10 月 20 日，王伍仁先生从香港九龙仓集团接过 Z15 地块设计团队的合同初稿，开启了中信集团独立开发建造 Z15 地块项目的新纪元。

• 12 月 2 日，中信集团常振明董事长率王伍仁等向北京市领导汇报 Z15 地块开发规划方案，获得了三项重大成果：（1）同意以中信集团 2011 年 12 月 2 日提交的方案一为基础进行大厦外形设计；（2）同意将大厦塔基底面积从原先的 70m×70m 放大到 82m×82m；（3）同意 Z15 地块红线按实际需要南移。

• 12 月 15 日，CBD-Z15 地块项目试桩工程动工。

2012 年

• 1 月 16 日，北京市政府召开第二次 CBD 规划建设联席会议决定：（1）Z15 地块受让方可改为中信和业投资有限公司；（2）同意 Z15 地块宗地红线南移方案。

• 2 月 2 日，Z15 地块项目成功完成 180m 深钻孔详勘，该详勘孔深度创造了北京市楼宇新纪录。

• 2 月 9 日，公司组织 9 位专家对 Z15 项目的基坑支护设计投标单位的技术标进行评审，北京市勘察设计研究院有限公司与上海岩土工程勘察设计研究院联合体中标。

• 2 月 29 日，集团批准启用"北京中信和业投资有限公司（后改为中信和业投资有限公司）"为 Z15 地块项目（以下简称中信大厦）的开发建设单位。

• 3 月 5 日，Z15 地块试验桩加载至 40,000kN，刷新了北京地区单桩静载试验最大加载纪录。

• 3 月 31 日～4 月 1 日，公司召开"Z15 大楼功能布局和竖向交通方案研讨会"及"北京 CBD-Z15 地块项目 50% 设计方案技术研讨会"。KPF 建筑设计事务所、北京市建筑设计研究院、中信建筑设计研究总院有限公司、奥雅纳、栢诚、弘达、三菱电梯等公司专家参与研讨。

• 4 月 17 日，公司邀请全国结构超限审查委员会八位专家参加"Z15 地块项目第二次结构超限预审会"。

• 4 月 26 日，启动 Z15 地块项目基坑围护工程，中信和业、CBD 国际、北京城建等单位参加（用中信大厦的地下连续墙替代公共管廊的抗浮桩获得 CBD 管委会的认可。此举不仅为政府节省了抗浮桩的费用，还使中信大厦的地下连续墙提前三个月施工，使中信大厦地下室面积增加了约 7,000m²）。

• 5 月 29 日，王伍仁总经理会见阿里巴巴代表，就其购买物业及功能需求等进行了深入沟通。

• 5 月 30 日，公司向北京市环保局提报了《Z15 地块项目环境评估报告》，向北京市发展改革委员会提报了《Z15 地块项目能源评估和项目申请报告》。

• 6 月 12 日，中信集团常振明董事长、田国立总经理听取汇报后，决定取消原设计中的公寓，中信大厦的业态调整为"办公 + 观光 / 多功能中心"。

• 7 月 1 日，完成了 Z15 地块北侧护坡桩的封闭施工（有效地防范了五十年一遇的北京 721 暴雨袭击）。

• 7 月 5 日，公司的 BIM 工作站安装完成，业主方、设计方及施工三方一体化的 BIM 工作室正式启用。

• 7 月 17 日，公司发布《Z15 地块项目工程总体进度计划》。

• 7 月 20 日，公司与 CBD 管委会和北京城建集团签订《基坑一体化工程建设合作协议》。

• 五十年一遇的"北京 721 暴雨"到来前一天公司与平安保险公司签订的《Z15 地块项目工程保险合同》生效。

• 7 月 24 日，公司召开"北京 CBD 核心区 Z15 地块项目桩基础设计专家论证会"，专家决定取消大厦核心筒与裙房间的混凝土后浇带，并优化了桩基设计（减少了约三分之一工程桩）。

• 7 月 25 日～27 日，公司召开"Z15 地块项目初步设计启动会"。KPF、BIAD、ARUP、PB、中信总院等 40 余名专家出席。

• 7 月 27 日，正式启动（Z15 红线外）地下连续墙施工。

• 7 月 30 日，完成 Z15 地块项目桩基础施工准备图，并对桩基础进行了优化设计（将原设计的 56.1m 桩长缩短至 46.1m，大幅降低了桩基造价并缩短施工期）。

• 8 月 8 日，公司与北京市勘察设计研究院有限公司签订了《北京朝阳区 CBD 核心区 Z15 地块项目基坑工程安全风险监测工程合同》。

• 8 月 7 日～15 日，公司聘请专家对 Z15 地块项目的强电、弱电、智能化、暖通、消防、给排水专业设计方案进行评审。

• 9 月 13 日，公司召开"Z15 地块项目施工塔吊设置方案研讨会"。

• 9 月 14 日，公司聘请国内知名设计、钢材生产、钢结构加工等方面的专家召开了"Z15 项目高性能钢材应用技术研讨会"。

• 9 月 21 日，公司与中信集团、北京市国土局签订"Z15 地块受让方由中信集团变更为中信和业投资有限公司协议书"。

• 9 月 26 日，公司获得了北京市规划委员会的《Z15 地块钉桩坐标成果通知书》，Z15 地块项目的轴线调整、红线南移申请获得批准。

• 10 月 19 日，在中国建筑科学研究院国家重点实验室的风洞试验室完成了 Z15 地块项目独立第三方风洞试验。

• 10 月 22 日，Z15 地块基坑挖至 -22m，累计出土 25 万 m³，实现了 Z15 地块项目地下连续墙内环封闭。

• 12 月 19 日，公司获得了中国地震局《北京朝阳区 CBD 核心区 Z15 地块发展项目工程场地地震安全性评价报告》，结论是：符合国家 GB17741-2005《工程场地地震安全性评价》要求。

• 12 月 31 日，五方设计联合体向公司提交了《Z15 地块项目 75% 初步设计成果》。

2013 年

• 1 月 13 日，完成了 Z15 地块项目 -22m 至 -27m 的土方开挖工程。

• 1 月 15 日，公司收到五方设计联合体提交的"100% 初步设计准备图（建筑部分）"。

• 1 月 18 日，中信集团发文成立"总部大楼项目指挥部"。中信集团副董事长、总经理田国立先生兼任总指挥。

• 1 月 23 日，公司获得了北京市发改委《关于 Z15 地块项目核准的请示》的核准批复。

• 同日，北京市住建委根据北京市住建委领导批转常振明董事长信函上"既要依法合规，又要加快推进"的批示召开专题会议，同意分段办理招投标及施工许可证手续；同意尽快启动 Z15 地块项目地下工程的招标程序；同意展开工程试验桩的施工作业。

• 2 月 7 日，五方设计联合体完成了 Z15 地块项目的 100% 初步设计图。

• 2 月 8 日，公司获得了北京市住建委发布的《Z15 地块项目结构超限设计通过审查通知书》。

- 栢诚提交了 Z15 地块项目机电工程的 100% 初步设计成果。

- 2 月 27 日，公司获得了北京市发改委《关于 Z15 地块项目节能专项审查意见》（节能策划获得认可）。

- 3 月 5 日 ~ 15 日，公司召开了 "Z15 地块项目建筑、结构专业 100% 初步设计评审会"、"Z15 地块项目初步设计概算初审会" 和 "Z15 地块项目机电专业 100% 初步设计评审会"。

- 3 月 18 日，公司收到北京市住建委颁发的《停工令》，Z15 项目红线内工程被迫停工。

- 4 月 9 号，中信和业公司与北京市土地储备中心签署 Z15 项目《土地移交书》。

- 4 月 10 日，在京城大厦举行 "北京市朝阳区 CBD-Z15 项目设计联合体六方合同签约仪式"，中信和业总经理王伍仁与北京建筑设计研究院董事长兼总建筑师朱小地、KPF 合伙人 Mr.Robert Whitlock、中信总院院长郭粤梅、ARUP 董事总经理萧锡才、PBA 董事总经理谢锦泉等签署了《Z15 地块项目联合体的设计合同》。

- 4 月 23 日，中信集团董事长常振明主持召开总部大楼项目指挥部第一次会议，听取了和业公司《Z15 项目近期运营情况》及《Z15 项目工程监理及地下结构工程施工招标方案》汇报，宣布指挥部成员调整决定。由中信集团副总经理居伟民接任总部大楼项目总指挥，王伍仁任中信和业投资有限公司总经理。

- 5 月 23 日，中信和业公司的办公地由京城大厦 F34 迁至万达广场 10 号楼。

- 6 月 6 日 ~ 9 日，董事长姚日波、总经理王伍仁、副总经理罗能钧等赴深圳、香港考察了深圳平安国际金融中心（在施）、香港环球金融中心（IFC）、香港环球贸易中心（ICC）等超高层建筑。

- 6 月 25 日，调整了公司组织构架：设立运营管理中心、成本管理中心、设计统筹中心、工程管理部、机电管理部、财务部、开发配套部、行政综合部。

- 7 月 9 日，中信和业公司与中建股份 / 中建三局联合体签署《CBD-Z15 地块地下结构工程施工承包合同》，仪式由王伍仁主持，中信集团董事长常振明、中国建筑集团董事长易军、中信总部大楼项目总指挥居伟民、董事长李庆萍等出席了签字仪式。

- 同日，公司与北京远达国际工程管理咨询有限公司签订《Z15 地块项目工程监理合同》。

- 7 月 11 日，公司获得北京市规委颁发的《Z15 地块建设用地规划许可证》。

- 7 月 15 日，完成北京市住建委的 "Z15 地块地下结构工程施工单位合同备案" 手续。

- 7 月 17 日，公司获得首都规划委员会颁发的《Z15 项目建设工程规划许可证（地下部分）》。

- 7 月 24 日，公司获得北京市住建委颁发的《Z15 项目建设工程施工许可证（地下部分）》。

- 7 月 29 日，举行 "中信集团 CBD-Z15 项目开工仪式"，北京市政府、朝阳区主要负责人、施工单位负责人，以及中信集团领导等出席，中信集团董事长常振明发布 Z15 地块项目正式开工令。

- 8 月 1 日，公司获得北京市交通委员会《关于北京 CBD 核心区 Z15 地块项目交通影响评价审查意见》。

- 8 月 9 日，公司向设计联合体下达《Z15 项目设计限额（概算控制）指标》，要求设计联合体按照上述造价指标调整、优化 Z15 地块项目的初步设计。

- 9 月 10 日，王伍仁总经理应邀代表中央企业出席 "北京市关于优化投资项目审批流程改革座谈会"，王伍仁在会上提出的 "关于特大型项目分段申办施工许可证的建议" 得到市领导的认可，随后北京市政府颁发了此项审批制度的改革决定。

- 9 月 24 日 ~ 25 日，公司召开第二次 Z15 地块项目初步设计复核评审会，聘请森大厦公司、仲量联行、中际北视、中信物业的专家对设计联合体编制的《Z15 地块项目的初步设计》进行评审。评审会对建筑、结构、机电三个专业提出了 268 项修订意见，设计联合体采纳了其中 241 条。

- 10 月 15 日，启动 Z15 地块项目北侧钢平台钢结构安装施工。

- 10 月 18 日，《Z15 项目的消防性能化设计》通过了国家消防总局委托北京市消防局组织的专家评审，这是 Z15 建设中重要的设计进度 "里程碑"。

- 11 月 1 日，集团领导常振明、王炯、居伟民、冯光、李庆萍等听取并肯定了王伍仁总经理所做的 CBD – Z15 地块项目开发建设的策划报告。

- 12 月 4 日，美国绿色建筑委员会（USGBC）向 Z15 项目颁发《LEED-CS 金级预认证证书》。

- 12 月 5 日，Z15 地块项目模型通过地震振动台的测试，测试结果证明中信大厦的结构设计达到了预定抗震性能目标（即遭遇 8.5 级巨震中信大厦也不会倒塌的性能目标）。

- 12 月 25 日，集团发文，集团副总经理李庆萍任北京中信和业投资有限公司董事长、王伍仁等三人任副董事长。

- 12 月 31 日，集团常振明董事长致函北京市领导，请求北京市政府协助解决 "零场地" 施工、办理地上部分《建设工程规划许可证》、交通运输保障、项目北侧光华路设置机动车出入口等影响项目建设的重大事宜。

2014 年

- 1 月 13 日，Z15 地块北侧大门正式启用，其与北侧钢平台一起构成 Z15 地块项目地下结构工程施工运输的 "生命通道"。

- 3 月 27 日，公司与中信银行签订《中信大厦部分写字楼购买意向协议书》。

- 4 月 23 日 10 时许，中信集团董事长常振明宣布 "中信大厦基础底板混凝土浇筑，现在开盘"，现场 16 台汽车泵、40 多台混凝土搅拌车同时启动大厦混凝土底板浇筑。中信股份副总经理、中信和业董事长李庆萍，中建股份副总裁王祥明，以及朝阳区政府、CBD 管委会领导，各参建单位负责人莅临现场。本次浇筑基础底板最厚达 6.5m，连续浇筑量约 5.6 万 m³，创北京市房建史混凝土浇筑新纪录。

- 4 月 28 日，董事长李庆萍，王伍仁一行赴幕墙公司视察幕墙样板，并召开董事长办公会听取了中信大厦首层大堂室内设计、楼体标识、楼体照明及幕墙视觉样板的汇报。会议决定采用铝合金板替代原设计的不锈钢板包袱玻璃单元板框（节省投资逾 6,000 万人民币）、会议还明确了装饰材料的研究方向，取消了大厦顶部的室外标识等。

- 5 月 15 日，中信大厦第一节巨柱和核心筒钢板墙吊装开始，工程进入钢结构全面展开吊装阶段。

- 7 月 31 日，公司取得了北京市规划委员会颁发的《建设工程规划许可证》（地上部分）（2014 规建字 0056 号）。

- 9 月 22 日，中信集团批准了中信大厦 F4 作为中信和业投资有限公司办公用房的申请。

- 9 月 26 日，现场施工如期完成了 B5 结构顶板混凝土浇筑（地下室结构施工进度控制点）。

- 9 月 27 日，中信大厦核心筒钢结构冲出正负零高度。

- 11 月 3 日，公司完成了中信大厦电梯供应及安装工程公开招标。

- 11 月 20 日，公司取得了《Z15 地块项目施工图审查合格书》（地上部分）房 -01103-14-1582 号。

- 12 月 8 日，公司取得了《Z15 地块项目建筑工程施工许可证》（地上部分）。

- 12 月 10 日，中信大厦地下结构工程施工达到正负零标高（实现了地下工程施工合同约定的目标）。

- 12 月 19 日，中信大厦施工总承包合同签约暨地上工程启动仪式在京城大厦举行。中信集团董事长常振明、总经理王炯、中信股份副总经理、中信和业董事长李庆萍等出席，王伍仁总经理主持。

2015 年

- 2 月 5 日，公司与通力电梯有限公司签订 "中信大厦电梯供应及安装合同"。中信和业董事长李庆萍、总经理王伍仁、通力集团执行副总裁兼通力电梯大中华区总裁姜威、芬兰驻华大使馆经贸事务公使、文化参赞等出席了仪式。中信和业董事、监事会代表及中信和业经营班子及相关人员等参加了仪式。

- 4 月 22 日，公司与中建安装工程有限公司签订 "中信大厦机电工程总承包合同"。中信和业董事长李庆萍、总经理王伍仁、副总经理罗能钧、董事严宁、徐翔、曹国强、王爱明，中建股份副总经理曾肇河、中建安装董事长罗能镇、总经理刘延峰等出席签字仪式。

- 6 月 8 日 ~12 日，王伍仁总经理率队赴日本专题调研。考察了虎之门新城、六本木新城、藤泽中心、晴海广场、涩谷项目、清水建设总部大厦等 6 个超高层商业综合体，着重对超高层的规划设计理念、节能设计经验及先进技术、材料、功能定位、流线规划、运营管理等进行调研。

- 9 月 28 日，中信大厦主体结构施工高度突破 100m，达到 103.4m。

- 9 月 30 日，罗能钧副总经理主持施工总承包单位与通力电梯公司签署《跃层电梯采购与安装运维合同》。

全程纪录
Full Record

- 10月9日，中信大厦获得了国家住建部颁发的《三星级绿色建筑设计标识证书》。
- 10月14日，与上海考克斯擦窗机设备工程有限公司/江苏建业建设集团有限公司联合体签署"中信大厦大厦擦窗机供应及安装合同"。
- 10月16日，中信集团董事长常振明会见了格力电器董事长董明珠一行，总经理王伍仁等陪同会见。
- 10月21日，公司举行了"中信大厦大厦机电专业分包合同签约仪式"。见证中建安装工程有限公司与深圳智宇实业发展有限公司、杭州源牌集团、中建一局集团建设发展有限公司签订的（1）《Z15地块项目暖通工程（高区）专业分包合同》、（2）《Z15地块项目暖通工程（低区）专业分包合同》、（3）《Z15地块项目楼宇智能化工程专业分包合同》和（4）《Z15地块项目冰蓄冷及暖通设备监控工程专业分包合同》。
- 12月25日，中信大厦完成幕墙工艺测试样板的安装，如期进入幕墙施工阶段。实现了公司年初制定的2015年大厦建设进度节点目标。

2016年

- 1月8日，中信大厦获得了中华人民共和国住房和城乡建设部颁发的《三星级绿色建筑设计标识证书》。
- 2月1日，集团发文，集团总经理王炯先生担任中信和业投资有限公司董事长、法定代表人。
- 3月31日，现场施工完成F40核心筒墙体混凝土浇筑，结构高度突破200m。
- 4月5日，集团发文罗能钧先生担任中信和业投资有限公司董事、副总经理。
- 4月15日，总经理王伍仁等会见西门子PLC有限公司一行，就中信大厦PLC电机变频等相关技术进行交流。
- 5月30日至6月3日，王伍仁、罗能钧等六人赴迪拜，对哈利法塔（828m）、帆船酒店（321m）、卡延塔（380m）、公主塔（413m）的观光平台及配套商业设施进行了调研，并与其开发和运营团队进行了深入的交流。
- 8月17日，全球首部超500m高度的跃层电梯（通力电梯的专利技术）投入使用。跃层电梯技术是当今化解超高层建筑垂直运力难题的最前沿的技术。中信大厦的跃层电梯最大行程达514m，电梯运行速度4m/s，单台跃层电梯的乘客运力约是同规格施工电梯的10倍，是全球速度最快、运载能力最大、行程最长的跃层电梯。
- 8月18日，在施工现场举行中信集团慰问施工人员暨见证"结构工程超越北京第一高度活动"，集团董事长常振明、总经理、中信和业公司董事长王炯，中国建筑股份公司总经理王祥明、朝阳区领导，以及业主方、施工方、监理方代表参加了慰问。
- 8月22日，通力电梯芬兰总部公司一行人员来访，就跃层电梯技术与公司王伍仁、罗能钧及相关专业人员进行了交流。
- 8月24日，德国罗德集团一行人员来访，就中信大厦精装修区域窗帘布置方案与公司王伍仁、罗能钧及相关专业人员进行了交流。
- 8月份，公司完成了冰蓄冷盘管、消防水炮两项重要机电设备的选型工作，完成12项机电设备材料审批工作。
- 9月6日~13日，应通力集团及法策集团的邀请，王伍仁率设计、机电等专业人员赴欧洲开展对通力电梯、法策钢索的实地考察。
- 9月22日，公司组织《冰蓄冷系统优化方案》、《蓄冰槽保温防水优化方案》评审，确认了上述方案。
- 9月26日，召开总经理专题会，审议了Z7区功能需求落实方案，并审议了拟向集团汇报的文件。
- 9月27日，向中信集团总经理王炯汇报了《中信大厦集团办公区功能需求落实方案》，会议审定了Z7区F89~F102各楼层功能定位，

以及集团对于办公、接待、会议、用餐及配套设施等需求。
- 9月29日，与阿里巴巴（中国）有限公司签署了《关于北京CBD核心区Z15项目的购买意向协议》，该协议涉及Z5区约40,750m^{2}的物业。
- 9月30日，北京建筑设计研究院正式提交成果文件（V5.3版）图纸。
- 9月份，完成了暖通设备板式换热器、监控子项智慧阀设备、给排水橡塑保温、油脂分离器、消防水自动喷洒系统与水喷雾灭火系统等5项重要机电设备选型，完成9项机电设备材料审批；完成了夜景照明实体样板安装；TA-02、03跃层电梯投入运行。
- 10月8日，TA-01和TX-01两台跃层电梯经北京市质量技术监督局验收，10月17日两台跃层电梯正式投入运行，至此，四台跃层电梯投入使用，在业主方的主导下，创新性地解决了困扰超高层建设中垂直运力不足的难题。
- 10月17日，清华大学、绿色三星顾问来访，就首层大堂及观光大堂气流组织模拟等事宜与王伍仁、罗能钧及相关专业人员进行了研讨。
- 10月20日，听取了黎设计（日）对标识设计阶段成果汇报并工作标识设计方案的完善提出了具体要求。
- 10月22日，听取Azbil对能源管理系统方案汇报后，明确了能源管理顾问的工作范围。
- 10月31日，听取了KPF和北京建筑设计研究院对车库入口、观光入口方案及相关夜景照明的汇报，并提出了具体修订、完善要求。
- 10月份，完成了消防专业自动扫描定位喷水灭火装置系统、冰蓄冷及空调控制系统化学加药装置、电气专业普通高压电缆、给排水专业玛钢件及暖通专业橡胶软接等7项机电设备材料审批；完成1项报价审核、3项设计变更费用审核；完成地上钢结构二区段结算工程量复核；完成窗边风机盘管、VAV-BOX、空调机组及新风机组等的询价工作。
- 11月1日~3日，受伦敦"2016年纵览基础设施建设大赛"邀请，齐燕妮率队参加了大赛，我司获得此次大赛最高奖项"特别荣誉奖"。11月3日公司召开样板区标识工程研讨会，确定月底前提交样板标识工程实施界面划分说明、标识施工图、技术规格书、材料实物样板及色板（色号）等文件。
- 同日，上海复荣环境科技有限公司专家来访，就消除臭氧的耦合催化技术；恩威德公司专家来访，就有害气体及二氧化碳吸附技术与王伍仁、罗能钧及相关专业人员进行了交流。
- 11月10日，公司召开《样板电梯轿厢设计方案汇报会》，要求北京建筑设计研究院对通力和弘高公司的轿厢精装方案提出审核意见。
- 11月10日，公司召开《热力施工图》设计成果汇报会，确定了热力检修期间地下区域生活热水热力站电加热负荷的设计方向，并就热力站自管、托管或移交热力集团三种管控模式进行了对比分析。
- 11月11日，听取了北京建筑设计研究院、KPF、金螳螂对首层大堂样板方案的汇报，审查通过了《首层大堂样板方案》。
- 11月17日，组织北京建筑设计研究院、清华大学、物业顾问完成了对《首层大堂空调优化方案》的审核。要求巨柱和核心筒分别送、回风，结合装修方案确定回风口的位置，确保首层大堂空调系统气流的均匀性，保证大厦舒适性要求。
- 11月28日，听取了北京建筑设计研究院、Azbil公司、物业顾问及机电总承包单位针对能源管理设计实施方案，审查通过了《能源管理系统实施方案》。明确了能源管理系统相关采集点的内容、相关计量点的数量及计量原则。
- 11月份，公司组织北京建筑设计研究院、物业顾问、机电总承包就动态UPS应用与荷兰Hitec、德国Piller、美国Active Power、北京泓慧能源等厂家进行了三次技术交流，基于动态UPS的优势，要求各家针对中信大厦的设计提出技术方案。完成了给排水专业减震类设备（不锈钢软接、橡胶软接、橡胶减震垫及弹簧减震器），消防专业

消防泵、自动扫描定位喷水灭火装置系统、应急广播及背景音乐系统、暖通专业组合式空调机组、风机盘管温控器、化学加药装置、橡胶软接、动态平衡电动两通阀、硅玻钛金耐高温软管等 10 项重要机电设备/材料的选型审批；完成了夜景照明灯具调试及测试，通过样板测试，确定了灯具固定方式、电源电压等级、出线方式，具备了灯具及工程商比选条件，下一步公司将开展灯具、电源、控制设备等的询价工作，要求机电总承包春节前完成招标工作。

• 12 月 1 日，公司召开专题会，审核了《F3、F4 层精装方案》；确定将首层大堂样板实施范围由东南角两个紧邻样板区变更为以中轴线对称的东南角和西南角；样板工程包含巨柱和雨篷室外区域。
同日，飞利浦公司专家来访，就夜景照明技术与王伍仁、罗能钧及相关专业人员进行了交流。

• 12 月 20 日，王炯董事长主持召开了中信和业 2016 年度第 4 次董事长办公会。会议审议了《中信和业投资有限公司 2016 年度工作汇报》等 4 项议题，董事长肯定了 2016 年中信和业的总体工作成果，并针对组建大厦运维管理团队、建立创新管理体制和奖惩机制、多功能中心业态定位及设计工作提出了具体要求。

• 12 月 22 日，公司召开《标识设计方案》审定会，就样板层标识实施及后续工作提出了要求；会议确定了 F10、F11 层标识工程的实施方案及实施单位。

2017 年

• 3 月 15 日，启动大厦顶部优化设计工作。

• 3 月 24 日，王伍仁与 Azbil 公司不破庆一先生代表双方公司签署了"中信大厦能源管理系统及空调顾问合同"。

• 4 月 17 日，集团常振明董事长、王炯总经理和中信和业王伍仁总经理应邀与北京市主要领导研讨中信大厦建设高度及安全两个问题。

• 4 月 30 日，公司启动中信大厦 F10、F11 十字区以及标准办公楼层的装修样板工程。

• 8 月 22 日，公司举行"中信大厦主体结构工程封顶暨慰问建设者活动"。中信集团领导常振明、王炯、蔡华相、冯光、李庆萍，中建总公司领导官庆、王祥明等，中信和业公司王伍仁，中信银行孙德顺，中信集团办公厅曹诚，财务部曹国强，稽核审计部舒扬，党务工作部杨林，中信和业董事罗能钧等出席了活动。施工总承包、机电总承包、设计单位、监理单位、顾问单位、专业分包、大型设备供应商等参建方的主要负责人也参加了活动。

• 自 2012 年 7 月 25 日正式启动初步设计起至 2018 年 9 月 6 日中信大厦的安防方案通过专家审查，中信和业公司组织外聘专家对五方设计联合体的设计成果组织了 72 次审查会，共提出审核意见 8,148 条，被采纳 7,403 条，采纳率达 91%。其中初步设计审核意见 1,126 条，被采纳 847 条，采纳率 75%；施工图审核意见 7,022 条，被采纳 6,556 条，采纳率 93%。

• 9 月 21 日，公司组织完成了中信大厦塔冠调整设计方案审定。

• 9 月 30 日，公司完成了大厦 Z7 区样板层（F094、F101）技术文件 V2.0 审查。

• 10 月 31 日，王伍仁总经理率队参加第十六届中国国际社会公共安全博览会（深圳），重点调研了国内外安全防护行业最新技术发展趋势及产品（视频监控、速通门、门禁系统、人脸识别及防爆安检设备等）。

• 11 月 23 日，公司组织造价顾问（QS）公司与公司机电、装饰、造价等部门及专业组长以上人员就如何从深化设计、施工、现场管理等各环节严控造价进行研讨。

• 12 月 13 日，公司与中信银行和阿里巴巴公司就《中信大厦中信银行与阿里巴巴区域行政、会议层实施方案》进行沟通。

• 12 月 26 日，公司完成了大厦 B1、Z4 区、Z6 区、Z7 区各区厨房设备清单及技术规格书审查。启动厨房设备采购比选工作。

2018 年

• 1 月 15 日，公司召开"保竣工计划实现 决战动员会"，总经理王伍仁、副总经理罗能钧及工程总监、结构、装饰、机电等各板块负责人、专业组长、施工总承包、机电总承包及专业分包负责人参会。

• 1 月 15 ~ 17 日，公司召开"中信大厦项目 2018 年重点工作计划分析、研讨会"，组织施工总承包单位、机电总承包单位及专业分包单位负责人共同研讨 2018 年保竣工重点工作计划。

• 1 月 28 日，应中信银行提前展开银行区的精装工程请求，公司决定将银行区域的装修工程列入施工总承包工程合同，并确保在国家消防总局改制前获得银行区域精装修的消防设计审批。

• 2 月 6 日，签署"Z7 区及 Z8 区室内精装修工程合同"，中信和业总经理王伍仁，中建三局副总经理汤才坤，苏州金螳螂董事长王汉林以及相关负责人参加签约仪式。

• 2 月 6 日，公司召开 2018 年工作会议，传达了中信集团 2018 年工作会议精神，回顾了 2017 年的工作。对 2017 年责任心、执行力、创新力上有突出表现的员工、管理者及优秀团队进行了表彰。会议"史无前例"历时近 8 小时，将直接关系到中信大厦 2018 年底实现竣工的 12 个主要专业倒排工期节点计划，从实物工程量、交安交装计划、产值影响、商务因素、专业搭接、难点排除等方面逐项、逐条、逐月梳理，形成了可落地的年度行动目标。

• 3 月 6 日，公司举行《2018 年岗位责任书》签字仪式。将 195 个年度工作的关键节点，依据责任体系，自总经理、副总经理、总经理助理、部门经理、专业组长逐级分解，签订了 28 份《岗位责任书》，将 2018 年的工作任务落实到人。

• 中信大厦消防性能化设计及防火分区审核历时三年余，通过了朝阳区消防支队审查（获得批复）。

• 3 月 19 日，公司针对中信大厦 Z1-Z4 十字区卫生间召开专题会议，决定对 Z1-Z4 区标准办公区的男卫生间恢复原设计。

• 4 月 2 日，公司组织两个总包单位负责人，及 20 余家专业分包法人单位负责人召开了"中信大厦项目 2018 年竣工决战誓师大会"。王伍仁指出，"行百里者半九十"，最后六个月的"决战"将决定中信大厦的总工期目标能否得以实现，合是否得到履行，要求参建单位继续拼搏争取决战的胜利。

• 4 月 12 日晚，总经理王伍仁带领相关单位技术人员，对大厦夜景照明工程的 F66~F72 线型灯具样板效果进行审查，通过在不同距离的观测，对中信大厦夜景照明效果、动画播放的流畅性、清晰度等关键指标进行实地察看。

• 4 月 17 日，公司召开中信大厦安防系统方案汇报会，听取速通门管理系统、防爆安全检查系统、视频监控系统、梯控系统、门禁管理系统、人脸识别系统、访客管理系统、安防通讯、停车场管理系统等十余项安防方案的汇报，就智能化工程的推进提出了方向性及工期要求。

• 4 月 23 日，针对自 3 月 27 日以来跃层电梯主机控制板、控制电源保护板多次烧毁，频繁跳闸，严重影响施工垂直运输的问题，召集两个总包技术负责人紧急研讨，为根本性解决垂直运力问题，王伍仁总经理决定提前启用大楼正式供电系统，替代施工总承包的临时供电系统，要求改造施工现场的临时供电系统，并就如何进行改造进行了部署。此举不仅解决了施工电源的可靠性问题，也彻底解决了因跃层电梯宕机导致垂直运输严重阻塞施工进度问题。

• 5 月 14 日，中信银行、行政楼层和中信集团 Z7 办公区取得消防支队的《建筑工程消防设计审核合格意见书》，此项《合格书》的取得为大厦整体消防验收和银行办公区早日启用奠定了坚实的基础。
从中信大厦 2013 年正式开工以来的五年间，公司先后取得了 8 份消防设计审核意见书，完成了多项方案优化（如：按国家颁布的消防新规范对防火分区进行调整、大厦南侧雨篷删除、添加等引发的消防设计变更）。

• 5 月 24 日，中信大厦地下室重要机房提前接入了中国移动和中国电信手机的无线通信，为地下室内的重要机房、机电系统联合调试等提供了通信保障。

• 5 月 30 日，经过参建单位一个多月夜以继日的奋战，中信大厦的制冷主机如期实现开机测试运行，这一目标的实现标志着中信大厦的空调进入系统调试及测试阶段。

• 6 月 1 日，中信和业携手国际建筑业主与管理者协会（Building Owners and Managers Association，简称 BOMA），联合中国 4 家国内商业地产领域的优秀企业，共同发布了《国际写字楼分级指南》汉化版（行业标准指导性文件）。中信和业在仪式上分享了中信大厦建造的创新经验。

• 6 月 6 日，中信大厦重要机房开通了中国移动 4G 网络；6 月 7 日相继开通中国电信 3G 及 4G 网络。中信大厦地下室等重要机房手机通讯系统较合同提前 3 个月开通，为中信大厦建设的"决战"之年提供了良好的通讯条件。

• 6 月 15 日，中信大厦完成了 M8R8（F104）及以下雨水系统工程施工（包括雨水管道系统连通、管道系统试验、减压水箱试验等），通过提前投用正式雨水系统，确保大厦雨季的排水安全。

• 6 月 20 日，经格力及参建单位的努力，中信大厦制冷主机实现了与格力珠海总部远程智能服务中心的衔接。远程智能监测的衔接，为中信大厦空调系统的调试和节能运行提供了可靠保障，提升了大厦运行安全性及智能化运维水平。

• 6 月 22 日，公司携手国际建筑业主与管理者协会（BOMA）、世界高层建筑与都市人居学会（CTBUH）、高层建筑人居环境委员会（CITAB）联合主办的"中信大厦"论坛——超高层建筑建设运营一体化研讨会。该论坛的宗旨是对标国际超高层建筑行业标准，从自持型超高层建筑业主的角度，以中信大厦的开发建设和未来运维的策划推动行业进步。第一期论坛对标了国际标准，交流了中信大厦开发建设实践。

• 6 月 26 日，经过 62 天的日夜奋战，中信大厦的最后一台跃层电梯（TX-01）完成从施工用电梯向正式电梯的转换，并通过政府验收。4 台跃层 / 消防电梯较计划提前 20 天完成转换，极大地缓解了施工垂直运输压力。

• 7 月 14 日，中信集团董事长常振明率王伍仁等参加北京市长专题会，请求市长帮助解决中信大厦正式电源接入问题。王伍仁就供电需求向参会的国家电网北京公司总经理进行了说明。

• 7 月 23 日，召开专题会，为防止北京 7 月 23 日晚~24 日因台风"安比"过境形成的风雨对中信大厦造成损害，组织施工总承包、机电总承包、监理以及幕墙、消防分包等单位部署了中信大厦的防汛工作。

• 8 月 7 日，Z15 地块项目被北京市优质工程评审委员会评为"2017~2018 年度结构长城杯金质工程"。

• 8 月 9 日，为实现集团领导提出的中信大厦从开发建设向运营管理"无缝、平稳过渡"要求，公司举行了中信大厦运营管理培训（第一期）。BOMA 中国执行总裁 Dominic Lau 以洛杉矶怡安中心和北京嘉铭中心为案例，介绍了 BOMA 制定的国际商业地产运营管理体系，以及助推中国商业地产提升运营管理水平的途径。

• 8 月 20 日，北京市发改委副主任、能源办主任王英建一行到中信大厦现场了解项目总体情况和电力设施建设。朝阳区代区长文献，副区长李国红，总经理王伍仁、副总经理罗能钧等陪同调研。

• 8 月 24 日，公司组织两个总包及各专业分包召开"总承包剩余工程销项计划汇报会"，审核剩余工程量销项计划，并对重要事项进行了协调及部署。公司将据此与施工、机电总承包签订《补充协议》，并调整奖惩节点设置。

• 9 月 4 日，公司召开中信大厦初步接收（PAC）证书颁发前置工作专题会，成立 PAC 证书颁发仪式筹备委员会、成立 PAC 证书颁发前验收领导小组、成立竣工资料验收小组，并提出了具体的工作要求。

• 9 月 5 日，国家电网发展策划部副主任张正陵，国电北京电力公司总经理万志军，发展策划部主任陈斌发等一行人员到中信大厦现场调研，并就解决中信大厦供电事宜与总经理王伍仁，副总经理罗能钧等进行了磋商。

• 9 月 6 日，召开《中信大厦安防方案》专家论证会，总经理王伍仁，副总经理罗能钧，机电总监以及中信大厦项目设计、监理、机电总承包、安防专业分包单位和供应商等各单位代表参加了会议。会议邀请公安部第一研究所、空军工程设计研究局以及中国公安大学的 5 位安防行业领域内的资深专家，对《中信大厦安防方案》进行评审。专家组认为：中信大厦的安防系统设计、施工标准远高于普通商业建筑，代表了国内当今超高层建筑的顶尖水平。

• 9 月 20 日，中信集团举行"中信大厦计划竣工倒计时 90 天座谈会"，中信集团及中国建筑领导王炯、王祥明，中信股份曹国强，王伍仁，罗能钧以及中信集团职能部门负责人，项目总包单位、分包单位主要负责人参加了会议。在双方领导的见证下，王伍仁、陈华元、田强分别代表中信和业、中建三局、中建安装签署了《确保中信大厦 2018 年年底竣工的补充协议》。

• 9 月 25 日，公司召开"关于 PAC 证书颁发前应完成工作梳理汇报会"，各专业组长对各专业 PAC 证书颁发前应完成工作进行汇报。

• 10 月 6 日，总经理王伍仁在现场发现因 10kV 施工临时供电设备无移动条件，导致无法形成消防环路的重大问题，王伍仁总经理在现场做出决定：委托机电总承包单位（1）改造施工用电的路由和临时供电、配电设备位置；（2）接管施工总承包的施工临时用电系统。要求两个总包要协同努力，确保中信大厦消防环路在 2018 年 11 月得以形成。

• 10 月 10 日，中信集团总经理办公会听取了中信和业关于中信大厦工程收尾及大厦运行维护方案的汇报后决定：中信大厦的建设要与大厦的运维无缝衔接，在 2020 年 12 月颁发工程最终接收证书（FAC）前，由中信和业公司组建专门运行维护部门，承担中信大厦的承接查验、运维管理等工作，中信和业要按照市场化、专业化和规范化要求运作，确保大厦运营管理安全、优质和高效，过渡期结束后组建中信大厦的专业运维公司。

• 11 月 14 日，中信集团副董事长、总经理王炯带队专程拜会新成立的国家应急管理部领导以及国家消防总局、北京消防局、朝阳消防分局负责人。王伍仁和罗能钧汇报了中信大厦的消防设计、施工进展和消防验收准备情况，应急管理部领导对中信大厦消防系统的设计和消防设备的配置，以及公司的重视程度给予了高度评价，他们将支持中信大厦的消防验收，要求中信大厦的消防系统设置必须严格按照"百分百满足法律、法规要求，百分百满足消防性能化设计的要求"。

11 月 14 日 中信大厦防雷接地系统通过第三方检验（评定为"优秀或极佳"）。

• 11 月 15 日，中信大厦冬季供暖工程正式投入运行。

• 11 月 27 日，举行中信大厦消防验收启动会。中信和业总经理王伍仁、北京市朝阳区消防支队队长高永路、副总经理罗能钧、两个施工总承包项目负责人汤才坤、曾运平、许立山、丁锐、刘庆海，监理公司张金涛等出席，朝阳区消防支队防火处约 20 名消防警官以及参建单位代表百余人参会。

• 11 月 28 日，中信大厦斩获亚洲地区地产领袖峰会（MIPIM ASIA）2018 年中国最佳未来大型项目金奖（Best Chinese Futura Mega Project）。

• 12 月 6 日，集团总经理王炯率王伍仁等拜访朝阳区委书记及区长等，请求帮助解决大厦供电、大厦命名、北侧红线外区域认养、消防环路审批等事宜。

• 12 月 14 日，公司取得北京市规划和国土资源管理委员会下发的《建设工程档案预验收意见书（合格）》。

• 12 月 25 日，中信集团召开党委会，会议决定按合同约定向中信大厦的总承包商颁发中信大厦的初步接收证书（PAC）。

• 12 月 28 日，中信大厦初步接受证书颁发仪式在中信大厦的 F11 举行，中信集团董事长常振明从中国建筑股份公司总经理王祥明手中接过金钥匙，并向中国建筑总经理王祥明颁发了中信大厦第一批初步接收证书。

通过近八年的"马拉松赛跑"，中信和业投资有限公司在中信集团党委的坚强领导下，率领八十余家设计及专业顾问公司的数百名设计师与工程专家、四百余家材料、设备制造厂及施工单位数千名工程师，以及数万名工人组成的"联合舰队"，披星戴月，日夜兼程，耗费了约 816 万工日的辛劳，于 2018 年底艰难地完成了中信集团交予的中信大厦建造任务。

主要设备及材料供应商（施工总承包）				
序号	主材或设备名称	制造厂	产地	备注（性能）
1	钢板	江阴兴澄特种钢铁有限公司	中国	-
2	钢筋	河北钢铁集团承德分公司	中国	
3	压型钢板	森特士兴集团股份有限公司	中国	
4	钢筋桁架板	北京多维联合集团香河建材有限公司	中国	-
5	防火涂料	四国化研有限公司	日本	厚型耐火 3h
		北京凌鹰科技发展有限公司	中国	厚型耐火 2.5h
6	高强螺栓	春雨（东莞）五金制品有限公司	中国	
7	栓钉	尼尔森植焊（天津）有限公司	中国	-
8	商品混凝土	北京市中超混凝土有限责任公司	中国	
		北京建工新型建材有限责任公司		
		北京新奥混凝土集团有限公司		
9	ALC 条板	天津天筑建材有限公司	中国	
10	吊顶矿棉板	德国可耐福爱孚管理有限责任公司	德国	吸音系数高达 0.76，具有超高吸声、洁净、超轻品质。与抗震龙骨搭配，既减轻了大楼的荷载，又给现代化办公提供了安静舒适的环境
11	抗震龙骨	大连舒心门业有限公司	中国	国家专利产品
12	软膜顶棚	广州朗域实业有限公司北京分公司	中国	防火性能 A 级
13	吊顶及墙面铝板	上海新大余氟碳喷涂材料有限公司	中国	6mm、12mm 蜂窝铝板薄型工艺
		广州金霸建材有限公司		
		萨克森建筑新型材料（廊坊）有限公司		
14	墙面岩板	北京申洋远航商贸有限公司（LAMINAM）	意大利	御火耐温、御击抗磨、御水防渗、御色不变
		佛山德赛斯建材有限公司	西班牙	
15	不锈钢板	北京康盛伟业科技发展有限公司	中国	双曲面、异形造型
16	壁纸	北京银河家墙纸有限公司	中国	防火性能 B1 级
		吴江优丽奇装饰材料贸易有限公司		
17	磁性漆	境洁环保科技（上海）有限公司	中国	美国 FDA 检测认证
18	成品隔断墙	驰瑞莱工业（北京）有限公司	中国	现场全装配化安装，隔声性能好 饰面种类多，满足防火性能要求
19	卫生间隔断	上海缘木家具有限公司	中国	
20	瓷砖	杭州诺贝尔陶瓷有限公司	中国	符合国家标准，吸水率优于国家标准
		东莞市唯美陶瓷工业园有限公司		
		信益陶瓷（中国）有限公司		
21	安哥拉棕石	厦门万里石股份有限公司	中国	环保性能通过国家检测
22	白色云朵拉灰石材	中成国泰石业有限公司	中国	吸水率 0.37%
23	GRC 网络地板	常州华通新立地板有限公司	中国	采用业内最先进的材料、科技与工艺，不仅荷载高、自重轻，而且防火等级达到 A1 级
24	地毯	美国美利肯公司	中国	Miliken Carpets

25	水泥自流平	圣戈班伟伯（北京）建材有限公司	中国	
		汉高黏合剂有限公司		
		北京诚成豪信建材有限公司		
26	钛瓷涂料	青岛普泰纳米新材料科技有限公司	中国	车库地坪耐磨涂料
27	屋面聚脲防水涂料	天津森聚科密封涂层材料有限公司	中国	技术领先
28	防火门	亚萨合莱天明（北京）门业有限公司	中国	
		无锡茂泰特种门有限公司		
29	防火卷帘	北京蓝盾创展门业有限公司	中国	
30	电动平开门	纳博克自动门（北京）有限公司	中国	-
31	折叠门	奢士拓（中山）智能科技有限公司	中国	SILENBLOCK 专利
32	气密平滑自动门（带应急功能）气密平开门、标准平开门	深圳市门老爷科技有限公司	中国	弱化超高层建筑的"烟囱效应"，改善门面产品节能环保性能，门面整体设计和智能系统解决方案定制
33	旋转门	北京宝盾门业技术有限公司	中国	
34	电动窗帘	望瑞门遮阳系统设备（上海）有限公司	德国	智能化新型遮阳系统
35	五金	亚萨合莱保安制品（深圳）有限公司	瑞典	-
36	卫生间一体化钢架	中建一局建设发展有限公司	中国	同层排水新型工艺
37	洁具	东陶（中国）有限公司	中国	节水环保性能世界领先
38	人造石	杜邦中国集团有限公司	中国	耐磨耐污染
39	橱柜及木饰面	天津祖阁汇致家具有限公司	中国	防火性能 B1 级

主要专业分包（施工总承包）		
序号	单位名称	合同责任（施工范围）
1	中建钢构有限公司	第二、四、六区段钢结构制作、钢结构安装
2	江苏沪宁钢机股份有限公司	第一、三、八区段钢结构制作
3	宝钢钢构有限公司	第五区段钢结构制作
4	武昌船舶重工集团有限公司	第七区段钢结构制作
5	北京江河幕墙系统工程有限公司	幕墙工程
6	苏州金螳螂建筑装饰股份有限公司	F003 层会议中心、首层大堂、Z7 区、中信银行 Z2 区精装修
7	北京华美装饰工程有限责任公司	Z7 区核心筒外的办公区标准装修、办公区核心筒内精装修
8	浙江亚厦装饰股份有限公司	Z3、Z5 空中大堂、地下大堂、中信银行 Z3 特殊层精装修
9	中国建筑装饰集团有限公司	Z1、Z4 区核心筒外的办公区标准装修、办公区核心筒内精装修
10	深圳市建筑装饰（集团）有限公司	Z2、Z5 区核心筒外的办公区标准装修、办公区核心筒内精装修
11	中建深圳装饰有限公司	Z3、Z6 区核心筒外的办公区标准装修、办公区核心筒内精装修，中信银行 Z3 区标准层精装修
12	北京市弘高建筑装饰设计工程有限公司	中信银行 Z1 区精装修
13	上海直玖机场设备有限公司	停机坪施工

MAJOR EQUIPMENT AND MATERIAL SUPPLIERS (CONSTRUCTION GENERAL CONTRACTORS)				
NO.	MAIN MATERIAL OR EQUIPMENT NAME	MANUFACTURING PLANT	PLACE OF ORIGIN	REMARKS (PERFORMANCE)
1	Steel Plate	Jiangyin Xingcheng Special Steel Works Co., Ltd.	China	-
2	Rebar	Hbis Group Company Limted. Chengde Branch	China	
3	Profiled Steel Sheet	Center Int Group Co., Ltd.	China	-
4	Steel Bar Truss Plate	Beijing Multidimensional United Group Xianghe Building Material Co., Ltd.	China	-
5	Fire Retardant Coating	SKK Co., Ltd.	Japan	Thick fire resistance for 3 hours
		Beijing Lingying Technology Development Co., Ltd.	China	Thick fire resistant 2.5 hours
6	High-Strength Bolt	ChunYu (Dongguan) Metal Products Co., Ltd.	China	-
7	Stud	Nelson®Stud Welding & Fastening Co., Ltd.	China	-
8	Ready-Mixed Concrete	Beijing Zhongchao Concrete Co., Ltd.	China	
		Bceg Advanced Construction Materials Co., Ltd.		
		Beijing Xinao Concrete Group Co., Ltd.		
9	Alc Panel	Tianjin tianzhu building materials co. Ltd.	China	
10	Mineral Fiber Ceiling Board	Germany Knauf Aifu Management Co., Ltd.	Germany	The sound absorption coefficient is as high as 0.76, with super high sound absorption, clean and ultra-light quality. Paired with the anti-seismic keel, it reduces the load on the building and provides a quiet and comfortable environment for modern office
11	Anti-Seismic Keel	Dalian Superego Door Products Co., Ltd.	China	National patent product
12	Stretch Ceiling	Guangzhou M-C Co., Ltd. Beijing Branch	China	Fire performance class A
13	Ceiling And Wall Aluminum Plate	Shanghai Xindayu Fluorocarbon Coating Material Co., Ltd.	China	6mm,12mm honeycomb aluminum plate thin process
		Guangzhou Gold Kings Building Material Co., Ltd.		
		Saxon Building New Materials (Langfang) Co., Ltd.		
14	Rock Wall Plate	Beijing shenyang yuanhang trading co., LTD. (LAMINAM)	Italy	Royal fire resistance, anti-wear, anti-wear, anti-seepage, and color
		Foshan Desais Building Material Co., Ltd.	Spain	
15	Stainless Steel Plate	Beijing Kangshengweiye Technology Development Co., Ltd.		Hyperboloid, Special shape
16	Wallpaper	Goodrich Global Pte Ltd.	China	Fire performance B1
		Beijing Galaxy Home Wallpaper Co., Ltd.	China	
17	Magnetic Paint	Jiejie Environmental Protection Technology (Shanghai) Co., Ltd.	China	US FDA testing certification
18	Finished Partition Wall	TRL Group (Beijing) Co., Ltd.	China	The site is fully assembled and installed, with good sound insulation performance and a wide variety of finishes to meet fire performance requirements
19	Toilet Partition	Shanghai Yuanmu Furniture Co., Ltd.	China	
20	Ceramic Tile	Hangzhou Nabel Group Co., Ltd.	China	In line with national standards, water absorption is better than national standards
		Dongguan Wonderful Croup Co., Ltd.		
		Champion-Tile (China) Co., Ltd.		
21	Angola Brown Granite	Xiamen Wanli Stone Stock Co., Ltd.	China	Environmental performance through national testing
22	White clouds pull grey stone	Zhongcheng Guotai Stone Industry Co., Ltd.	China	Water absorption rate 0.37%
23	GRC Network Floor	Changzhou Huatong-Floor Co., Ltd.	China	Huatong GRC floor use the most advanced materials, containing the most advanced technology and technology in the industry, It's not only high load and light weight, but also reached A1 fire grade

24	Carpet	Miliken Co.,Ted.	China	Miliken Carpets
25	Cement Self Leveling	SAINT-GOBAIN Weber (Beijing)Building Materials Co.,Ltd.	China	
		Hangao Adhesive Co., Ltd.		
		Beijing Cheng Cheng hao Xin Building Materials Co., Ltd.		
26	Titanium Ceramic Coating	Qingdao Putai Nano materials Technology Co., Ltd.	China	Wear resistant paint for garage floor
27	Polyurea Coating for Roof Waterproof	Tianjin Shinjuker Coating & Sealing Co., Ltd.	China	Leading technology
28	Fire Door	ASSA ABLOY Tian Ming (Beijing) Door Industry Co., Ltd.	China	
		Wuxi Moritec Special Door Co.,Ltd.		
29	Fire Shutter Door	Beijing Lan Dun Chuangzhan Door Industry Co., Ltd.	China	
30	Electric Side-hung Door	Nabco Autodoor (Beijing) Co.,Ltd.	China	-
31	Accordion Door	Cexito (Zhongshan) Intelligent Technology Co., Ltd.	China	SILENBLOCK patent
32	Airtight and smooth automatic door (with emergency function), airtight swing door, standard swing door	Lafaya Technology Co.,Ltd.	China	Weaken the "Chimney Effect" of super high-rise buildings, improve the energy-saving and environmental protection performance of facade products, and design the overall design of the facade and intelligent system solutions
33	Revolving Door	Shenzhen Lafaya Technology Co.,Ltd.	China	
34	Electric Curtain	Warema Sun Shading Systems Shanghai Co., Ltd.	Germany	Intelligent new sunshade system
35	Hardware	ASSA ABLOY Security Products(Shenzhen) Co. Ltd	Sweden	-
36	Toilet Integrated Steel Frame	China Construction First Group Construction & Development Co., Ltd.,	China	New technology for the same layer drainage
37	Sanitary Appliance	Toto (China) Co.,Ltd	China	Water-saving and environmental performance leading the world
38	Artificial Stone	Dupont China Holding Co., Ltd.	China	Wear and stain resistance
39	Cabinet And Timber Finish	Tianjin Zuge Huizhi Furniture Co., Ltd.	China	Fire performance B1

MAIN PROFESSIONAL SUBCONTRACTORS (CONSTRUCTION GENERAL CONTRACTORS)		
NO.	COMPANY NAME	CONTRACTUAL LIABILITY (CONSTRUCTION SCOPE)
1	China Construction Steel Structure Corp.ltd	Manufacture and Installation of the Z2, Z4 & Z6
2	Jiangsu HNGJI Machinery Co., Ltd.	Manufacture of Steel Structure of the Z1, Z3 & Z8
3	Baosteel Co., Ltd.	Manufacture of Steel Structure of the Z5
4	Wuchang Shipbuilding Industry Group Co., Ltd.	Manufacture of Steel Structure of the Z7
5	JANGHO GROUP Co.,Ltd.	Curtain Wall Engineering
6	Suzhou Gold Mantis Construction Decoration Co.,Ltd.	Fine Decoration of F003 Floor Conference Center, First Floor Lobby, Z7, Z2 of China CITIC Bank
7	Beijing Huamei Decoration Engineering Co., Ltd.	Standard Decoration of the Office Area outside the Core Tube and Fine Decoration of the Office Area in the Core Tube in Z7
8	Zhejiang YASHA Decoration Co., Ltd.	Fine Decoration of Z3 & Z5 Sky Lobby, Underground Lobby, Special Floors in Z3 of China CITIC Bank
9	China State Decoration Group Co., Ltd.	Standard Decoration of the Office Area outside the Core Tube and Fine Decoration of the Office Area in the Core tube in Z1 and Z4
10	Shenzhuang Group Co., Ltd.	Standard Decoration of the Office Area outside the Core Tube and Fine Decoration of the Office Area in the Core Tube in Z2 and Z5
11	CSCEC Shenzhen Decoration Co., Ltd.	Standard Decoration of the Office Area outside the Core Tube and Fine Decoration of the Office Area in the Core Tube in Z3 and Z6; Fine Decoration of Standard Floors in Z3 of China CITIC Bank
12	Beijing Honggao Architectural Decoration Design Engineering Co., Ltd.	Fine Decoration of Z1 of China CITIC Bank
13	Shanghai Heliport-9 Airport Equipment Co., Ltd.	Construction of Parking Apron

主要设备及材料供应商（机电总承包）				
序号	合同责任	供应商或制造厂	产地	备注
1	制冷机组供应	珠海格力电器股份有限公司	中国	双工况永磁同步变频离心机，COP 为 6 .46，IPLV 为 9 .57
2	空调机组、风机盘管供应	江苏风神空调集团股份有限公司	中国	"风神"超薄型窗式风盘、超静音组合式空气处理机组、新风热回收机组、干工况风盘实现了中信大厦的构建的一体化集成空调机组，节省空间、消声效果好；大温差超薄静音风机盘管，安全、舒适、环保、节能的目标
3	高压垂吊电缆供应	远东电缆有限公司	中国	世界最长高压垂吊电缆，安装方便，节约空间
4	综合能源管理系统及智慧阀供应	阿自倍尔自控工程（上海）有限公司	日本	综合能源管理系统提供动态的能耗分析与能效评估，智慧阀具备动态平衡、电动调节和冷热量计量等功能
5	VAV、FASU、低温风口供应	北京协力东方科技发展有限公司	日本	第二代 VAV 测量数据稳定且精度高，气流稳定，整流效果好，节能。FASU 单元工厂化生产，各支路风量平衡性好
6	柴油发电机组供应	合肥康尔信电力系统有限公司	中国	国际顶尖技术采用奔驰 MTU 高压共轨式电喷柴油发机组，具有节能、环保、高可靠性特点
7	LED 超薄集成灯盘供应	江苏新广联光电股份有限公司	中国	超薄集成化创新设计，光效达到 110 lm/W
8	隔离开关、断路器等电气开关变频器、电力监控系统供应	施耐德电气（中国）有限公司	法国	智能化电力监控设备 Smart Panel 实现低压系统中的配电设备信息互通，动态监控
9	网络设备、程控电源供应	华为技术有限公司	中国	无线架构技术领先，板卡式 AC 有效解决无线数据流量的瓶颈；支持业务随行功能；AP 实测速率高
10	数字安防设备供应	杭州海康威视数字技术股份有限公司	中国	一级平台、安防设备管理网和建筑设备管理网核心交换机采用 CLOS 架构，业界领先；支持数据中心特性的跨三层迁移技术，提升数据迁移效率；安全设备兼容性好，具备高级别的安全联动功能
11	夜景照明灯具供应	欧司朗（中国）照明有限公司	德国	满足国际标准的 DMX512 协议，具备标准的 RDM 自动巡检功能
12	火灾自动报警及消防联动系统供应	霍尼韦尔消防安防系统（上海）有限公司	美国	NFS2 -3030 为霍尼韦尔最高端系列产品，生产工艺自动化程度高，性能成熟稳定
13	电能质量产品供应	帝森克罗德集团有限公司	中国	电能质量产品技术全国领先，核心产品寿命可达 15 年以上
14	高、低压变配电开关柜供应	上海柘中电气有限公司	中国	核心主件采用施耐德元器件
15	变压器供应	江苏华鹏变压器有限公司	中国	
16	配电箱供应（部分）	江苏恒凯电气有限公司	中国	
17	配电箱供应（部分）	江苏弘历电气有限公司	中国	
18	PLC 控制器接触器及 VAV 控制器供应	西门子（中国）有限公司	德国	西门子产品处理速度快，网络接口多，存储器容量大，可在 -20 ℃ 至 60 ℃ 的温度下稳定运行
19	空调水泵给水泵组污水提升装置供应	格兰富水泵（上海）有限公司	丹麦	
20	冷却塔供应	东莞菱和宝德冷热设计有限公司	中国	引风式冷却塔体内负压，降低飘水率，提高冷却塔的热力性能，有效防止细菌滋生
21	消防风机供应	浙江双阳风机有限公司	中国	
22	风机供应	台州华德通风机有限公司	中国	
23	冰盘管供应	杭州源牌环境科技有限公司	中国	采用聚合物基纳米复合材料，结冰层厚度更薄为 18-20mm，结冰、融冰好性好；耐腐蚀，使用寿命达 30 年以上；重量轻，韧性好
24	水阀供应	广东永泉阀门科技有限公司	中国	空调系统的防结露电控一体阀，不仅节能显著，还实现了智慧泵房；具有防止"水锤"的安全性能
25	耦合催化板供应	上海复荣环境科技有限公司	中国	可以降解静电产生的臭氧，高效分解 TVOC
26	DUPS 飞轮储能系统	Euro-Diesel	比利时	采用 DUPS 飞轮储能系统替代传统 UPS 蓄电池，更加安全可靠、环保
27	光伏系统供应	北京汉能薄膜太阳能电力工程有限公司	德国	采用进口的 145 Wp CIGS 薄膜太阳能电池，光电转换效率为 15 .43 %，在同类产品中转换效率最高
28	电梯的供应及安装	通力电梯有限公司	芬兰	跃层电梯大大提高施工阶段的垂直运力；穿梭电梯采用碳纤维曳引绳，使用寿命可达到 15 年，减少能耗及钢丝绳维修更换时间
29	擦窗机的供应及安装	上海考克斯擦窗机设备工程有限公司	西班牙	

30	内保温风管	山东品通机电科技有限公司	中国	消声保温一体，抗菌防霉

主要专业分包（机电总承包）			
序号	合同责任	单位名称	备注
1	低区空调工程	中建三局安装工程有限公司	
2	高区空调工程	中建一局集团建设发展有限公司	
3	暖通设备监控工程	杭州源牌环境科技有限公司	
4	冷源工程	杭州华电华源环境工程有限公司	
5	热源工程	北京东方中远市政工程有限责任公司	
6	消防工程	北京四海消防工程有限公司	
7	夜景照明工程	豪尔赛科技集团股份有限公司 / 北京良业环境技术有限公司（联合体）	
8	综合布线、机房工程	中建电子信息技术有限公司	
9	智能化工程	深圳市智宇实业发展有限公司	
10	中区电气工程	上海唯中建设有限公司	
11	窗台一体化系统 一体化钢制门	中建五洲工程装备有限公司	

MAJOR EQUIPMENT AND MATERIAL SUPPLIERS (MECHANICAL AND ELECTRICAL GENERAL CONTRACTOR)				
NO.	CONTRACTUAL LIABILITY	SUPPLIER OR MANUFACTURER	PLACE OF ORIGIN	REMARKS
1	Supply of Refrigeration Unit	Zhuhai Gree Electric Appliances, Inc.	China	Dual-mode permanent magnet synchronous frequency conversion centrifuge with COP of 6.46 and IPLV of 9.57
2	Supply of Air Conditioning Unit and Fan Coil	Jiangsu Fengshen Air Conditioner Group Co., Ltd.	China	"Aeolus" ultra-thin window wind disk, ultra-quiet combined air handling unit, fresh air heat recovery unit, dry working condition wind disk realizes the integrated air conditioning unit constructed by CITIC Tower, saving space and eliminating sound effects; Huge temperature difference ultra-thin silent fan coil, safety, comfort, environmental protection, energy saving goals
3	Supply of High-voltage Hanging Cable	Far East Cable Co., Ltd.	China	The world's longest high-voltage hanging cable is easy to install and saves space.
4	Supply of Intelligent Valve & Integrated Energy Manage-ment	Azbil Control Solutions (Shanghai) Co., Ltd.	Japan	Integrated energy management systems provide dynamic energy analysis and energy efficiency assessments, The smart valve has functions such as dynamic balance, electric adjustment and cold heat metering
5	Supply of VAV, FASU and Cold Air Diffuser	Beijing Xieli Dongfang Technology Development Co. Ltd.	Japan	The second-generation VAV measurement data is stable and has high precision, stable airflow, good rectification effect and energy saving. The FASU unit is factory-produced, and the air volume balance of each branch is good.
6	Supply of Diesel Generator Sets	Hefei Calsion Power System Co. Ltd.	China	The international top technology adopts Mercedes-Benz MTU high-pressure common rail EFI diesel generator unit, which has the characteristics of energy saving, environmental protection and high reliability
7	Supply of LED Ultra-thin In-tegrated Light Panel	Jiangsu XGL Opto. Co., Ltd.	China	Ultra-thin integrated and innovative design, the light efficiency reaches 110lm/W.
8	Supply of Isolation Switch, Circuit Breaker and Other Electrical Switch, Inverter, Power Monitoring System	Schneider Electric (China) Co., Ltd.	France	Intelligent power monitoring equipment (Smart Panel realizes information exchange and dynamic monitoring of power distribution equipment in low-voltage systems)
9	Supply of Network Equip-ment, Program-controlled Power	Huawei Technologies Co., Ltd.	China	Leading wireless architecture technology, board-type AC effectively solves the bottleneck of wireless data traffic; supports business mobility; AP measured rate is high
10	Digital Security Equipment Supply	Hangzhou HIKVISION Digital Technology Co. Ltd.	China	Level 1 platform, security equipment management network and construction equipment management network The core switch adopts the CLOS architecture and is leading the industry;Support cross-Layer 3 migration technology for data center features to improve data migration efficiency; The safety equipment has good compatibility and has a high level of safety linkage.
11	Nightscape Lighting Fixtures Supply	OSRAM (China) Lighting Co., Ltd.	Germany	Meet the international standard DMX512 protocol, With standard RDM automatic inspection function.
12	Supply of Automatic Fire Alarm and Fire Linkage Sys-tem	Honeywell Fire Security system (Shanghai) Co., Ltd.	United States	NFS2-3030 is Honeywell's highest-end product line. The production process is highly automated and the performance is mature and stable.
13	Power quality product supply	TYSEN-KLD Co., Ltd.	China	Power quality product technology is leading the country, and the life of core products can reach more than 15 years.
14	Supply of High and Low Voltage Power Distribution Switchgear	Shanghai Tuozhong Electric Co., Ltd.	China	The core main parts adopt schneider components
15	Transformer Supply	Jiangsu Huapeng Transformer Co., Ltd.	China	
16	Distribution Box Supply (Part)	Jiangsu Hengkai Electric Co., Ltd.	China	
17	Distribution Box Supply (Part)	Jiangsu Hongli Electric Co., Ltd.	China	
18	Supply of PLC Controller, Contactor and VAV Control-ler	Siemens (China) Co., Ltd.	Germany	Siemens products have fast processing speed, many network interfaces and large memory capacity. Stable operation at temperatures from -20°C to 60°C.
19	Supply of Air Conditioning Pump, Sewage Lifting Device for Feed Pump Unit	Grundfos Pumps Shanghai Co.ltd.	China	
20	Cooling Tower Supply	Dongguan Linghe Baode Cooling & Heating Equipment Co. Ltd.	China	The negative pressure inside the induced cooling tower tower reduces the drift rate.Improve the thermal performance of the cooling tower to prevent bacterial growth.
21	Fire Blower Supply	Zhejiang Shuangyang Fan Co., Ltd.	China	
22	Fan Supply	Taizhou Wolter Ventilator Co., Ltd.	China	
23	Ice-on-coil Supply	Hangzhou RUNPAQ Environmental Technology Co. Ltd.	China	Using polymer-based nano composites, the thickness of the icing layer is thinner 18mm-20mm, with good icing and melting properties; corrosion resistance,The service life is more than 30 years; light weight and good toughness.
24	Water Valve Supply	Guangdong Yongquan Valve Techonology Co. Ltd.	China	The air-conditioner system's integrated anti-condensation valve not only saves energy significantly, but also realizes the intelligent pump room. It has the safety performance of preventing "water hammer"
25	Coupled Catalytic Plate Sup-ply	Shanghai Furong Environmental Technology Co. Ltd.	China	Can degrade the ozone generated by static electricity and efficiently decompose TVOC

26	DUPS Flywheel Energy Stor-age System	Euro-Diesel	Belgium	Replace the traditional UPS battery with DUPS flywheel energy storage system.More safe, reliable and environmentally friendly.
27	Photovoltaic System Supply	Beijing Hanergy Thin Film Solar Power Engineering Co., Ltd.	Germany	Imported 145Wp CIGS thin film solar cells, The photoelectric conversion efficiency is 15.43%, and the conversion efficiency is the highest among similar products.
28	Supply and Installation of Elevator	Kone Elevator Co., Ltd.	Finland	The hop elevator greatly improves the vertical capacity during the construction phase;The shuttle elevator uses a carbon fiber hoisting rope and has a service life of 15 years. Reduce energy consumption and wire rope repair and replacement time
29	Supply and Installation of Window Cleaning Machine	Shanghai COX Window Cleaning Device Engineering Co., Ltd.	Spain	
30	Internal Insulation Duct	Shandong PIM Technology Co.,Ltd.	China	Sound insulation and insulation, antibacterial and mildewproof

MAIN PROFESSIONAL SUBCONTRACTORS (ELECTROMECHANICAL GENERAL CONTRACTOR)

NO.	CONTRACTUAL LIABILITY	COMPANY NAME	REMARKS
1	Lower Zone Air Conditioning Engineering	China Construction Third Bureau Installation Engineering Co., Ltd.	
2	Higher Zone Air Conditioning Engineering	China Construction First Group Construction & Development Co., Ltd.	
3	HVAC Equipment Monitor-ing Engineering	Hangzhou RUNPAQ Environmental Technology Co. Ltd.	
4	Cold Source Engineering	Hangzhou Huadian Huayuan Environmental Engineering Co., Ltd.	
5	Heat Source Engineering	Beijing D.F.Z.Y Civicism Engineering Co., Ltd.	
6	Fire Protection Engineering	Beijing Sihai Firefighting Engineering Co., Ltd.	
7	Nightscape Lighting Engi-neering	Joint venture of HES Technology Group Co. Ltd. Beijing Liangye Environmental Technology Co. Ltd. (Consortium)	
8	Engineering of Integrated Wiring, Machine Room	CSCEC Electronic Information Technology Co., Ltd.	
9	Intelligent Engineering	Shenzhen Zhiyu Industry Development Co., Ltd.	
10	Medium Zone Electric Engineering	Shanghai Willzone Construction Co., Ltd.	
11	Window Sill Integration Sys-tem Integrated Steel Door	China Construction Equipment & Engineering Co., Ltd.	

梁传新
Liang Chuanxin

参与地标性建筑开发建设，几乎是每一个工科专业人的梦想。我在中信建设工作期间，就对中信大厦项目颇为关注，建筑体量大、投资额巨大、开发建设周期漫长、建设难度大，中信和业有一支综合能力突出，高效敬业的专业化队伍。在中信集团的坚强领导下，七年来，中信大厦在安全、品质、造价受控的前提下，如期竣工（初步接收），创造了国内超高层开发建设的纪录。

2018 年下半年，中信集团主要领导根据中信大厦开发建设及运营管理的需要，要求我到中信和业接棒已经圆满完成开发建设任务的王伍仁同志，继续推进中信大厦项目。这之后，我进行了认真的思考，作为一名有着 30 余年工程建设从业经验的专业管理人员，执掌中信大厦的工程收尾和运维是一件既充满诱惑力又极具挑战的工作，但"编框编篓，重在收口"，我清晰地认识到自己将面临何等复杂而艰巨的工作形势、肩负多么重要而光荣的使命。

2019 年 1 月 3 日，中信集团党委正式调任我加入中信和业，出任副董事长职务，开始熟悉工作。通过 4 个月的深入调研与广泛交流，我深刻体会到了中信大厦开发建造成果的来之不易，了解了和业公司的发展变迁，也感受到了和业团队的专业能力和拼搏精神。那些业界难以想象的难题和挑战，那些栉风沐雨、忘我拼搏的峥嵘岁月，那些欢笑与泪水、求索与奋进的攀登足迹……如今全都沉淀在《中国建造 中信大厦建造纪实》这本图文并茂的书中，厚重而详实，带着沁人心脾的墨香，向世人讲述中信大厦开发建造的故事。

作为中信大厦开发建设的"船长"，王伍仁同志对本书的编撰倾注了大量心血，连续几个礼拜，他都废寝忘食地亲自校验稿件、核对照片、检视版面，是想用一版精品的图录书册展现精品工程。

翻阅手中这本《中国建造 中信大厦建造纪实》样稿，我心生敬意，中信大厦不平凡的建造历程跃然眼前，七年筑梦岁月，编者如数家珍、娓娓道来，文字与图片交相辉映，承载着开发建设者的激情与梦想，诉说着七年来中信人的拼搏与创新。

2019 年 5 月 13 日，中信集团党委决定由我接任中信和业副董事长、总经理职务，在集团党委的信任和领导下，在王伍仁同志的支持下，我结合中信大厦进入工程收尾和运维展开的基本形势，调动各种资源积极推动中信大厦供电事宜、未完工程销项及缺陷整改、设备调试、竣工验收及第三方承接查验等工作，为迎接中信集团、阿里巴巴等客户的入驻做好"硬件"准备。

"硬件"品质一流，"软件"更要精良，系统才能高效运转。承担中信大厦"软件"角色的运维管理对长期在项目建设中摸爬滚打的我来说是一个全新的课题。中信大厦作为未来中信集团的总部大楼不仅承载着中信集团的品牌形象，更关系到北京的政治安全，中信大厦工程收尾和运维管理必须做到万无一失，对此，我深感责任重大、使命光荣。

在"爱与包容"的良好氛围中，在"责任心、执行力、创新力"的企业核心价值观的指引下，如今，中信和业团队的每一名成员都打起了十二分的精神。在中信集团的正确领导下，我们将以开发建设、运维全生命周期管理为基础，以实现开发建设向运维"无缝转换"为目标，团结协作、不负重托，实现中信大厦舒适、安全、低成本的高效运营，为中国超高层建筑提供一个可借鉴的样本，缔造中国超高层建筑的新高度，为中信大厦的开发建设及运维管理再立新功！

It is the dream of almost everyone with engineering majors to participate in the development and construction of landmark buildings. During my work in CITIC construction, I paid close attention to CITIC Tower, characterized by large building volume, huge investment, and long development and construction cycle. Fortunately, CITIC Heye boosts its professional team with outstanding comprehensive ability and high efficiency and dedication. Firmly led by CITIC Group over the past seven years, CITIC Tower has been completed (preliminary acceptance) on schedule in the principle of safety, quality and cost control, setting a record for the development and construction of super high-rise buildings in China.

In the second half of 2018, to manage the development, construction and operation of the CITIC Tower, I was appointed to take over the project from Mr. Wang Wuren, who has successfully completed the development and construction task of the CITIC Heye, and continue to push forward the project. Then I pondered over that it is both tempting and challenging for a professional manager with more than 30 years' experience in engineering construction to take charge of the CITIC Tower's project closing, operation and maintenance. However, as the old saying goes, "the ending work counts most". I was aware of the complicated and arduous work situation, and the important and glorious mission in front of me.

On January 3, 2019, the Party Committee of CITIC Group formally appointed me to join CITIC Heye as Vice Chairman of the Board of Directors and began to familiarize myself with my work. Through four months of in-depth research and extensive exchange, I deeply realized the hard-won development and construction achievements of CITIC Tower, understood the development and changes of Heye Company, and was also impressed by the professional abilities and hard-working spirits of the Heye Team. Those difficult problems and challenges in the industry, those eventful years, selfless struggle, those laughing and tears, seeking and striving… now all are written into the pages in the BUILT BY CHINA Construction Record Of CITIC Tower. With refreshing ink, it will tell the world the story of CITIC Tower development and construction.

As the "captain" of the development and construction of CITIC Tower, Comrade Wang Wuren devoted a great deal of effort to the compilation of this book. For several weeks in a row, he proofread the manuscripts, checked the photographs and the layout with a wish to show the excellent project with an edition of fine illustrated book.

Thumbing through the sample manuscript of BUILT BY CHINA Construction Record Of CITIC Tower, we can see with respect that the extraordinary construction process of CITIC Tower leaps into our mind. For the seven years of dream-building, the editors have enumerated a wealth of stories, and the words and pictures reflect each other, carrying the passion and dream of the developers and builders, and telling the hard work and innovation of CITIC people over the past seven years.

On May 13, 2019, the Party Committee of CITIC Group has decided that I took over the posts of Vice Chairman and General Manager of CITIC Heye. Under the trust and leadership of the group's Party committee, and with the support of Comrade Wang Wuren, I, in accordance with the basic situation when CITIC Tower was in the closing and operation and maintenance stage, mobilized various resources to actively promote CITIC Tower power supply, unfinished project items and defects rectification, equipment commissioning, completion acceptance and third-party acceptance inspection, and other tasks, so as to make "hardware" preparation for the arrival of CITIC Group, Alibaba and other customers.

With first-class "hardware" and more sophisticated "software", the system can operate efficiently. For me, who has been working in the project construction for a long time, it is a brand-new subject to bear the role of operating and managing "software" of CITIC Tower. The CITIC Tower, as the headquarters building of CITIC Group in the future, not only embodies the brand image of CITIC Group, but also relates to the political safety of Beijing. The project closing, operation and maintenance management of CITIC Tower must be absolutely safe, which is of major responsibility and honorary mission.

In the good atmosphere filled with "love and tolerance", guided by the corporate core values of "responsibility, execution and innovation", each member of CITIC Heye team remains very vigorous and positive. Under the correct leadership of CITIC Group, based on the whole life cycle management of development and construction, operation and maintenance, we will follow the objective of "seamless conversion" from development and construction to operation and maintenance, cooperate with one another, realize effective operation of CITIC Tower in such aspects as comfort, safety and low cost, provide a referable sample for the super high-rise building in China, create the new achievements and make new contributions to the development and construct ion, operation and maintenance of CITIC Tower.

图书在版编目（CIP）数据

中信大厦建造纪实 ：汉英对照 / 王伍仁主编 ；中信和业，
中建三局，中建安装编. － 北京 ：中国建筑工业出版社，2019.9（2023.1重印）
（中国建造）
ISBN 978-7-112-24147-7

Ⅰ．①中… Ⅱ．①王… ②中… ③中… ④中… Ⅲ.
①超高层建筑－建筑实录－朝阳区－汉、英 Ⅳ.
①TU972

中国版本图书馆CIP数据核字(2019)第179735号

责任编辑：徐 纺 滕云飞 张振光
责任校对：王 烨

艺术顾问：Steve Mok
工艺顾问：周小琴
视觉策划：胡文杰
建成摄影：胡文杰
视觉设计：胡文杰 王静雯 华烯宇 胡兰丽
　　　　　高 清 王者风 刘昊星 周 琦
图表设计：胡兰丽 华烯宇 王 黎 张玲玲

中国建造
BUILT BY CHINA
中信大厦建造纪实
CONSTRUCTION RECORD OF CITIC TOWER
王伍仁 主编
WANG WUREN EDITOR IN CHIEF

中信和业 | 中建三局 | 中建安装 编
EDITED BY CITIC HEYE INVESTMENT CO., LTD. I CHINA CONSTRUCTION THIRD
ENGINEERING BUREAU CO., LTD. I CHINA CONSTRUCTION INDUSTRIAL & ENERGY
ENGINEERING GROUP CO.,LTD.

*
中国建筑工业出版社出版、发行（北京海淀三里河路9号）
各地新华书店、建筑书店经销
北京富诚彩色印刷有限公司印刷
*
开本：889×1194毫米 1/8 印张：42 插页：1 字数：1048千字
2019年10月第一版 2023年1月第二次印刷
定价：360.00元
ISBN 978-7-112-24147-7
　　　（34632）